SOLANO COMMUNITY COLLEGE
3 7045 00023 5772

D0602865

# Descriptive Astronomy

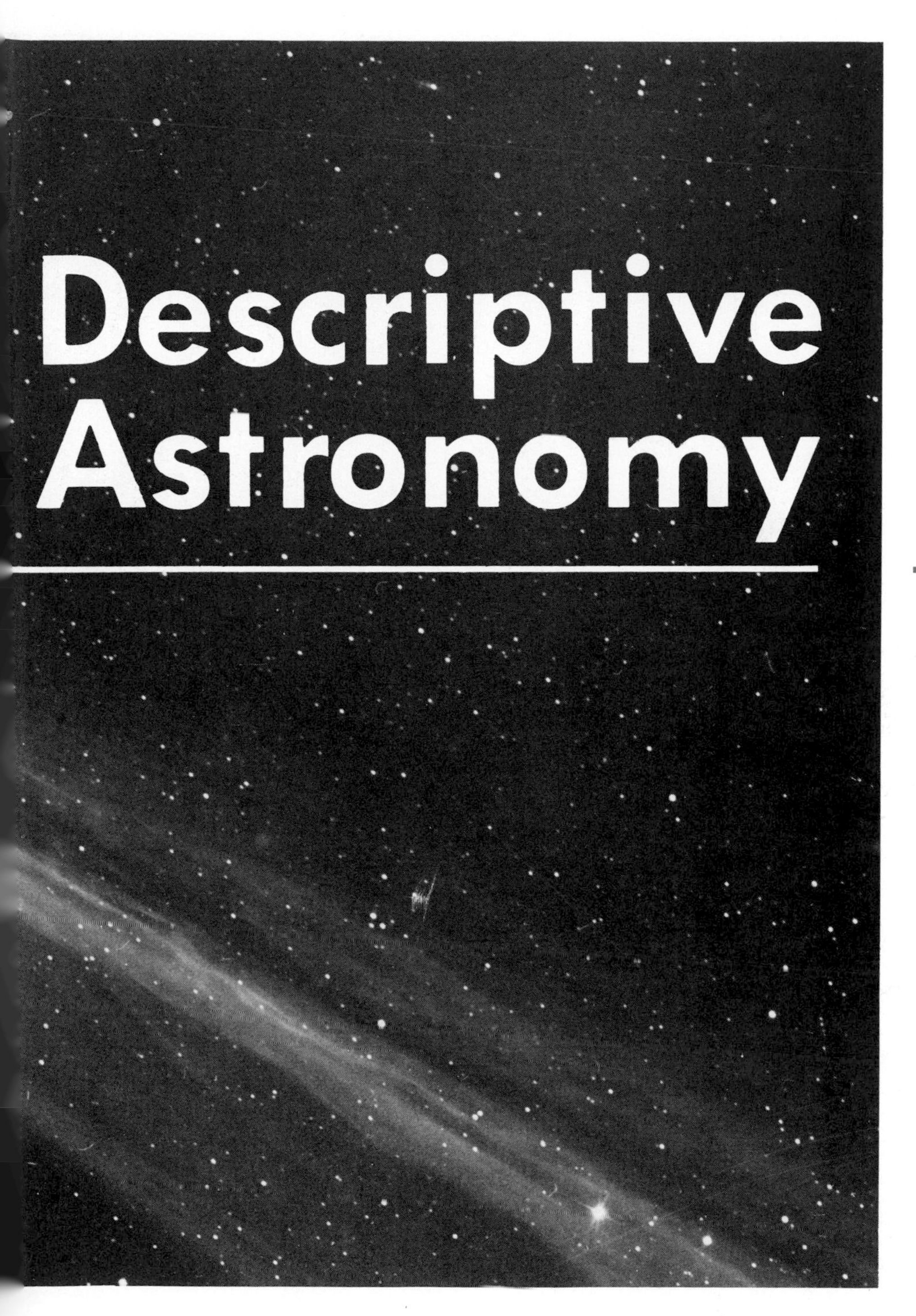

# Descriptive Astronomy

**Rebecca M. Berg**
Strasenburgh
Planetarium
Rochester, New York

**Laurence W. Fredrick**
Director,
Leander McCormick
Observatory
University of Virginia

D. Van Nostrand Company

New York • Cincinnati • Toronto
London • Melbourne

Cover illustration by Michael Freeman (from Bruce Coleman, Inc.): The illustration is an interpretation of the view of Jupiter from the barren surface of its largest moon, Ganymede. Compare this artist's conception with the recent spacecraft photographs of Jupiter and Ganymede in Chapter 8.

D. Van Nostrand Company Regional Offices:
New York   Cincinnati

D. Van Nostrand Company International Offices:
London   Toronto   Melbourne



Library of Congress Catalog Card Number: 77-89365
ISBN: 0-442-25472-5



Published by D. Van Nostrand Company
135 West 50th Street, New York, N.Y. 10020

10 9 8 7 6 5 4 3 2 1

# PREFACE



Modern astronomy is an exciting science whose laboratory is the entire Universe. Unlike other scientists, astronomers cannot handle or manipulate the objects in their laboratory. To study the remote objects of the Universe, they call upon diverse, creative, and complex tools of science. Every technological advance and new discovery triggers a better understanding of our Universe.

The frontiers of astronomy are constantly changing. Before World War II, most astronomers measured positions and distances of the planets, stars, and galaxies and patiently and laboriously cataloged the properties of the different objects in space. Hints of the spectacular changes to come were few, and those few hints were largely ignored. Now, in the second half of the twentieth century, astronomy, with its remarkable new discoveries, is setting the pace for the physical sciences. Astronomers discuss with confidence conditions in the deep interiors of stars. They even discuss with reasonable certainty the conditions in the Universe during the first few minutes of its existence.

Over the past three decades, new x-ray, optical, and radio telescopes and remarkable spacecraft have made discoveries almost unparalleled in the history of science. Quasars, pulsars, black holes, and extraterrestrial life have become topics of conversation for everyone. People want to know what is up there. Today's college students, the leaders of the future bent upon nonscientific careers, want to know and need to know. They are the people who will make decisions affecting the course of astronomy over the years to come. More importantly, they will be making decisions affecting planet earth, and a knowledge of the variety and physical conditions of other celestial objects may help them with their decisions.

DESCRIPTIVE ASTRONOMY presents modern astronomy to the nonscience college student. We use an informal writing style and avoid the use of mathematics. Outlines at the beginning and summaries at the end of each chapter help crystalize the material. There are no prerequisites to reading and enjoying DESCRIPTIVE ASTRONOMY. The great problems faced by astronomers and solved over a period of time, sometimes many years, are presented by giving the accepted interpretation first and then telling how that interpretation was reached. In many ways, knowing the result first can intensify our appreciation of the work that goes into deriving scientific conclusions.

In Chapter 1 we set the grand scene by telling the story of the development of the Universe from the beginning. In Chapters 2 through 13 we trace the observations and conclusions that lead to the view presented in Chapter 1. First, we discuss some essential background information on the appearance of the sky and the properties of light. We continue with the developments of the 17th, 18th, and 19th centuries, which set the stage for modern astronomy. Next we treat the contents and structure of the Milky Way galaxy and deal with the formation and evolution of stars. The development of the solar system follows in its natural place as a result of the formation of the sun. Then we deal with the details of the sun, the death of stars, and the properties of other galaxies. Finally, in Chapter 13, we address the questions of cosmology and of life elsewhere in the Universe.

An extensive glossary assists the student with astronomical terms. Words that do not appear in the text are included because they may be used in lectures or found in outside readings. A number of tables related to the text are in the Appendix. For example, there is a Brief Chronology of Astronomy, some Basic Astronomical Data, information about The Brightest Stars and The Nearest Stars, and the Characteristics of Members of the Solar System. We have also listed most of the observatories located in the United States, Canada, and Mexico. Since most observatories have scheduled visitor nights, our list could prove to be a useful resource for the interested student.

We would like to acknowledge the help of many collegues from around the world in providing photographs, a few of which appear for the first time in this book. We also want to thank Richard A. Berg for his useful, critical readings and suggestions; the D. Van Nostrand editorial and production staff for assistance and for independently obtaining professional manuscript reviewers, including John B. Winters, Glendale Community College, Gladwyn Comes, Broward Community College, Stephan Hill, Michigan State University, whose suggestions were extremely helpful; Patrick Seitzer for the computer program that generated the sky map grids; and Frances Fredrick for typing and retyping the manuscript many times.

Rebecca M. Berg
Laurence W. Fredrick

# CONTENTS

# 1 The Universe: From the Beginning



## BEFORE TIME BEGAN

Before time had any meaning, the Universe consisted entirely of photons and exotic particles. Photons are bundles of energy. There were x-ray photons, visible light photons, and photons of every other possible energy in the Universe. The exotic particles of the Universe were minute subdivisions of matter, even smaller than a single atom. In order for them to exist, extreme conditions of high temperature, great density, and intense pressure prevailed. The photons had the same energy as the exotic particles in the original mixture. Thus the Universe was in equilibrium.

The temperature of the mixture of photons and particles exceeded 100 billion degrees. The photons moved randomly through the mixture at the speed of light.* The speed of the particles was only slightly less than the speed of light. Under such circumstances, nothing was happening that was distinguishable from anything else.

*The speed of light is almost 300,000 kilometers per second (km/sec).

FIGURE 1.1
The explosion of a hydrogen bomb. The violence and fury of the original primeval fireball far exceeded that of an H bomb. (U.S. Air Force photograph)

## THE FIRST THIRTY MINUTES

Then, billions upon billions of years ago, something happened. The equilibrium was disturbed. The disturbance caused the chaotic but balanced motions of the photons and particles to expand or explode (Figure 1.1). With a sudden violence, the Universe exploded. Its temperature was 100 billion degrees, and at that temperature such an explosion was indescribable in its fury. We call this the primeval fireball. At the moment of the explosion there began an orderly sequence called time. After the explosion, the Universe—its space and time—was expanding.

Before the first second of time had passed, the entire Universe was cooling dramatically due to its expansion. The Universe cooled enough to allow the initial exotic particles to form stable particles called electrons, protons, and neutrons. These stable particles were the building blocks of all matter. After only one second of time, the temperature of the Universe cooled down to 10 billion degrees (see Figure 1.2). The Universe consisted of basic stable particles and an intense photon energy, called radiation, left over from the violent explosion only a second ago. The particles and photons had the same temperature.

The Universe continued to expand and cool uniformly. During the next few seconds, the stable particles tried in vain to bind together and form atoms. The temperature of the Universe was still hot enough that collisions occurred continuously between particles or between particles and radiation. If any atoms formed in these first instants of time, they were immediately disrupted by collisions. Before primitive atoms could survive, the temperature of the Universe had to cool to a billion

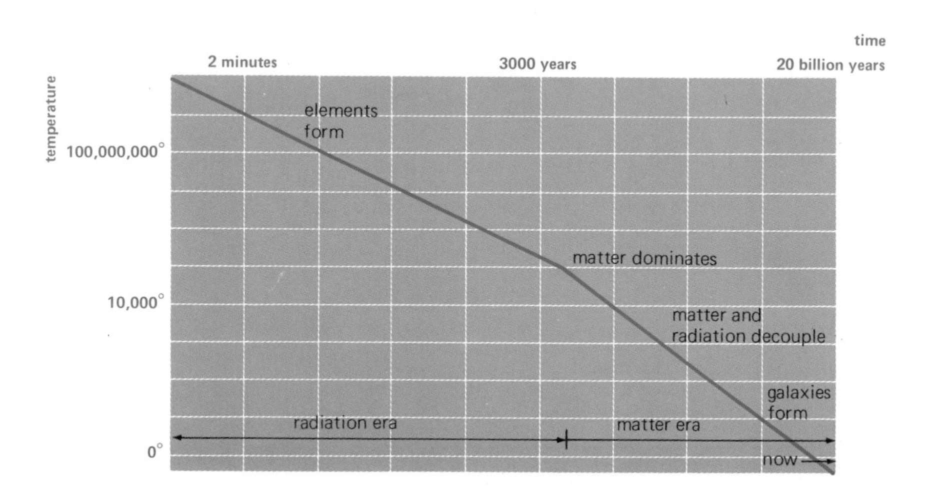


FIGURE 1.2
How the universe cooled with time. The various events are marked.

degrees. That temperature was reached in only a few seconds. Then protons and neutrons combined to form atoms of deuterium (also called heavy hydrogen). The deuterium atoms in turn combined to form helium atoms. When the Universe was only thirty minutes old, the atom-building process ended. The matter of the Universe was then 75% hydrogen and almost 25% helium. A small fraction of deuterium also remained.

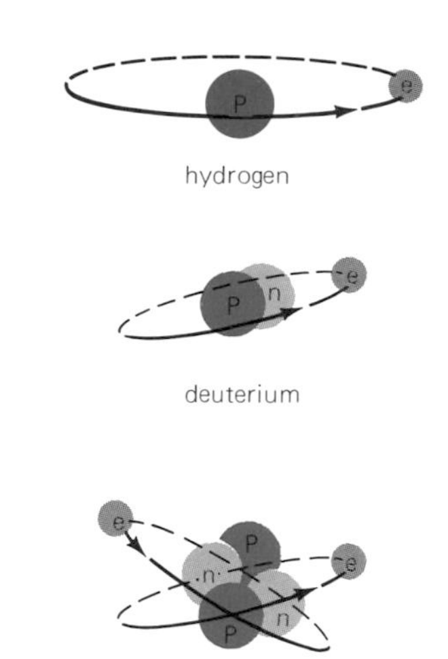


**FIGURE 1.3**
A schematic idea of how atoms are built. The nuclei of deuterium and helium contain a neutron (n) for each proton (p). The proton has a positive electrical charge which is balanced by an orbiting electron (e) with a negative charge. Neutrons have no electrical charge.

## MATTER AND RADIATION SEPARATED

For the next two thousand years nothing of great significance happened. The Universe continued to cool and expand. Matter and radiation cooled together in equilibrium. Then one day another change occurred. The space between particles of matter became large enough that the matter retained some heat, while the radiation continued to cool at the same rate as before. Expansion and cooling continued, but the matter and radiation would never again be in equilibrium. The slightly higher-temperature matter became the dominant component of the Universe.

Now the time between events lengthened. A million years of relentless expanding and cooling passed. The temperature of the Universe cooled to 3,000 degrees. At that temperature, matter and radiation decoupled and evolved independently. From this point on the radiation cooled uniformly as the Universe expanded. A more interesting future was in store for the expanding matter.

## GALAXIES FORMED

Matter began to interact with matter. Slowly, atom by atom, the matter collected into clouds. For millions of years, only the Universe knew that this cloud-building process was going on. Finally, great gas clouds of hydrogen and helium existed (Figure 1.4). These clouds were thousands to millions of light years* across. Each atom was attracted to every other atom in the cloud. This mutual attraction, called gravitation, kept the clouds from dispersing into space. Under certain circumstances, gravitation also caused the vast clouds to collapse. In time, the largest clouds broke up into a group of hundreds of smaller clouds. These clouds were the early stages of galaxies.

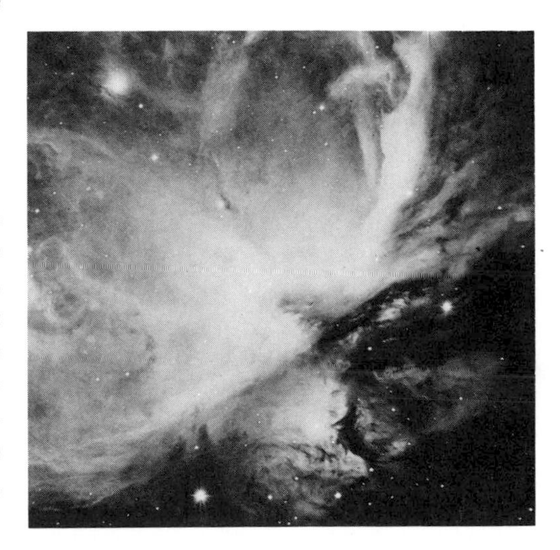

**FIGURE 1.4**
A portion of the great gas and dust clouds in the constellation of Orion.

*One light-year (lt-yr) is the distance covered in one year by a photon moving at the speed of light. It equals nearly 10 trillion kilometers.

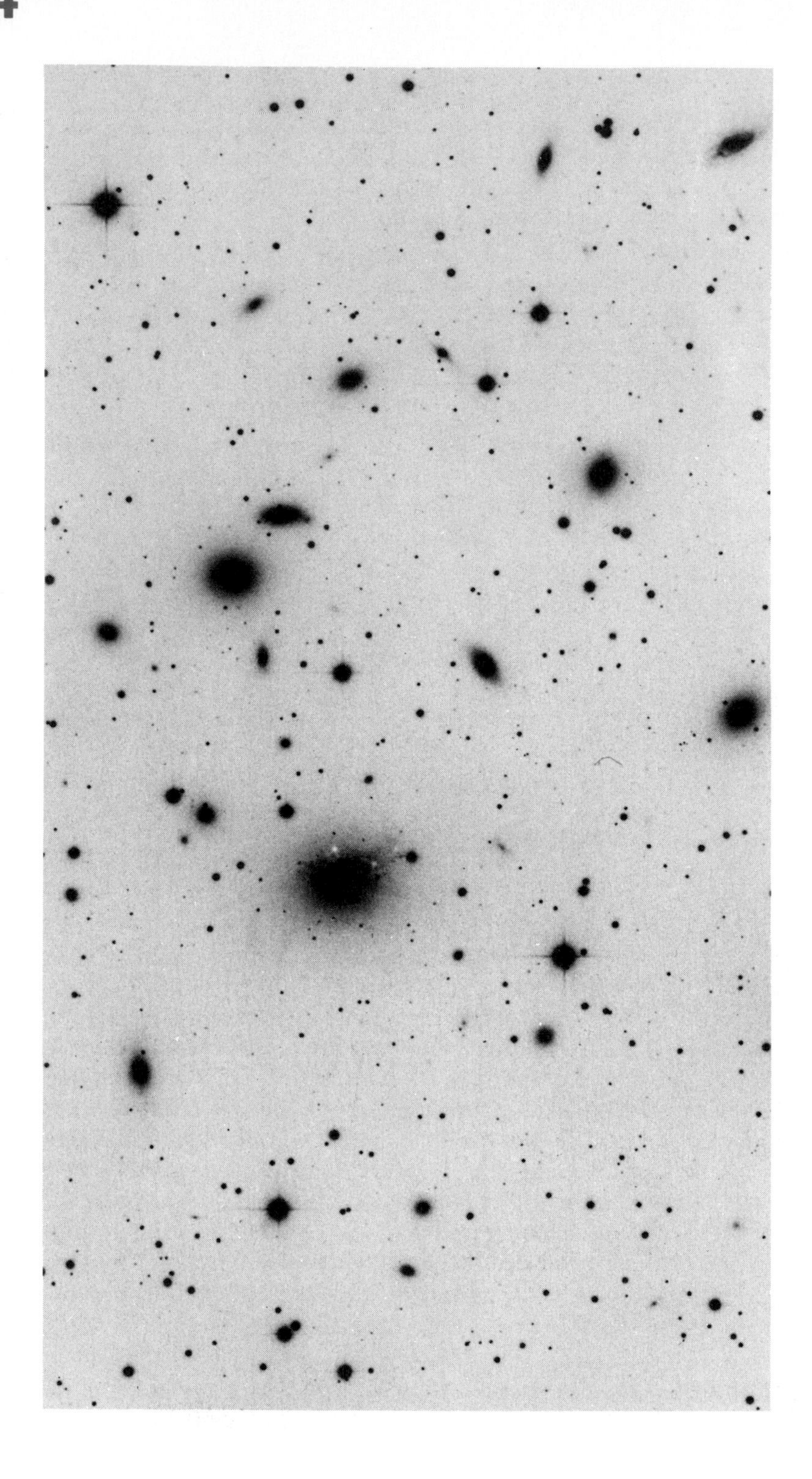

**FIGURE 1.5**
A portion of a cluster of galaxies in the constellation of Perseus. Each galaxy formed from a fragment of the original cloud. (Hale Observatories photograph)

## THE MILKY WAY GALAXY FORMED

At this point, it had been a billion years since the cataclysmic beginning of the expanding Universe. In one place in the Universe, an immense cloud of hydrogen and helium fragmented into several dozen clouds. One of these occupied over one hundred thousand light years of space and became the Milky Way galaxy. Each galaxy followed a similar development. Smaller cloud condensations formed within the larger galaxy clouds. Over millions of years, these condensations gave birth to stars. In some galaxies the stars formed efficiently: All the matter was locked up in stars. In other galaxies, stars formed inefficiently: A lot of hydrogen and helium remained between stars.

**FIGURE 1.6**
A typical galaxy resembling in many ways the Milky Way galaxy as seen from above. (Kitt Peak National Observatory photograph)

## STARS FORMED, AGED, AND DIED

Condensations within the Milky Way galaxy slowly collapsed, crushing matter against matter. The temperatures and pressures at the center of the condensations increased dramatically. When a critical central temperature of about 13 million degrees was reached, the conversion of hydrogen into helium began producing energy. A star was born. Throughout the Milky Way (Figures 1.6 and 1.7), billions of

**FIGURE 1.7**
A typical galaxy resembling the Milky Way galaxy as seen from the side. (Hale Observatories photograph)

stars began to light up the darkness of space. The collapse of large condensations produced stars more quickly than the collapse of smaller condensations. Often groups of many stars formed almost at the same time. Even after billions of years of star formation, the Milky Way galaxy had gas left over. This gas would never be completely used in forming stars (Figure 1.8).

The first stars were composed entirely of hydrogen and helium, and their energy came mainly from converting hydrogen into helium. As the stars used up their energy supply, they aged. The aging process was slow and constant. For millions of years, changes in the stars were inconspicuous. But after a vast amount of time, the centers of some stars grew exceedingly hot. At temperatures reaching over 100 million degrees, helium was converted into heavier elements such as carbon and nickel. The heaviest, most massive stars produced all the elements up to iron, in their interiors. These stars aged fastest and often came to spectacular ends.

The most massive stars exploded. In an instant, with unimagin-

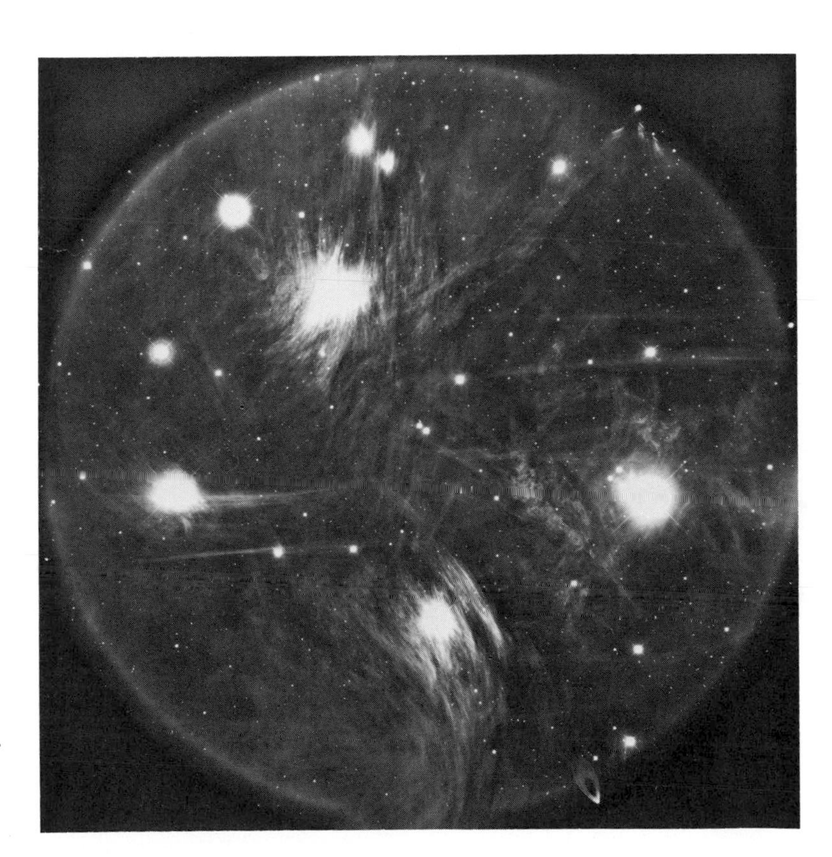

**FIGURE 1.8**
An open cluster of stars called the Pleiades. Some of the original material that formed the stars can be seen. (Kitt Peak National Observatory photograph)

able violence, they spewed their contents back into space (Figure 1.9). The heavy elements they had contained mixed with the hydrogen and helium left over in space after the initial star formation phase. More years passed; then a cloud of the new mixture formed. The ends of some stars signaled the beginning of others. Condensations in the new cloud slowly collapsed and formed stars. The most massive stars formed first from the new mixture and then followed the cycle of their ancestors. They aged more quickly than the other stars. After several million undramatic years, they abruptly burst forth, sending more heavy elements into space.

After fifteen billion years, the Milky Way contained several generations of stars. New stars formed from a mixture of 70% hydrogen, 27% helium, and 3% heavier elements. By this time the galaxy contained dust and molecules as well as atomic elements. Some of the molecules were quite complex. Many of them were organic molecules, the kind on which life as we know it is based.

10 March 1935

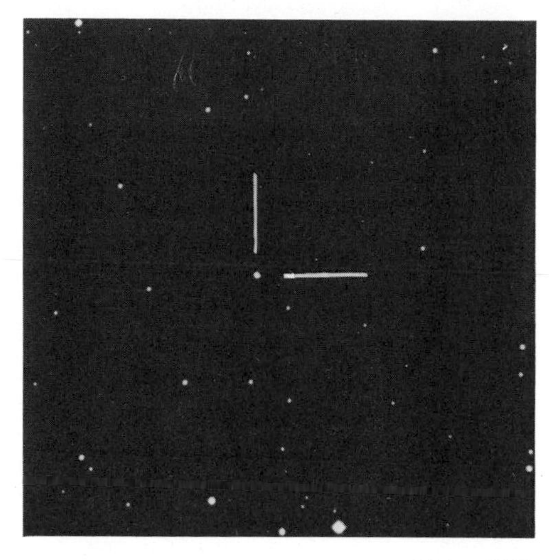

6 May 1935

**FIGURE 1.9**
Two pictures of Nova Herculis, 1935. An explosion made the star extremely bright on 10 March 1935. The result of such a cataclysmic event is to return enriched material to interstellar space. By 6 May 1935, the star had returned to more normal brightness. (Lick Observatory photograph)

## THE SUN AND PLANETS FORMED

The billions of galaxies continued to ride the wave of the expanding Universe. Then one day, inside the Milky Way, a special cloud began forming. It was off on one side of the galaxy and formed from material containing the debris of dead stars. This cloud would determine the course of our history (Figure 1.10).

As the cloud collapsed, its rotation increased. The center of the cloud grew denser and hotter. But in the outer regions of that cloud, clumps of material stuck together to form bigger clumps. Slowly, the clumps swept up more material. Out of the spinning, collapsing cloud emerged planets. They were circling a new medium-sized star, our sun (Figure 1.11).

The earth, the third planet away from the sun, accumulated enough material and gases so that its interior heated up due to compression. The center of the earth was compressed to a hot molten fluid from the weight of all that overlying material. It was surrounded by an atmosphere rich in hydrogen and helium.

Through the clouds of gas and dust and the clumps of material pervading the solar system, the young sun was barely visible as a dull, glowing sphere. Finally the sun became a self-sustaining star. Deep in its interior, hydrogen was being converted into helium. When that nuclear energy first turned on, a blast of intense radiation swept outward through the solar system. It cleared out the gas and dust between the

**FIGURE 1.10**
If the Milky Way galaxy were this galaxy, the sun and solar system would be located at the arrow. (Kitt Peak National Observatory photograph)

**FIGURE 1.11**
Sketches of the early formative phases of the solar system. Proceeding down the diagrams, the sun becomes visible as an enlarged, low-density spheroid and the planets accrete material.

planets, stripped the earth of its primitive atmosphere, and melted its surface.

The surface of the earth cooled quickly and formed a thin, hard crust. Because the earth was turning, pressures built up in the crust and caused cracks to appear. At the same time, the large pieces of material that could not be swept away by the blast of the sun's radiation continued to be collected by the earth. Great and small craters were blasted out on earth, intensifying the cracking and triggering volcanic activity. For eons, the earth was in a formative upheaval. Volcanoes brought molten rock up from the interior. This rock contained water vapor and molecules of all types. Some of the molecules floated above the earth. Very gradually, a second atmosphere was built up around the earth. Then water began to collect on the surface (Figure 1.12).

Finally, after nearly two billion years, the bombardment of debris from interplanetary space ceased. Now the earth's features were determined by the action of drifting land masses, the growing quantity of water, and the atmosphere. Still more time passed. Somewhere on earth a protected pool of water formed and was warmed by its surroundings. The water running into the pool leached minerals and organic molecules from the surface material. Under the influence of the sun, life began on earth.

**FIGURE 1.12**
The planet earth as viewed from space. (National Aeronautics and Space Administration)

## TODAY

Twenty billion years after time began, intelligent life exists. The human mind looks back into space and time and wonders. When, how, why did the Universe begin? These are questions we have sought to answer in one way or another since before recorded history. Most ancient peoples simply accepted the existence of the Universe (their world), and went on from there. More recently, people have tried to answer these questions analytically because their observations seem to say there was a beginning. If there was a beginning, we should be able to say when. If we can say when, then we should be able to say how. But it does not follow that if we can say when and how, we should also be able to say why.

## SUMMARY

We have recounted one story of the development of the Universe. This story is not the result of wild imagination. The temperatures, times, and events are based on some of the latest astronomical theories.

We mark the beginning of time from the violent explosion that occurred when the Universe contained exotic particles and photons at a temperature of 100 billion degrees. From that time on, all the matter and radiation in the Universe began to expand and cool. During the first 30 minutes of time, all the initial atoms of hydrogen and helium had formed. About a million years later, the matter and radiation separated. The radiation continued to expand into space and cool. The matter also continued to expand into space, but accidents of gravity caused the matter to collect into the familiar constituents of the Universe: galaxies, stars, planets, clouds of gas and dust.

The galaxies formed from great clouds of gas in space. Often they formed in clusters. The Milky Way galaxy began forming about a billion years after the Universe exploded. The stars formed out of smaller clouds of gas in the galaxies. The stars, too, often formed in clusters. The stars shine by producing nuclear energy. This energy is produced when stars convert hydrogen into helium in their centers. Heavy, massive stars can continue to produce energy after their hydrogen is used up by converting helium and other light elements to heavier elements, all the way to iron. The most massive stars used up all their fuel and died first. They exploded their contents back into space to mix with the original hydrogen and helium. New stars formed from this enriched mixture. One of these was the sun, born nearly 15 billion years after the Universe exploded. The earth and other planets formed from the outer parts of the cloud that collapsed to produce the sun. Life slowly developed on earth. It is now 20 billion years since the Universe exploded.

The following terms appeared in Chapter 1. They will all be used and defined in detail in the remaining chapters of this text.

| *Matter Terms* | *Energy Terms* | *Large Collections of Matter* | *General Terms* |
|---|---|---|---|
| exotic particles | photons | clouds of gas<br>dust | Universe |
| atoms made of electrons<br>protons<br>neutrons | radiation | galaxies | temperature |
| | light | Milky Way galaxy | density |
| atomic elements like hydrogen<br>helium<br>deuterium | | stars | pressure |
| | | sun | equilibrium |
| molecules | | planets | speed of light |
| | | earth | light year |
| | | | gravitation |
| | | | rotation |
| | | | nuclear energy |
| | | | time |

## REVIEW QUESTIONS

_____ 1. Which began first, formation of galaxies, stars, or planets?

_____ 2. Was the sun one of the first stars formed? Explain.

_____ 3. Did all the gas and dust in the Milky Way galaxy form into stars? Explain.

_____ 4. About how old is the sun today?

_____ 5. What force kept the clouds of collapsing gas and dust from dispersing into space?

_____ 6. Describe what happened to the original radiation after the matter and radiation separated. If you were looking for the original radiation today, where would you look?

_____ 7. What caused the interior of the earth to be hot?

_____ 8. What happened to the earth's original atmosphere?

# 2 The Sky: From Legend to Science



## THE CONSTELLATIONS

### Origins

If you have ever watched the sky, the motions of the sun and moon, or the stars that make up the constellations, you may consider yourself an astronomer. Unlike a modern astronomer, who uses giant telescopes and sophisticated electronic equipment, you are imitating the first astronomers who watched and remembered and eventually recorded the patterns in the sky.

Since the stars are so distant, their relative positions in our sky have been the same for thousands of years. This single fact, that we see the same patterns as our ancient ancestors, makes constellation figures and legends intriguing. Easily recognizable pictures of constellations have been found carved on walls of certain caves. References to the constellations are found in Assyrian tablets dating from 3000 B.C. to 1000 B.C., in the Bible, in the writings of Homer, and in many other early works. The earliest known star catalog was compiled by the Chinese about 350 B.C. It contained the positions of 800 stars.

**Constellations** are arbitrary groupings of stars that just happen to lie in the same direction in space. The Chinese mapped the sky into several hundred constellations (Figure 2.1). There is evidence that most of our present constellations were mapped and named by Mediterranean, Mesopotamian, and Egyptian cultures about five thousand years ago. The Greeks later renamed some of the sky figures after the heroes and creatures of their mythology. By 270 B.C., the Greeks had divided the sky visible to them into 48 constellations. A poet, Aratus, described nearly all these constellations in his *Phenomena*, a manual for sailors. Nearly three centuries later, Ptolemy published his *Almagest*, which contained descriptions of the stars and constellations based on an extensive star catalog compiled by Hipparchus.

## Magnitudes of Stars

Hipparchus, an able Greek astronomer and mathematician, is sometimes called the father of astronomy. In addition to constructing a star catalog around 140 B.C., Hipparchus sorted the stars that could be seen by the naked eye into six brightness categories (**magnitudes**). His first-magnitude stars were the brightest that could be seen with the naked eye, and his sixth-magnitude stars were the faintest. We still use Hipparchus' system, although the magnitude scale has now been extended at the bright end to zero and even minus one (−1) to account for some extraordinarily bright stars.

After the invention of the telescope, fainter stars could be seen. In keeping with the system, these were assigned higher numbers for magnitudes. Stars as faint as twenty-third (+23) magnitude can be seen with our largest telescopes. The magnitudes of stars from zero to plus four are indicated by different symbols on the sky maps in this chapter. Notice that Polaris and most of the stars in the Big Dipper are only second-magnitude stars.

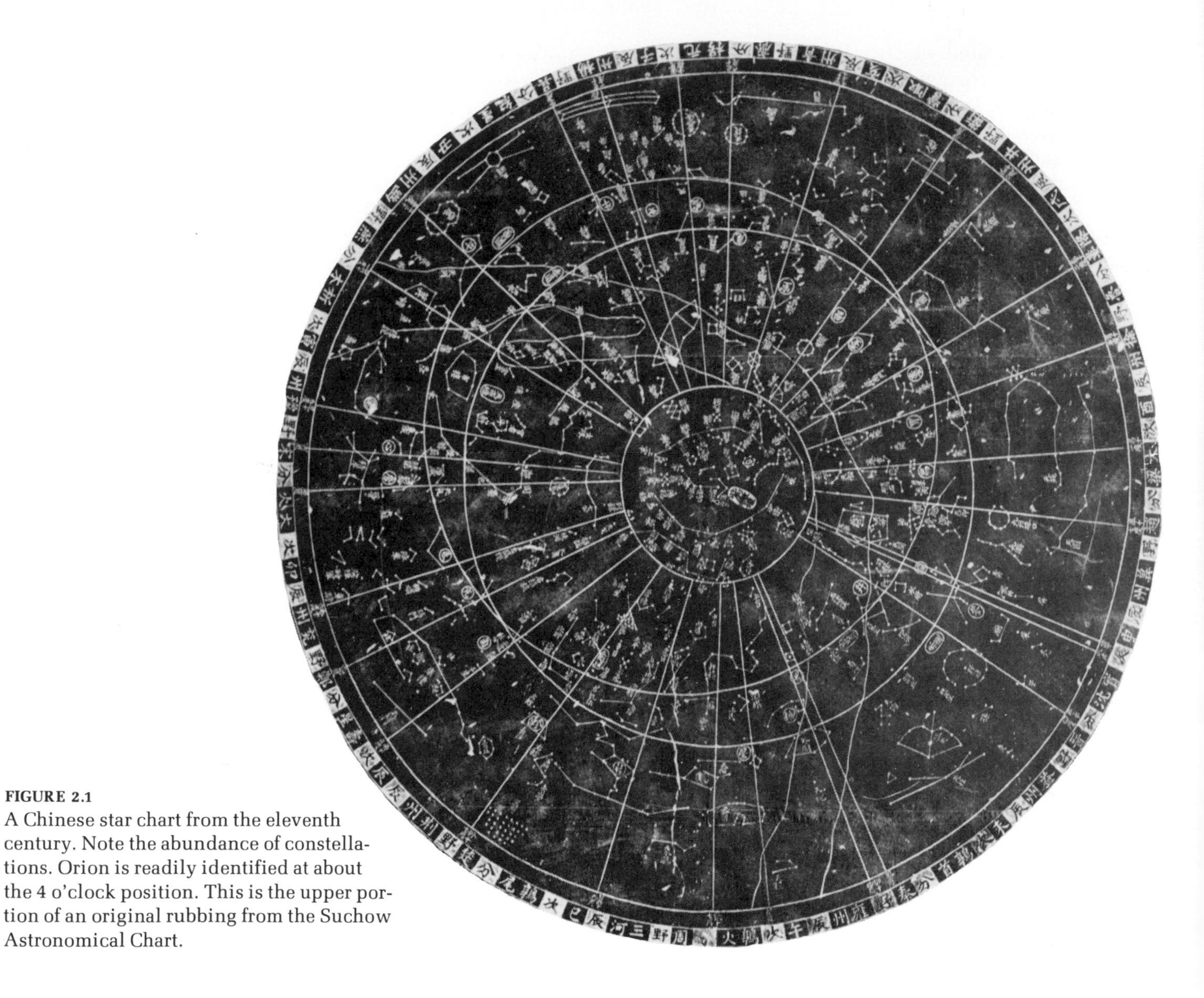

**FIGURE 2.1**
A Chinese star chart from the eleventh century. Note the abundance of constellations. Orion is readily identified at about the 4 o'clock position. This is the upper portion of an original rubbing from the Suchow Astronomical Chart.

## Official Constellations

Since the 48 Greek constellations did not cover the whole sky, more were added by 17th- and 18th-century mapmakers. The new constellations were named after inanimate objects such as telescopes, microscopes, and clocks. Today there are 88 constellations covering the entire sky (a complete list is given in Table 2.1, pp. 16–17). The boundary lines of these constellations were changed to straight north-south or east-west lines by an international agreement in 1928. The new straight boundaries for the constellations still zigzag to keep parts of the mythological characters together (Figure 2.2).

The Big Dipper, which many people regard as the most famous constellation, is really not one of the official 88. It is a subgrouping of stars within Ursa Major, the Great Bear (Figure 2.3). Such patterns within constellations are called **asterisms**. The Little Dipper is another asterism.

Astronomers now use constellation names merely to indicate a general area of the sky in much the same way that state names indicate a region of the United States. If you say you are from Virginia, that immediately brings to mind the East Coast, mid-Atlantic region of the United States. If an astronomer says there is an interesting object in Perseus, that brings to mind a specific region of the sky. Historically, there is some reason to believe that constellations or groups of constellations marked out reference points or circles in the sky. Hydra, a staggering collection of 25 faint stars, may have marked the equator of the sky about 3000 B.C. The most famous circle of constellations is the one made of the twelve zodiacal constellations, discussed on page 22.

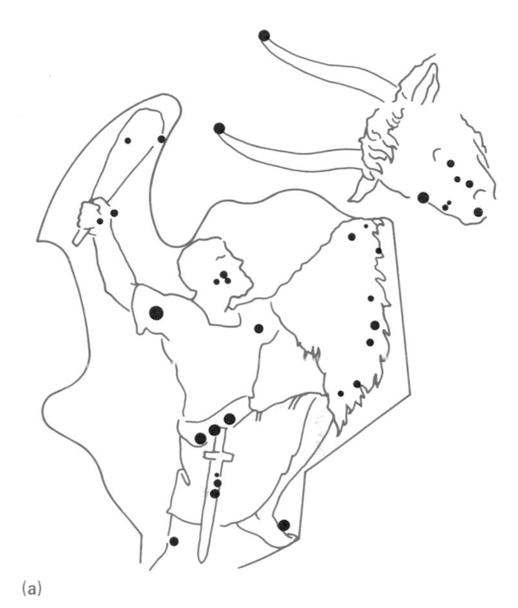

(a)

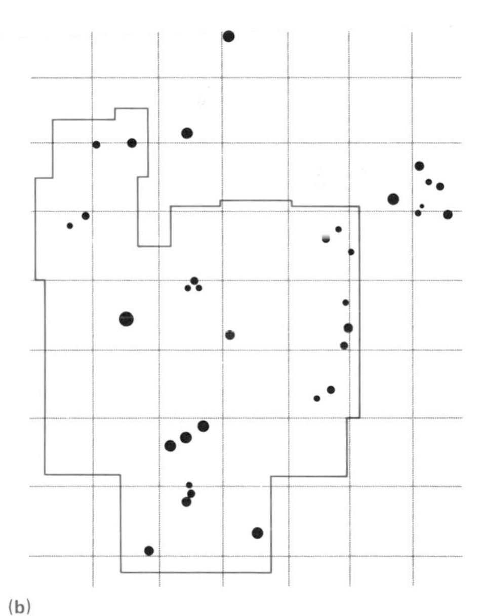

(b)

**FIGURE 2.2**
The old and new boundaries of Orion.

## Star Names

In addition to constellation names, there are personal names for the brightest stars. These names, handed down through the ages, were modified in ancient times by the Greeks and later by the Arabs. The Arabs were responsible for preserving much of Greek knowledge during the decline and fall of the Roman Empire.

Here are a few examples of how some of the stars were designated. The star Procyon, meaning "before the dog," precedes Sirius the Dog Star in its rising. Aldebaran means "the follower"; it rises after the Pleiades. Antares is the "rival of Mars" because it is red. Other star names have come to us in different ways.

In the earliest catalogs, such as Ptolemy's, the stars were distinguished by their positions in the imagined figures of heroes and animals. One star was the "mouth of the Fish"; another was the "tail of the Bird." Transcribed later into the Arabic, some of these expressions finally degenerated into single words. Betelgeuse, the name of the bright red star in Orion, was originally three words perhaps meaning the "armpit of the Central One."

The plan of designating the brighter stars by letters was introduced by J. Bayer, a Bavarian attorney, in 1603. In a general way, the stars of each constellation are denoted by the lowercase letters of the Greek alphabet in order of their brightness, and the Roman alphabet is used for further letters. If there are several stars of nearly the same brightness in the constellation, they are likely to be lettered in order from the head to

**TABLE 2.1 The Constellations**

| Latin Name | Possessive | English Equivalent | Map†† |
|---|---|---|---|
| *Androm'eda | Androm'edae | Andromeda | 4, 5 |
| Ant'lia | Ant'liae | Air Pump | |
| A'pus | A'podis | Bird of Paradise | |
| *Aqua'rius | Aqua'rii | Water Carrier | 4 |
| *Aq'uila | Aq'uilae | Eagle | 3, 4 |
| *A'ra | A'rae | Altar | 6 |
| *A'ries | Ari'etis | Ram | 4, 5 |
| *Auri'ga | Auri'gae | Charioteer | 5 |
| *Boö'tes | Boö'tis | Herdsman | 2, 3 |
| Cae'lum | Cae'li | Graving Tool | |
| Camelopar'dus | Camelopar'dalis | Giraffe | |
| *Can'cer | Can'cri | Crab | 2, 5 |
| Ca'nes Vena'tici | Ca'num Venatico'rum | Hunting Dogs | 2 |
| *Ca'nis Ma'jor | Ca'nis Majo'ris | Larger Dog | 5 |
| *Ca'nis Mi'nor | Ca'nis Mino'ris | Smaller Dog | 5 |
| *Capricor'nus | Capricor'ni | Sea-Goat | 4 |
| †Cari'na | Cari'nae | Keel | 6 |
| *Cassiope'ia | Cassiope'iae | Cassiopeia | 1, 4 |
| *Centau'rus | Centau'ri | Centaur | 2, 6 |
| *Ce'pheus | Ce'phei | Cepheus | 1, 4 |
| *Ce'tus | Ce'ti | Whale | 4, 5 |
| Chamae'leon | Chamaeleon'tis | Chameleon | |
| Cir'cinus | Cir'cini | Compasses | |
| Colum'ba | Colum'bae | Dove | 5 |
| Co'ma Bereni'ces | Co'mae Bereni'ces | Berenice's Hair | 2 |
| *Coro'na Austra'lis | Coro'nae Austra'lis | Southern Crown | |
| *Coro'na Borea'lis | Coro'nae Borea'lis | Northern Crown | 3 |
| *Cor'vus | Cor'vi | Crow | 2 |
| *Cra'ter | Crater'is | Cup | 2 |
| Crux | Cru'cis | Cross | 6 |
| *Cyg'nus | Cyg'ni | Swan | 3, 4 |
| *Delphi'nus | Delphi'ni | Dolphin | 4 |
| Dora'do | Dora'dus | Dorado | |
| *Dra'co | Draco'nis | Dragon | 1, 3 |
| *Equu'leus | Equu'lei | Little Horse | |
| *Erid'anus | Erid'ani | River | 5, 6 |
| For'nax | Forna'cis | Furnace | |
| *Gem'ini | Gemino'rum | Twins | 5 |
| Grus | Gru'is | Crane | 4 |
| *Her'cules | Her'culis | Hercules | 3 |
| Horolo'gium | Horolo'gii | Clock | |
| *Hy'dra | Hy'drae | Water Snake | 6 |
| Hy'drus | Hy'dri | Sea Serpent | 2 |
| In'dus | In'di | Indian | |
| Lacer'ta | Lacer'tae | Lizard | |
| *Le'o | Leo'nis | Lion | 2 |

**TABLE 2.1 The Constellations (con't)**

| *Latin Name* | *Possessive* | *English Equivalent* | *Map††* |
|---|---|---|---|
| Le'o Mi'nor | Leo'nis Mino'ris | Smaller Lion | |
| *Le'pus | Le'poris | Hare | 5 |
| *Li'bra | Li'brae | Scales | 3 |
| *Lu'pus | Lu'pi | Wolf | 3 |
| Lynx | Lyn'cis | Lynx | |
| *Ly'ra | Ly'rae | Lyre | 3, 4 |
| Men'sa | Men'sae | Table Mountain | |
| Microsco'pium | Microsco'pii | Microscope | |
| Monoc'eros | Monocero'tis | Unicorn | |
| Mus'ca | Mus'cae | Fly | 6 |
| Nor'ma | Nor'mae | Level | |
| Oc'tans | Octan'tis | Octant | |
| *Ophiu'chus | Ophiu'chi | Serpent Holder | 3 |
| *Ori'on | Orio'nis | Orion | 5 |
| Pa'vo | Pavo'nis | Peacock | 6 |
| *Peg'asus | Peg'asi | Pegasus | 4 |
| *Per'seus | Per'sei | Perseus | 4, 5 |
| Phoe'nix | Phoneni'cis | Phoenix | 4 |
| Pic'tor | Picto'ris | Easel | |
| *Pis'ces | Pis'cium | Fishes | 4 |
| *Pis'cis Austri'nus | Pis'cis Austri'ni | Southern Fish | 4 |
| †Pup'pis | Pup'pis | Stern | 5 |
| †Pyx'is | Pyx'idis | Mariner's Compass | |
| Retic'ulum | Retic'uli | Net | |
| *Sagit'ta | Sagit'tae | Arrow | 3, 4 |
| *Sagitta'rius | Sagitta'rii | Archer | 3 |
| *Scor'pius | Scor'pii | Scorpion | 3 |
| Sculp'tor | Sculpto'ris | Sculptor's Apparatus | 4 |
| Scu'tum | Scu'ti | Shield | |
| *Ser'pens | Serpen'tis | Serpent | 3 |
| Sex'tans | Sextan'tis | Sextant | |
| *Tau'rus | Tau'ri | Bull | 5 |
| Telesco'pium | Telesco'pii | Telescope | |
| *Trian'gulum | Trian'guli | Triangle | 4, 5 |
| Trian'gulum Austra'le | Trian'guli Austra'lis | Southern Triangle | 6 |
| Tuca'na | Tuca'nae | Toucan | 6 |
| *Ur'sa Ma'jor | Ur'sae Majo'ris | Larger Bear | 1, 2 |
| *Ur'sa Mi'nor | Ur'sae Mino'ris | Smaller Bear | 1, 3 |
| †Ve'la | Velo'rum | Sails | 2, 6 |
| *Vir'go | Vir'ginis | Virgin | 2 |
| Vo'lans | Volan'tis | Flying Fish | |
| Vulpec'ula | Vulpec'ulae | Fox | |

*One of the 48 constellations recognized by Ptolemy.
†Carina, Puppis, Pyxis, and Vela once formed the single Ptolemaic constellation Argo Navis.
††See appropriate accompanying map for location in the sky.

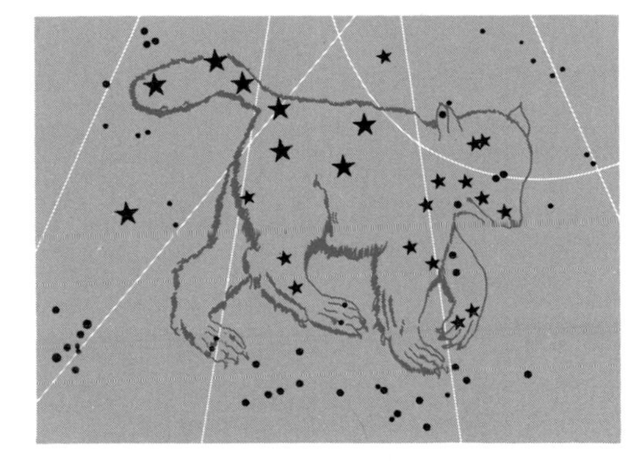

**FIGURE 2.3**
The Great Bear. The Big Dipper makes up part of the rump and tail of this famous constellation.

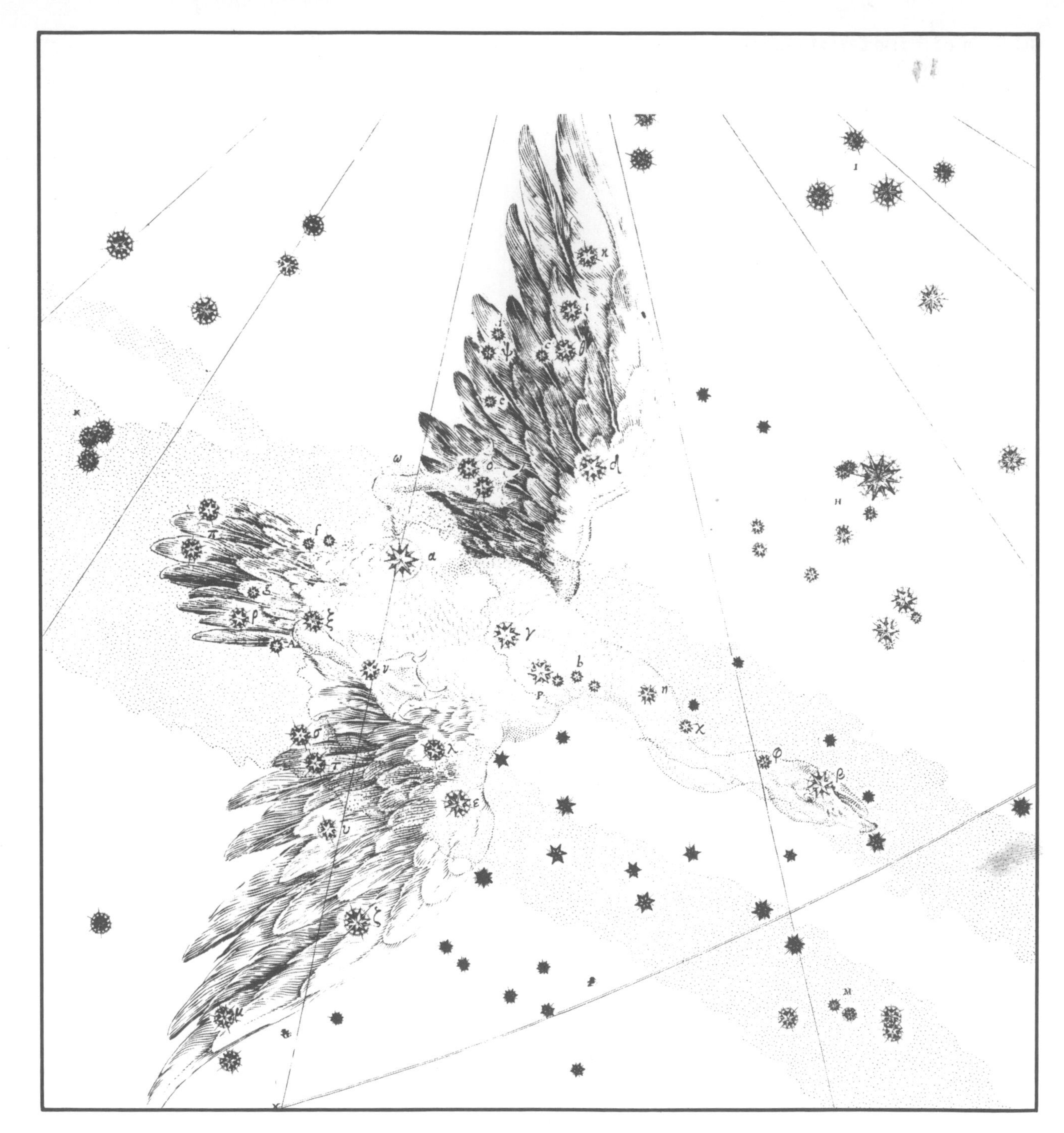


**FIGURE 2.4**
Cygnus the Swan from Bayer's *Uranometria*.

the foot of the legendary creature. The full name of a star in the Bayer system is the letter followed by the possessive of the Latin name of the constellation. Thus Capella, the brightest star in Auriga, is alpha ($\alpha$) Aurigae (the letters for some of the brighter stars are shown in the maps that follow). Bayer also produced one of the most famous illustrated star catalogs of the 17th century. Some illustrations from Bayer's *Uranometria* appear in Figures 2.4 and 2.5. The engravings were originally made by Albrecht Dürer.

**Greek Alphabet**

| | | | |
|---|---|---|---|
| $\alpha$ | alpha | $\nu$ | nu |
| $\beta$ | beta | $\xi$ | xi |
| $\gamma$ | gamma | $o$ | omicron |
| $\delta$ | delta | $\pi$ | pi |
| $\epsilon$ | epsilon | $\rho$ | rho |
| $\zeta$ | zeta | $\sigma$ | sigma |
| $\eta$ | eta | $\tau$ | tau |
| $\theta$ | theta | $\upsilon$ | upsilon |
| $\iota$ | iota | $\phi$ | phi |
| $\kappa$ | kappa | $\chi$ | chi |
| $\lambda$ | lambda | $\psi$ | psi |
| $\mu$ | mu | $\omega$ | omega |

## THE CELESTIAL SPHERE

### A Description

"Millions of stars shone down from the celestial vault." That is the way a poet might react to a clear night sky. Though millions of stars exist, only about 2,500 are seen by the naked eye, at any one time, on the clearest, darkest night. The stars seem to be fixed to the inside of a large inverted bowl or dome. The phrase we use to describe this, **celestial sphere,** refers to the imaginary sphere completely surrounding the earth, but infinitely far away. Everything in the sky can be located on this imaginary sphere. The sun, moon, planets, and all the stars are

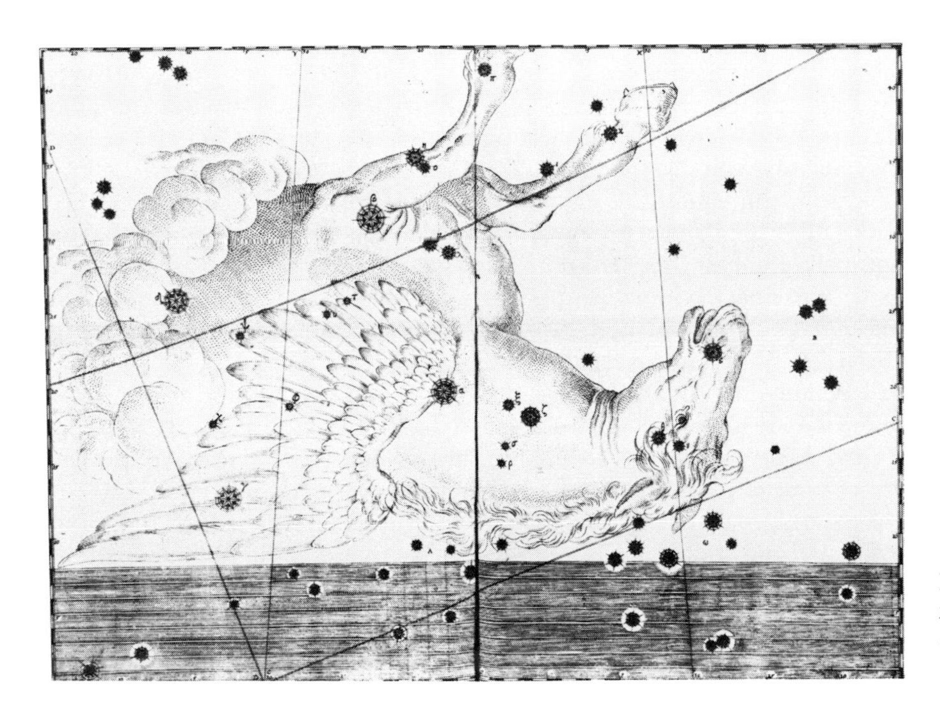

**FIGURE 2.5**
Pegasus the horse from Bayer's *Uranometria*. Pegasus, along with several other constellations, appears upside down on star maps.

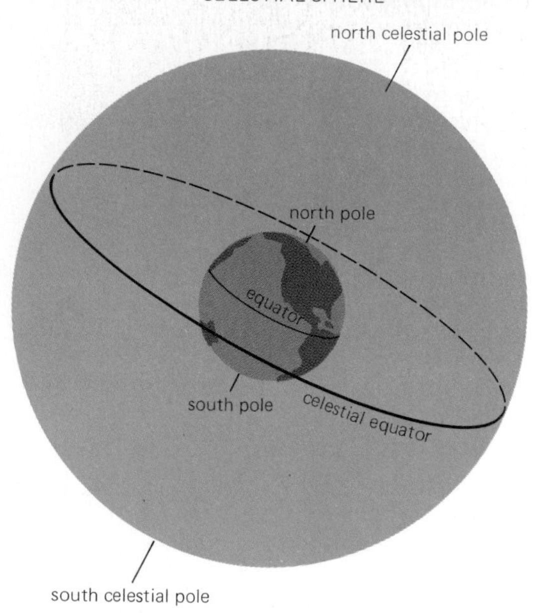


**FIGURE 2.6**
The celestial equator and poles are the projection of the earth's equator and poles to the imaginary celestial sphere.

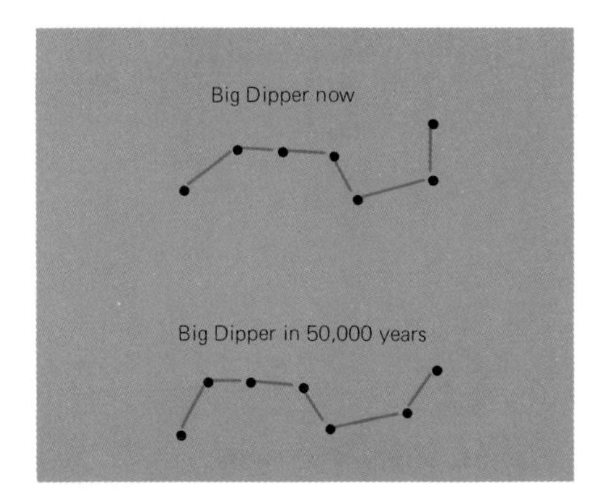


**FIGURE 2.7**
The stars of the Big Dipper as they appear from the earth now and 50,000 years from now.

at many different distances from us. The real positions of these objects in space have nothing to do with a sphere. It is only as a convenience that we still use the concept of the celestial sphere. We may assign to the celestial sphere a north pole, a south pole, and an equator. All these positions lie directly above the corresponding positions on the globe of the earth (Figure 2.6).

## Apparent Motion of the Celestial Sphere

The next step after picturing the constellations is to watch and see if they change. The space motions of individual stars, which might change the outlines of the constellations, are ordinarily not detectable for thousands of years. The stars are just too far away for their motions to be noticeable (Figure 2.7). After several hours of looking at the constellations, however, it becomes obvious that they move slowly in unison. The entire celestial sphere seems to turn around the earth. In most ancient cultures, this turning of the celestial sphere was believed to be real. When people finally accepted the fact that the earth could turn without our falling off, the correct explanation surfaced. The motion of the celestial sphere is only apparent and results from the earth spinning on its **axis** or **rotating** once each day. As the earth rotates from west to east, the stars in space, including the sun, seem to drift across the sky from east to west.

Picture yourself standing at the north pole of the earth. Look up, and you will see the north pole of the celestial sphere. The North Star, Polaris, marks the north celestial pole almost exactly. As you turn with the earth, the North Star remains overhead, but other stars seem to move in counterclockwise circles around the North Star. A time-exposure photograph shows the circling of the stars around the pole (Figure 2.8). Viewed from the north pole, the stars do not rise or set; they just trace circles parallel to the horizon (Figure 2.9).

From the equator of the earth, the motion of the celestial sphere looks different. The equator of the sky passes through the **zenith,** the point right above your head. All the stars seem to rise in the east as the earth turns east. The stars appear to drift westward in arcs parallel to the equator. Here all the stars rise and set, and only half their daily circle is visible (Figure 2.10).

Somewhere in between the north pole and the equator, where most of us live, the celestial sphere seems to be tilted in the sky (Figure 2.11). As the earth turns, the North Star still seems stationary. We see the entire counterclockwise circles of some stars in the northern sky, but we see only part of the daily circles of other stars. There is a region of stars near the south celestial pole that is never visible to us. Stars that never set in the northern region are called **circumpolar stars** (i.e., they circle the pole). Some famous circumpolar patterns for most United States locations are the Big Dipper, the Little Dipper, Cassiopeia, Cepheus, and Draco. These appear on Map 1.

## Seasonal Positions of the Constellations

As they circle Polaris, the circumpolar stars do not end up in the same position at the same time every night. To see this effect, turn Map 1 until October is at the top. That is the way the stars will appear at 9 P.M. Standard Time in October. Now advance the chart three months by putting January at the top. Notice that the Big Dipper will be higher in the sky in January at 9 P.M. than it was in October at the same time.

**FIGURE 2.8**
Star trails around the north pole on a long-exposure photograph. Stars that are never obscured by the horizon are called circumpolar stars. (Yerkes Observatory photograph)

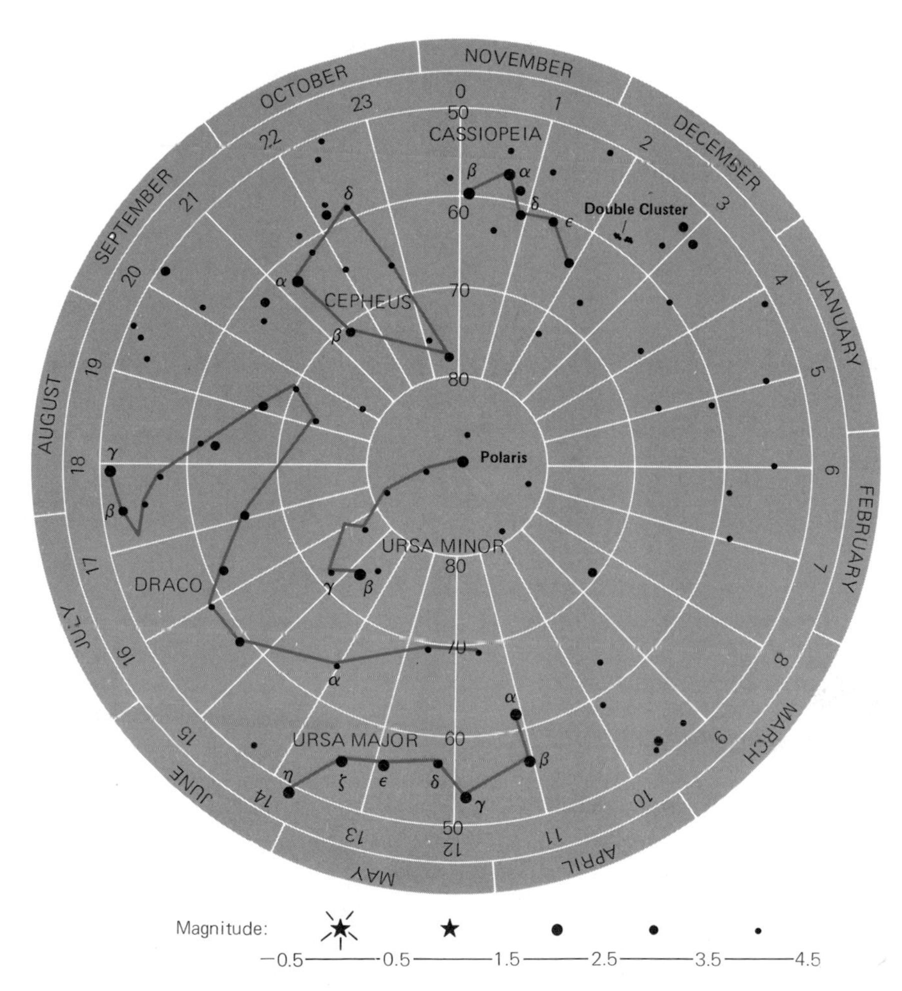


Map 1. The northern stars and constellations

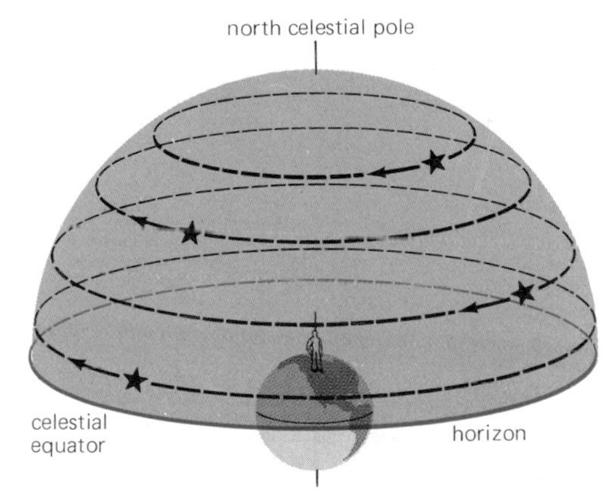


**FIGURE 2.9**
Star trails as observed from the north or south pole appear parallel to the equator and horizon.

**Relation between Hours and Degrees**

| | |
|---|---|
| $1^h$ | = 15° |
| $2^h$ | = 30° |
| $6^h$ | = 90° |
| $12^h$ | = 180° |
| $18^h$ | = 270° |
| $24^h$ | = 360° |

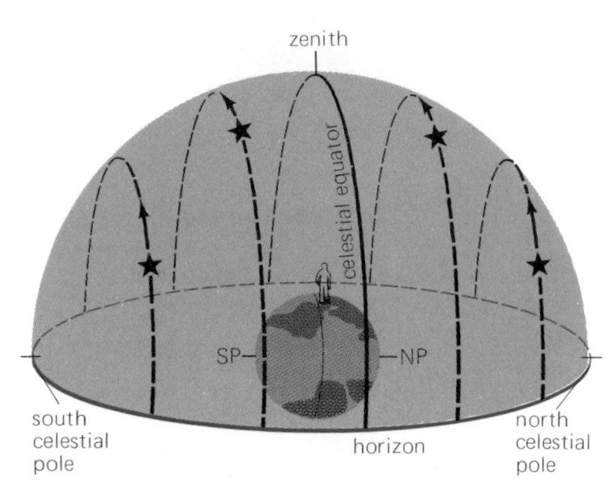


**FIGURE 2.10**
Star trails as observed from the equator appear parallel to the equator, but rise and set vertical to the horizon.

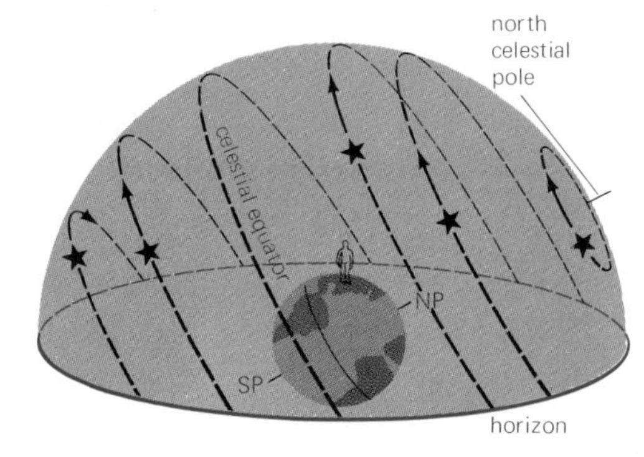


**FIGURE 2.11**
Star trails as observed from intermediate latitudes appear parallel to the equator, but rise and set obliquely to the horizon.

The reason for this change from season to season is another motion of the earth, **revolution.** The fact that the earth revolves around the sun once each year combines with the earth's rotation to make the celestial sphere appear to turn around once in only 23 hours 56 minutes. The 4-minute difference between 23 hours 56 minutes and our 24-hour day produces a change in the positions of the stars each night. In one month, 4 minutes per night adds up to a 2-hour shift in position. In terms of angles, a 2-hour shift is 30°. In six months, the shift would be 12 hours, or 180°. In one year, the shift would be 24 hours or 360°, and the cycle is complete. The position of the Big Dipper at 9 P.M. at the beginning of each of the four seasons is shown in Figure 2.12.

The part of the celestial sphere where stars rise and set for us is much larger than the circumpolar region. It takes four maps to show it. For convenience, Maps 2, 3, 4, and 5 each represent a different season. The same 4-minute per night shift in position is seen in these stars. As the earth revolves around the sun, the stars rise 4 minutes earlier each night. The net effect is to give us different constellations in the night sky at different seasons (Figure 2.13). The part of the celestial sphere directly behind the sun is blocked from our view by the bright sunlight. As the earth revolves around the sun, different parts of the celestial sphere are blocked out. We do not see Orion on a summer night because it is behind the sun then. We do not see Boötes in the winter for the same reason. The constellations that appear in the different seasons are shown in Maps 2, 3, 4, and 5. Map 6 shows the southern circumpolar stars that never rise for us living in middle northern latitudes. Notice that there is no bright star to mark the south pole of the sky the way Polaris marks the north celestial pole.

## The Zodiac

The band of constellations that lie directly behind the earth's orbit are the **zodiacal constellations** [Figure 2.14 (a)]. One of these twelve constellations is always directly behind the sun's position in space. The expression "the sun is in Virgo" means the sun is in front of the stars that form Virgo. The sun is "in" each zodiacal constellation for about one month each year. Two thousand years ago, at the beginning of spring, the sun was in Aries. At the beginning of spring today, March 21, the sun is in the constellation Pisces and will not enter Aries until the end of April. The reason for this change is a very slow wobbling motion of the earth's axis, called **precession.**

The axis of the earth is tilted with respect to the plane of its orbit. The earth's orbit plane intersects the celestial sphere in a circle called the **ecliptic.** The ecliptic runs right through the center of the zodiac and may be thought of as the apparent yearly path of the sun in the sky.

Map 2. The spring stars and constellations

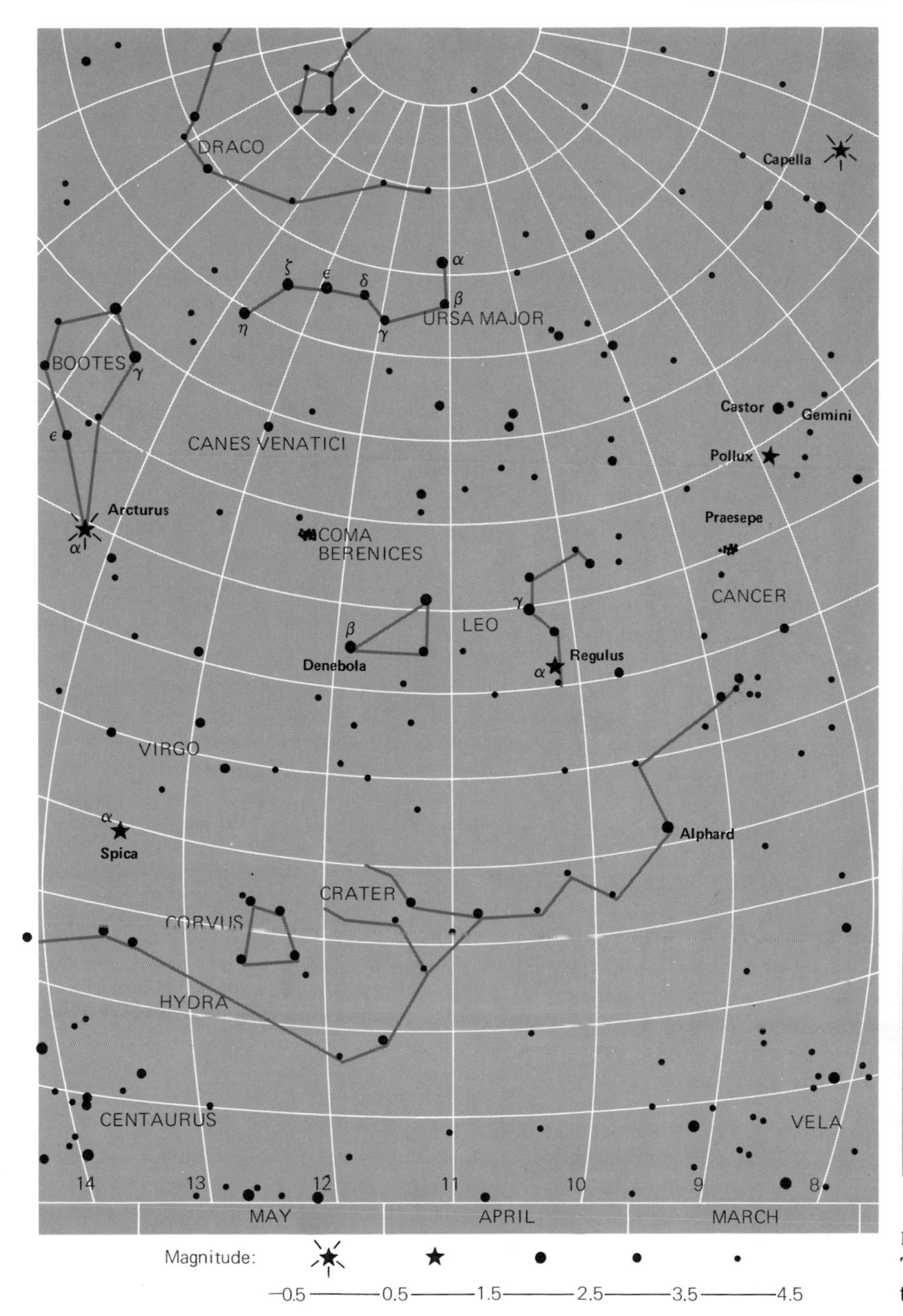


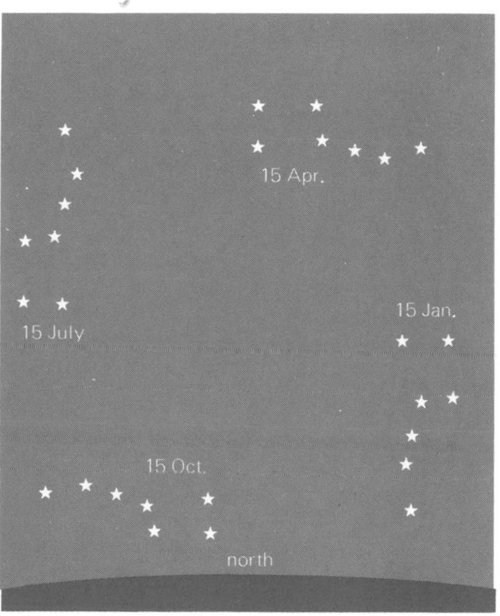


**FIGURE 2.12**
The Big Dipper as observed at 9 o'clock in the evening at different seasons.

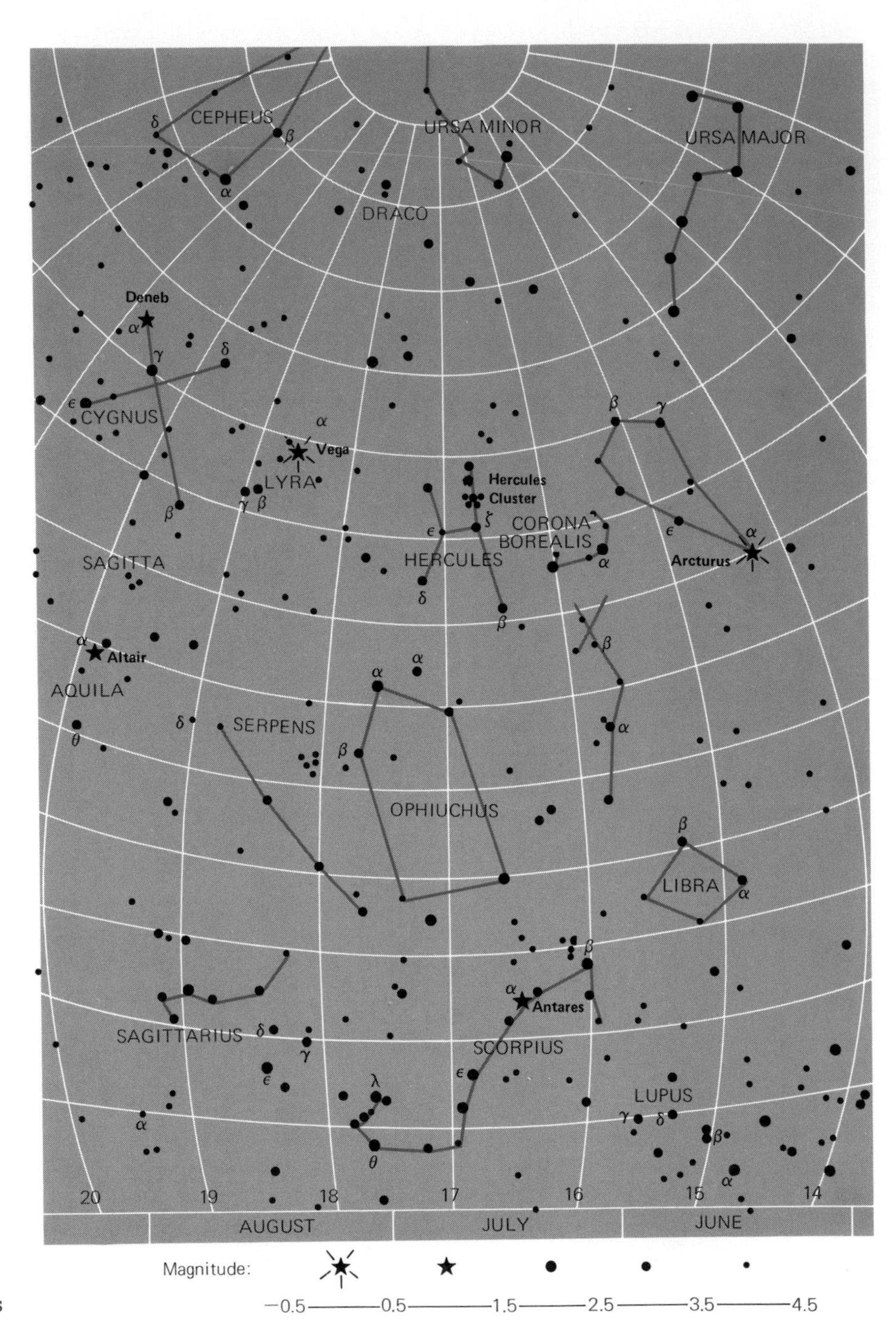


Map 3. The summer stars and constellations

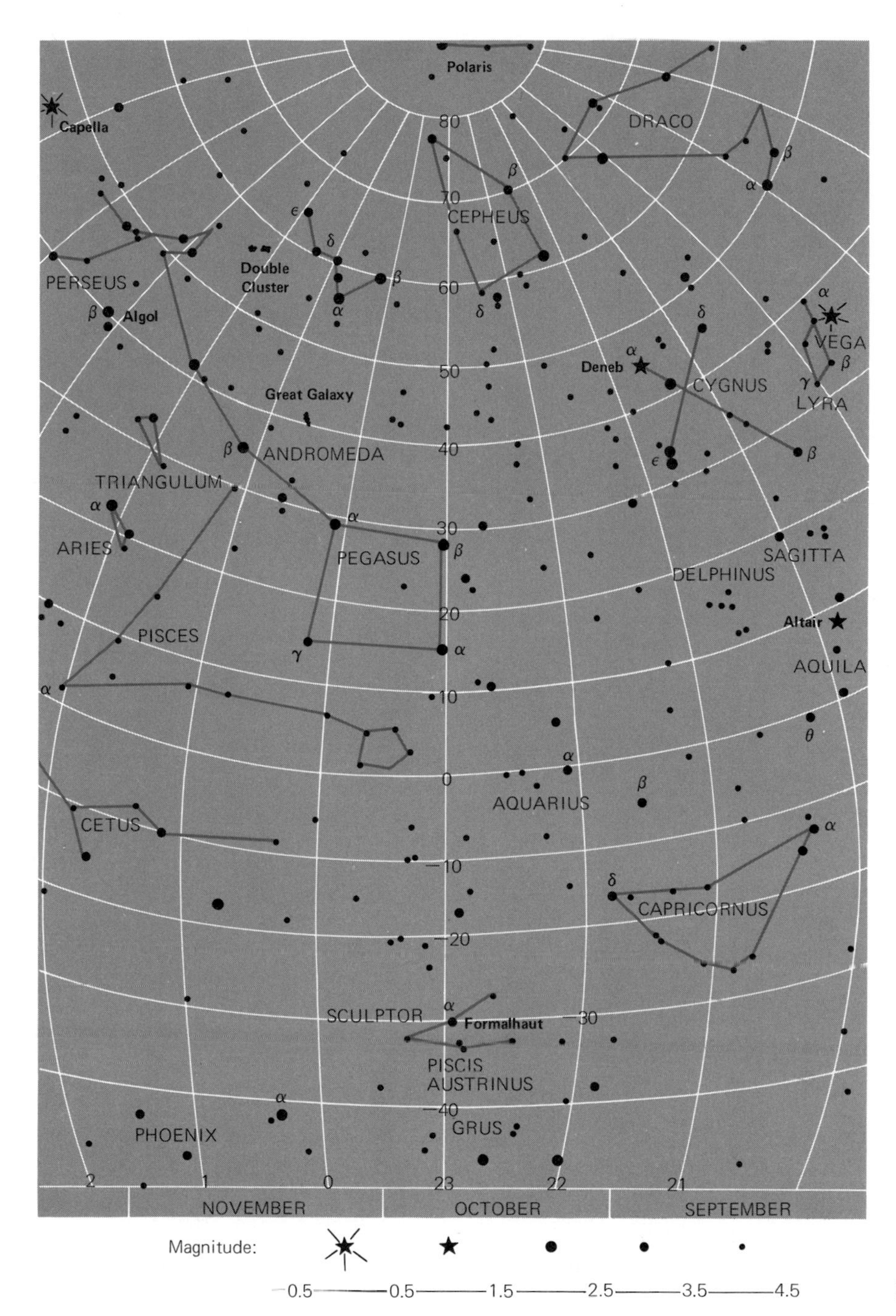


Map 4. The fall stars and constellations

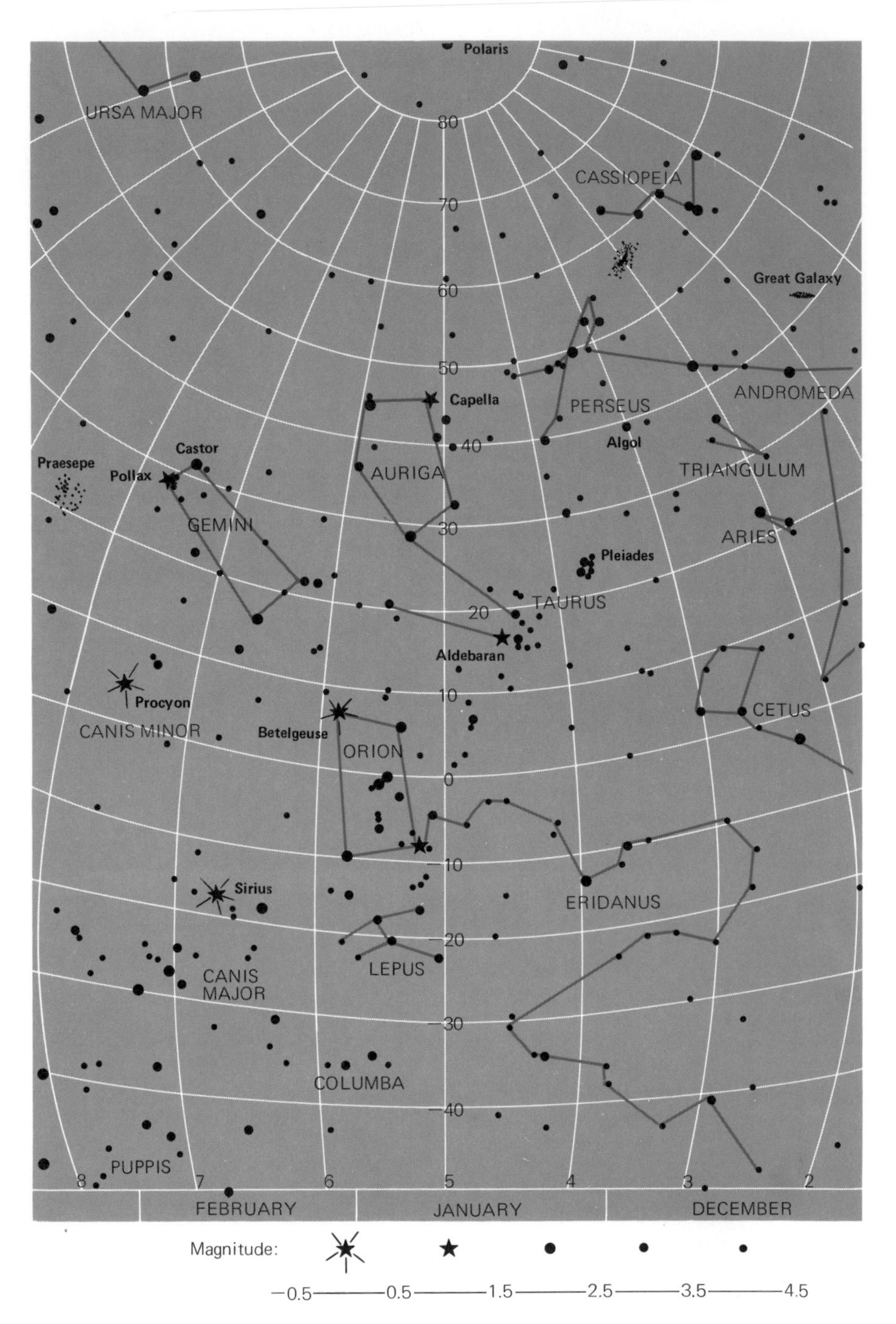


Map 5. The winter stars and constellations

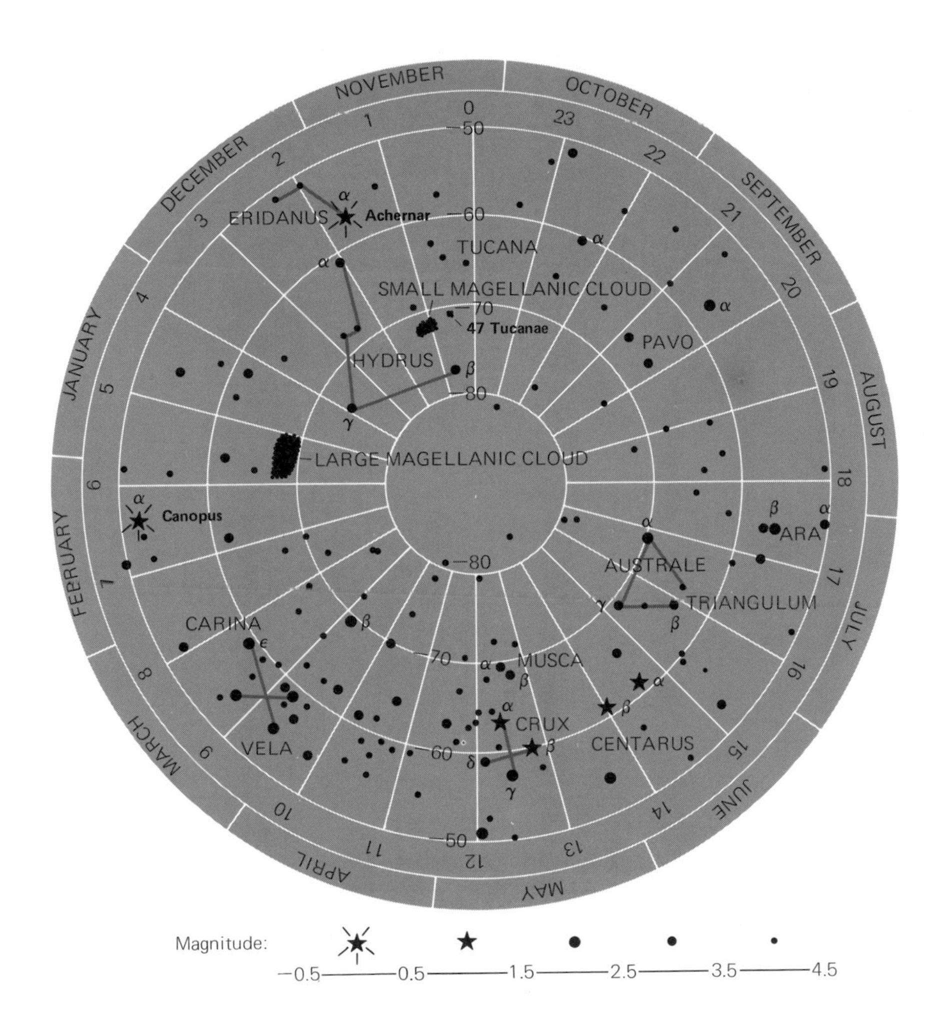


Map 6. The southern stars and constellations

Figure 2.14(b) shows that the ecliptic crosses the celestial equator in two places: the **spring equinox** and the **fall equinox.** About half way in between these points are the **summer solstice** and **winter solstice** positions of the sun. As the earth spins, the direction of its axis in space slowly precesses causing the equinox points to move. They move all the way around the ecliptic, through the zodiacal constellations, in about 26,000 years.

The motion of the equinoxes has caused some difficulty for astrology. In **astrology,** the **signs of the zodiac** are fixed to the spring equinox. The signs of the zodiac are 12 equal divisions, each 30° long, marked off eastward beginning at the spring equinox. Because of the precession described above, the signs are now far from the actual

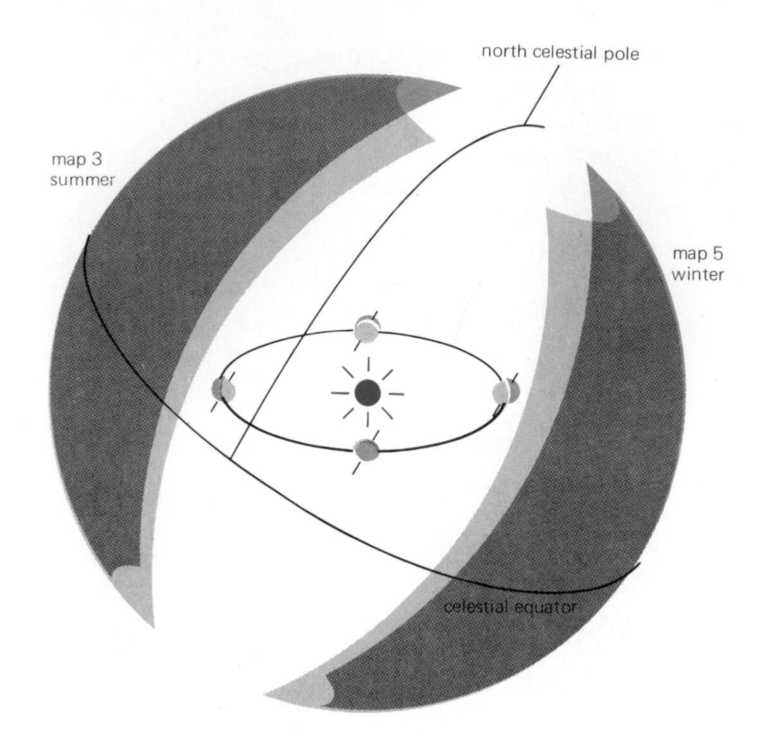


**FIGURE 2.13**
In the summer the sun hides the stars beyond it, and at night we see the summer sky. In the winter, the opposite is the case.

constellations they were named for nearly two thousand years ago. This, along with the arbitrariness in choosing the moment to assign an individual a sign, should cause one to pause before putting much faith in astrology. At your moment of birth, the sun was located in one of the signs of the zodiac. According to astrologers, the position of the sun at your birth determines your sign and your destiny. The real moment of destiny is the moment of conception, but that would be a rather difficult moment to assign.

Beyond describing the positions of the sun, moon, and planets in the celestial sphere, astrology and astronomy have almost no common ground. Astronomy employs deduction and the scientific method to reach conclusions about the Universe; astrology employs interpretation and personal beliefs to relate celestial events to everyday occurrences in the lives of individuals.

## The Days of the Week

The **constellations of the zodiac** were special not only because the sun appeared in front of one of them each month, but because they contained a total of seven objects that moved: the sun, the moon, and the five visible **planets** (also called wandering stars by the Greeks). All seven objects were always found in front of the zodiacal constellations. The number of days in a week and the names of the days go back to these seven ancient wanderers. The Babylonians regarded them as deities. Each deity ruled certain hours of the day.

If you list the seven "wanderers" in order of the amount of their daily motion, Saturn (which moves least) will be first, then Jupiter, Mars, the sun, Venus, Mercury, and finally the moon. If you repeat this list over and over again (24 times), you will end up with a table similar

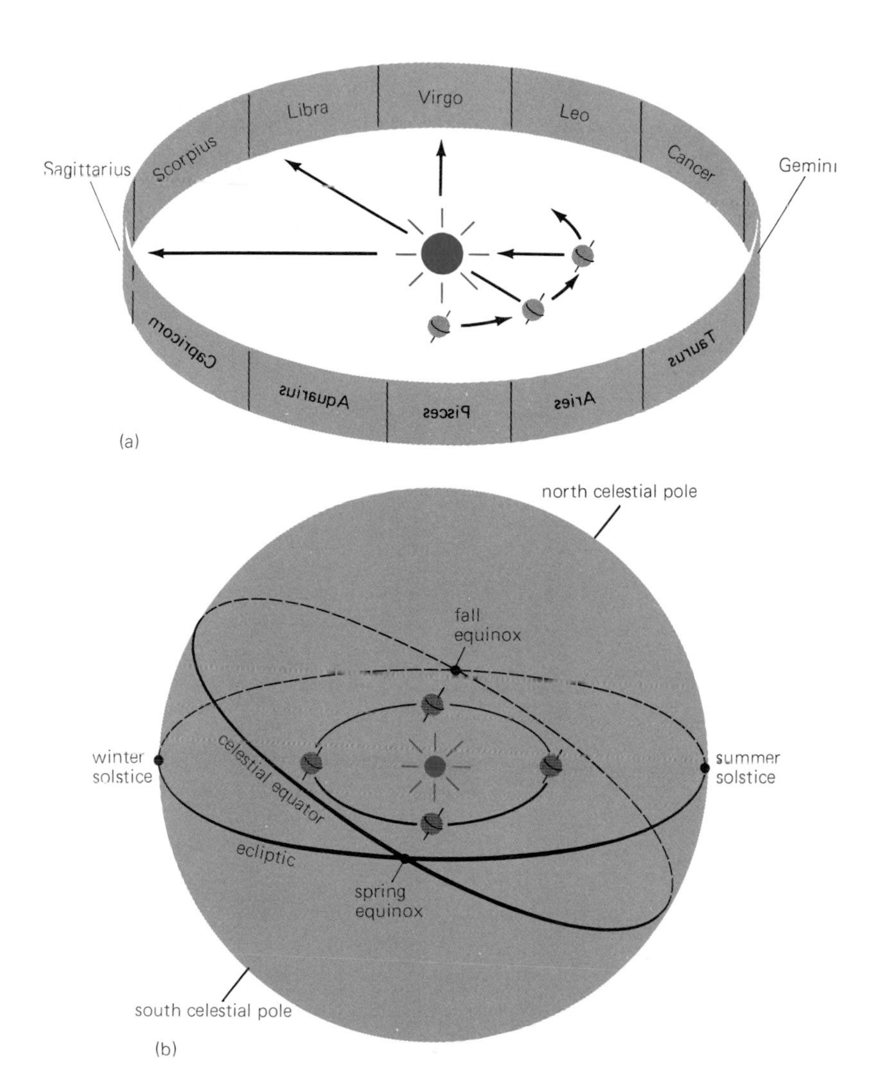


**FIGURE 2.14**
(a) The band of 12 constellations called the Zodiac. (b) The celestial sphere showing the relationship of the ecliptic to the celestial equator.

to Table 2.2. Saturn ruled the first hour of the first day, Jupiter rules the second hour of the first day, and so on. Each day was named for the deity that rules the first hour of that day. In English, Saturday, Sunday, and Monday correspond well with Saturn's day, Sun's day, and Moon's day. The other days of the week retain their associations better in other languages. In Italian, Tuesday or Mars' day is Martedi, Wednesday or Mercury's day is Mercoledi, Thursday or Jupiter's day is Giovedi (Jove), and Friday or Venus' Day is Venerdi.

## THE PLANETS

### Motions

The apparent daily motions of the stars are fairly simple and easy to describe using the concept of the celestial sphere. They rise in the east and set in the west. The daily motions of the sun and moon are

**TABLE 2.2 Days of the Week**

| Hour | *First Day* | *Second Day* | *Third Day* | *Fourth Day* | *Fifth Day* | *Sixth Day* | *Seventh Day* |
|---|---|---|---|---|---|---|---|
| 1st | Saturn | Sun | Moon | Mars | Mercury | Jupiter | Venus |
| 2nd | Jupiter | Venus | Saturn | Sun | Moon | Mars | Mercury |
| 3rd | Mars | Mercury | Jupiter | Venus | Saturn | Sun | Moon |
| 4th | Sun | Moon | Mars | Mercury | Jupiter | Venus | Saturn |
| 5th | Venus | Saturn | Sun | Moon | Mars | Mercury | Jupiter |
| 6th | Mercury | Jupiter | Venus | Saturn | Sun | Moon | Mars |
| 7th | Moon | Mars | Mercury | Jupiter | Venus | Saturn | Sun |
| 8th | Saturn | Sun | Moon | Mars | Mercury | Jupiter | Venus |
| 9th | Jupiter | Venus | Saturn | Sun | Moon | Mars | Mercury |
| 10th | Mars | Mercury | · | · | · | · | · |
| 11th | Sun | Moon | · | · | · | · | · |
| 12th | Venus | Saturn | · | · | · | · | · |
| 13th | Mercury | Jupiter | · | · | · | · | · |
| 14th | Moon | Mars | · | · | · | · | · |
| 15th | Saturn | Sun | · | · | · | · | · |
| 16th | Jupiter | Venus | · | · | · | · | · |
| 17th | Mars | Mercury | · | · | · | · | · |
| 18th | Sun | Moon | · | · | · | · | · |
| 19th | Venus | Saturn | · | · | · | · | · |
| 20th | Mercury | Jupiter | · | · | · | · | · |
| 21st | Moon | Mars | · | · | · | · | · |
| 22nd | Saturn | Sun | · | · | · | · | · |
| 23rd | Jupiter | Venus | · | · | · | · | · |
| 24th | Mars | Mercury | · | · | · | · | · |

more complicated, but regular enough to be easy to describe. They also rise in the east and set in the west, but their intervals are different from those of the stars and from each other's. The five wandering stars or planets are quite another matter. Their motions appear very complex, and explaining their wanderings was the main preoccupation of astronomers for centuries. The motions of the planets, in fact, were not understood until well into the 16th century.

The great complication in describing the motions of the planets comes from our position on a constantly moving earth. An observer must separate the earth's motions from the planets' motions. Ancient observers and astronomers assumed the earth was stationary and at the center of things. Over several weeks or months, the planets trace out looping paths through the zodiacal constellations (Figure 2.15). All this motion was assumed to belong to the planets themselves.

In a planetarium, where motions of the planets can be demonstrated, we can make time-lapse photographs. The photograph in Figure 2.16 shows the annual path of Mars in 1979–80 as seen from the earth. When a planet seems to back up or loop westward in its path, it is said to be **retrograding.** Elaborate schemes were developed to account for this retrograde motion.

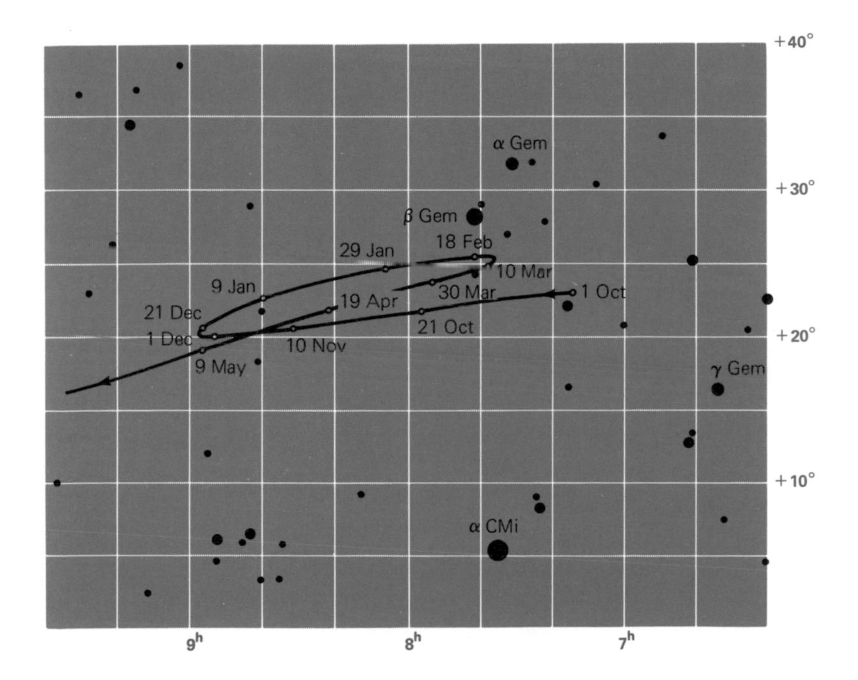


**FIGURE 2.15**
Predicted path of Mars from October 1977 through May 1978.

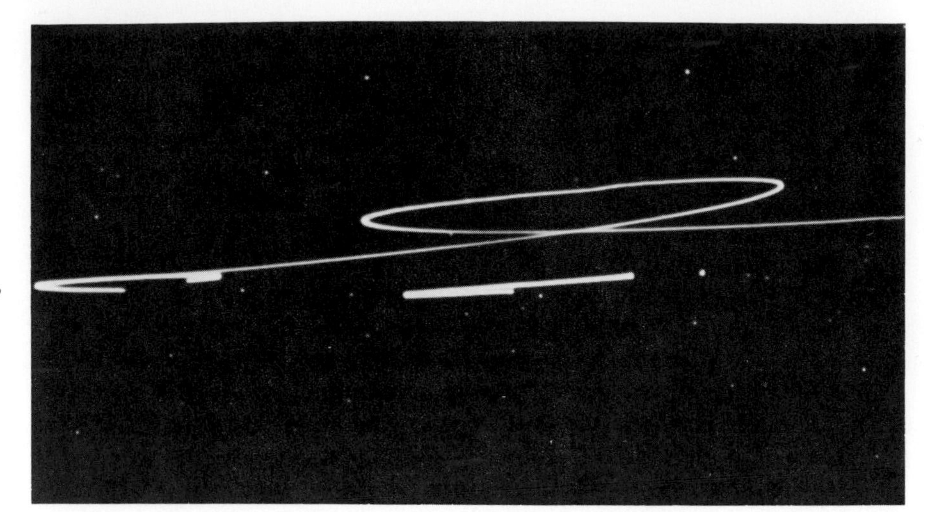

**FIGURE 2.16**
During the winter and spring of 1980, Mars, Jupiter, and Saturn will be retrograding in the constellation of Leo. The large loop is Mars's path through the sky. Jupiter's path is below that of Mars. Saturn's path is farther to the east (left). Regulus is the brightest star in this photo. (Photograph courtesy of Jack Gross)

## Ptolemy's System

The most enduring early plan for describing the planetary motions was proposed by Ptolemy in the 2nd century A.D. Ptolemy was the last of the important Greek astronomers, and he worked at the school of Alexandria in present-day Egypt. The Ptolemaic system is a geometric solution using epicycles for planet motions. In its simplest form, each planet moved uniformly in a circle, the **epicycle,** while the center of this circle was in uniform motion on a circle, the **deferent,** around the earth (Figure 2.17). A circle has the greatest symmetry and was

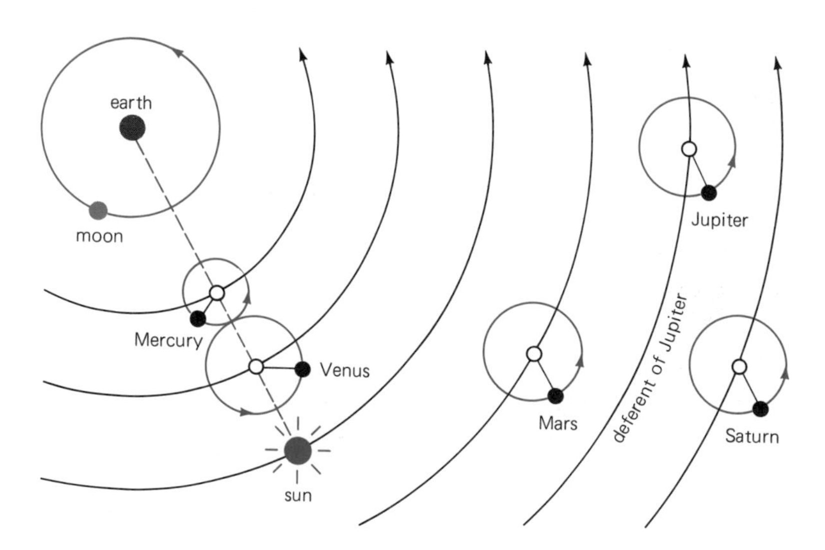


**FIGURE 2.17**
The Ptolemaic system for explaining planetary motions. Note that Mercury and Venus are aligned with the sun and that Mercury is closer to the earth than is Venus.

assumed to be the perfect orbit for a planet. The Greeks therefore used only circles or combinations of circles in their planetary orbits.

Basically, Ptolemy had done what all the Greek philosophers aimed to do. He described the observations and was able to predict future positions of the planets. The Greeks were not concerned with describing reality; that was left for the gods.

## The Way It Is

Today, we can state a few simple rules that describe the real motions of the planets.

*Rule 1.* The planets move in **elliptical orbits** around the sun. The sun is not at the center, but rather at a point called the **focus of the ellipse** (Figure 2.18). Ellipses are flattened circles varying in shape from slightly flat to cigar-shaped. The planet orbits are very nearly circles. In the figures representing planet orbits in this book, the flatness of the ellipses is exaggerated.

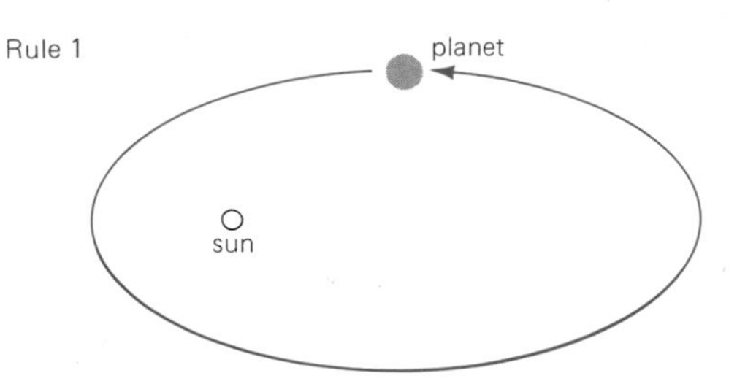


**FIGURE 2.18**
A planet orbits the sun on an ellipse with the sun at one focus. This illustrates Rule 1.

*Rule 2.* The planets move at variable speeds around their orbits. They travel fastest when they are nearest the sun. This point of closest approach is called the **perihelion** of the orbit. They travel slowest when they are farthest from the sun, at a pointed called the **aphelion** (Figure 2.19). Sometimes this is called the **rule of equal areas,** because a line joining the sun to a given planet would sweep out equal areas in equal amounts of time for that planet.

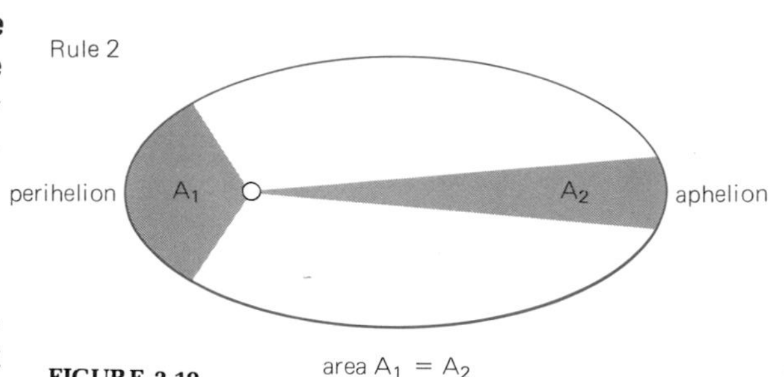


**FIGURE 2.19**
The areas swept out by a line joining the planet to the sun are equal for equal intervals of time. This illustrates Rule 2.

*Rule 3.* The length of a planet's year (the time it takes to orbit the sun once) is related to the distance of the planet from the sun. The closer the planet is to the sun, the shorter its year is. For example, a year on Mercury is only 88 of our days. The Jupiter year is nearly 12 of our years. Pluto's year is over 247 of our years. A more explicit statement of this rule appears in Chapter 3; a table of planet periods appears in the margin and in the appendices.

**TABLE 2.3 Planetary Periods**

| *Planet* | *Orbit Period* |
|---|---|
| | days |
| Mercury | 87.969 |
| Venus | 224.701 |
| Earth | 365.256 |
| Mars | 686.980 |
| | years |
| Jupiter | 11.862 |
| Saturn | 29.458 |
| Uranus | 84.013 |
| Neptune | 164.794 |
| Pluto | 247.686 |

These three rules are called **Kepler's laws of planetary motion.** They were first formulated by Johannes Kepler, a 17th-century astronomer. Kepler used the careful and extensive records kept by the Danish astronomer Tycho Brahe (1546–1601) to develop these laws of motion. Tycho's many observations of the position of Mars over a twenty-year period were particularly valuable to Kepler. In Chapter 3 we will examine Kepler's technique to determine the orbit of Mars.

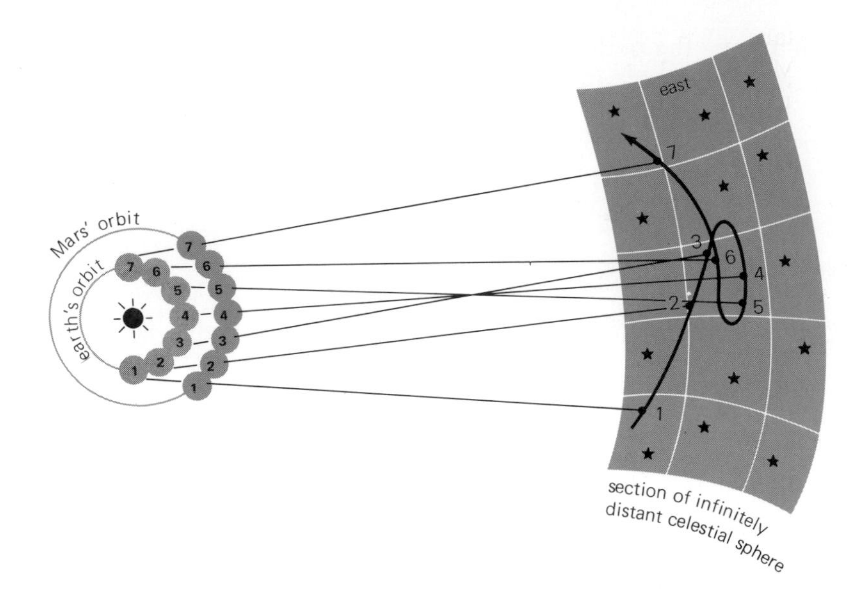


**FIGURE 2.20**
The retrograde motion of Mars is explained by the faster-moving earth overtaking and passing Mars.

Explaining retrograde motion is just a question of taking the right point of view. We view the planets from a moving platform, the earth. When we look at Mars, for example, it is also moving. According to rule 3 above, Mars takes longer to orbit the sun since it is farther from the sun than we are. Every two earth years, the earth passes Mars in space. When this happens, Mars seems to retrograde or back up from our point of view. In Figure 2.20, as the earth moves from position 1 to 2 to 3, and so on, Mars is also moving from its position 1 to 2 to 3, and so on. As the earth passes Mars, the line of sight shifts, causing Mars to retrograde compared with the background stars.

## PREDICTIONS

### Lunar Calendars

The study of the sky has been prompted more than anything else by the need to make accurate predictions of celestial events. Astrologers in particular pressed for more and more accurate past, present, and future positions of the sun, moon, and planets. And as soon as primitive people turned from a nomadic existence to an agricultural one, predictions became critical. Knowing when to plant and when to harvest could mean the difference between life and death. So calendars were developed to record past events and predict future ones. Most calendars are based on the moon, the sun, or a combination of both.

**Moon** or **lunar calendars** are simple to set up and were probably the earliest kind used. Each month begins with the first appearance of the month of the crescent moon after sunset. The main trouble with the lunar calendar is that its 12-month year is about 11 days shorter than the year of the seasons. An error of 11 days per year is too great for an agricultural society. For practical purposes, then, the sun began to be used to calculate the seasons of the year.

But the moon did not disappear completely from calendars. The complicated **luni-solar combination** calendar of the Hebrews for example, is still used today for religious purposes. The computation of the date of Easter is also based on the moon and the sun. Easter is the first Sunday after the fourteenth day of the ecclesiastical moon (nearly the full moon) that occurs on or immediately after the vernal equinox. Thus if the fourteenth day of the moon occurs on Sunday, Easter is observed one week later. Unlike Christmas, Easter is a movable feast because it depends on the moon's phases. Its dates can range from 22 March to 25 April (Table 2.4); in the year 2000, it will fall on 23 April.

**TABLE 2.4 Dates of Easter Sunday**

| Year | Date | Year | Date |
|---|---|---|---|
| 1978 | 26 Mar | 1989 | 26 Mar |
| 1979 | 15 Apr | 1990 | 15 Apr |
| 1980 | 6 Apr | 1991 | 31 Mar |
| 1981 | 19 Apr | 1992 | 19 Apr |
| 1982 | 11 Apr | 1993 | 11 Apr |
| 1983 | 3 Apr | 1994 | 3 Apr |
| 1984 | 22 Apr | 1995 | 16 Apr |
| 1985 | 7 Apr | 1996 | 7 Apr |
| 1986 | 30 Mar | 1997 | 30 Mar |
| 1987 | 19 Apr | 1998 | 12 Apr |
| 1988 | 3 Apr | 1999 | 4 Apr |

## Solar Calendars

The ancient Egyptians had determined that the year was nearly 365.25 days long. As early as 4236 B.C., they had a 360-day solar calendar of 12 months of 30 days each. They added 5 feast days to the end of the year to keep the calendar in step with the seasons.

When Julius Caesar became ruler of Rome, a luni-solar calendar was in use by the Romans. The calendar was so complicated and poorly managed that Caesar set about correcting it. With the advice of the astronomer Sosigenes of Alexandria, Caesar had the change made. The year 46 B.C. was made 445 days long to correct errors. Small wonder it was called "the year of confusion." The new **Julian calendar** began on 1 January 45 B.C. It had 3 years each of 365 days followed by a fourth year having 366 days. This is the familiar system of the **leap year.** Unfortunately, the year of the seasons is not exactly 365.25 days long. It is really 11 minutes 14 seconds shorter. That does not seem like much, but by A.D. 1582, the calendar was in error by 10 days.

Pope Gregory XIII then ordered the Julian calendar corrected. This was done by dropping 10 days from the calendar. The date 4 October 1582 was followed by 15 October. England and its colonies, including America, did not make the change until 1752, when 11 days had to be dropped. The date 2 September 1752 was followed by 14 September.

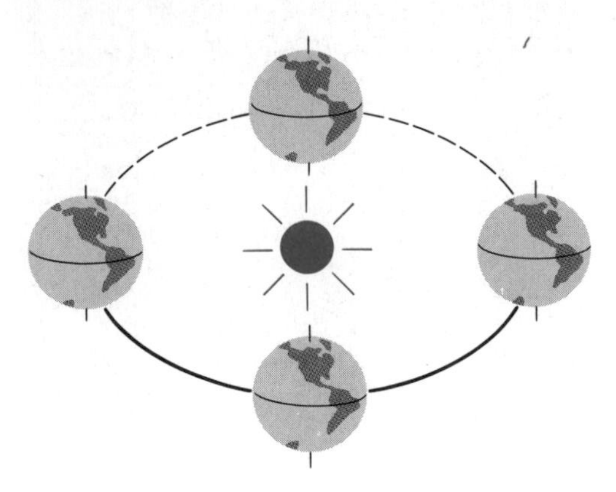

FIGURE 2.21
If the axis of rotation of the earth were perpendicular to its orbit, there would be no seasons and day and night would each last 12 hours.

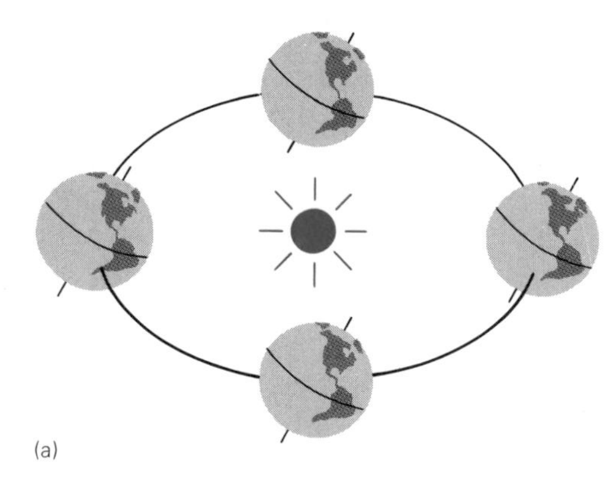


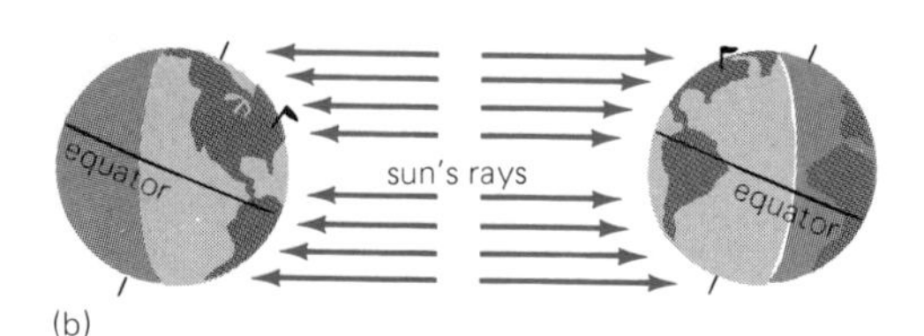


FIGURE 2.22
The earth's tilt causes the sun to warm the northern and southern hemispheres alternately. More energy falls on the northern hemisphere when it is tilted toward the sun than when it is tilted away from the sun.

Riots broke out at the time; people thought they had been cheated out of 11 days' wages.

To make sure the calendar did not get out of step again, Pope Gregory proposed that century years (1800, 1900, 2000, 2100, and so on) be leap years only if they were evenly divisible by 400. The year 1900 was not a leap year. The **Gregorian calendar** as described above is now in use at least for civil purposes in most nations. It will not get out of step with the seasons by even a day for at least three thousand more years.

## A Reason for Seasons

When a planet is at **perihelion** in its orbit, it is closest to the sun for its year (Figure 2.19). At **aphelion** it is farthest away from the sun. The changing distance to the sun has little, if any, effect on the seasons. The earth, for example, is closest to the sun in January and farthest away in July. The reason for our seasons is that the **axis of the earth** is not perpendicular to the direction of the sun. The axis is an imaginary line through the north and south poles of the earth. If the axis were at right angles to the sun, the seasonal changes would be negligible and it would be light for 12 hours of every day (Figure 2.21).

In reality, the axis of the earth is tipped 23 1/2° from the vertical to the plane of the earth's orbit. This produces four distinct seasons (Figure 2.22). The earth's north pole always points toward the North Star in space. But during the year, the tilt of the north pole is alternately in the direction toward and away from the sun. When the north pole leans toward the sun, it is summer in the northern hemisphere. At that time, the sun appears high in our sky; we get more direct sunlight; and the days are longer. Six months later, the north pole leans away from the sun. It is then winter; the sun is low in the sky; we get less direct sunlight; and the days are shorter. During spring and fall, the axis is not leaning either toward or away from the sun. Everything we have said about the northern hemisphere also applies in the southern hemisphere, only there the times of the seasons are reversed.

## Eclipses

Predictions of eclipses of the sun and the moon have been made far back into history. It is said that in 2159 B.C. Hsi and Ho, two Chinese astronomers, were executed for failing to predict an eclipse. Whether or not this is a true story, it does point to the high value placed on predictions of this sort. We know Chinese astronomers were predicting eclipses by A.D. 25. At that time, the Astronomical Bureau of China was an important government office. The bureau's astronomers not only

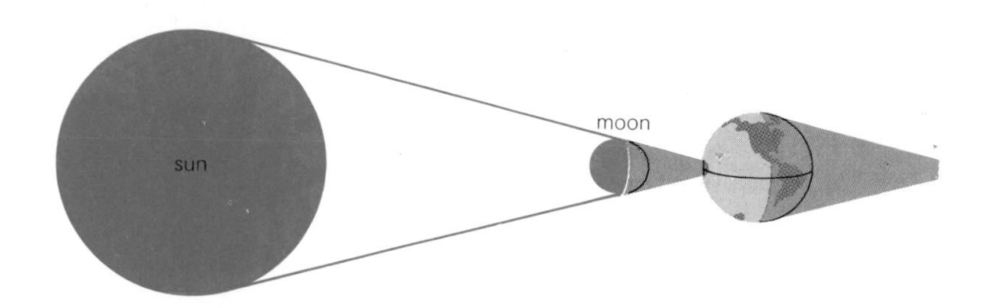


**FIGURE 2.23**
The condition for a solar eclipse occurs when the moon is between the earth and sun. In the drawing the sizes and distances are not to scale.

predicted eclipses, but also interpreted them as good or bad omens for the kingdom.

The sun is about 400 times larger and 400 times farther away from us than the moon is. These two facts combine to make the sun and moon appear to be the same size when viewed from the earth. When the earth, moon, and sun line up in space, the moon can completely block the sun from our view. This event is called a **solar eclipse** (Figure 2.23). At other times, the earth, moon, and sun may be lined up so that the moon passes through the shadow of the earth. This configuration is called a **lunar eclipse** (Figure 2.24). During any one year, seven eclipses are possible, five of the sun and two of the moon, four of the sun and three of the moon, or three of the sun and four of the moon. There must be at least four eclipses, two solar and two lunar. When three celestial bodies are lined up in space, such as during a solar or lunar eclipse, the configuration is called **syzygy.**

Predicting the times of eclipses requires some knowledge of the positions of the sun and moon. After sufficient experience we learn that

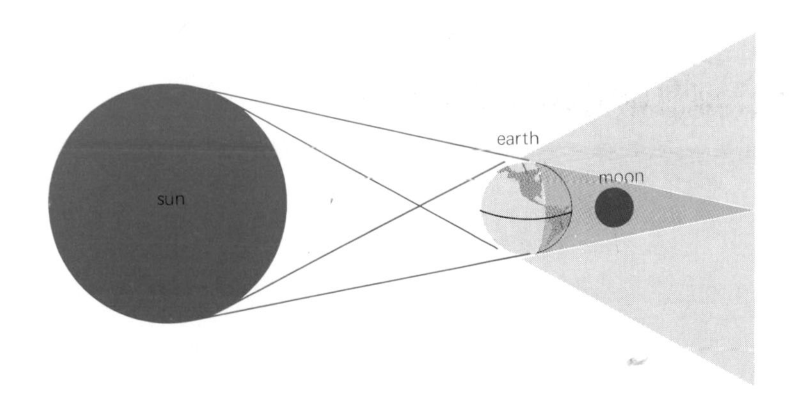


**FIGURE 2.24**
When the earth is between the sun and moon, a lunar eclipse can occur. Often the moon is dimly visible during the eclipse because the earth's atmosphere refracts sunlight into the normally dark shadow. The figures and distances are not to scale.

**FIGURE 2.25**
A ground view of several of the principal stones at Stonehenge. (Photograph by G. Hawkins)

eclipse conditions repeat themselves at certain intervals. It takes many generations of observation to deduce the necessary intervals. Once this is done, it is possible to construct a device that will predict an eclipse. Such a device can also indicate when the sun crosses the spring equinox or the summer solstice.

The great ring of stones in England called Stonehenge (Figure 2.25) is one of the most recent testimonies to the knowledge and sophistication of ancient astronomers. Stonehenge was used to predict the beginning of each season and probably eclipses as well. The building of Stonehenge was begun about 3500 years ago. It took place in three phases over a period of 400 years. There was obviously a plan. Each significant stone lines up with another to point to some extreme position of the sun and moon. A set of holes around the outside probably helped predict and keep track of eclipses (Figure 2.26). Stonehenge was a calendar, first, and perhaps a place of worship also. To be able to predict special risings and settings of the sun as well as eclipses must have accorded someone devine power and a permanent high place in that early culture. Stonehenge was completed a few hundred years before the fall of Troy, but one thousand years after the great pyramids of Egypt.

The pyramids were not used to predict eclipses, but they have long interested astronomers for other reasons. The pyramids bear witness to the fine engineering and orientation abilities of the early Egyptians. All the large pyramids except the first are oriented within a few degrees of true north, south, east, and west. The principal descending corridor which opens on the north face of the Great Pyramid must have been engineered to point to the north pole of the sky. Other alignments of passages to stars have been suggested.

## COSMOLOGY

### Old Ideas

The world view or **cosmology** of ancient people was governed by legend and mythology. Astronomers were usually priests in charge of timing religious occasions and keeping the calendar. Their fanciful views of the world are all egocentric.

The Egyptian universe, for example, was considered to be a large rectangular box with Egypt at the center. The dome of the heavens was supported at the north, south, east, and west points by mountain peaks.

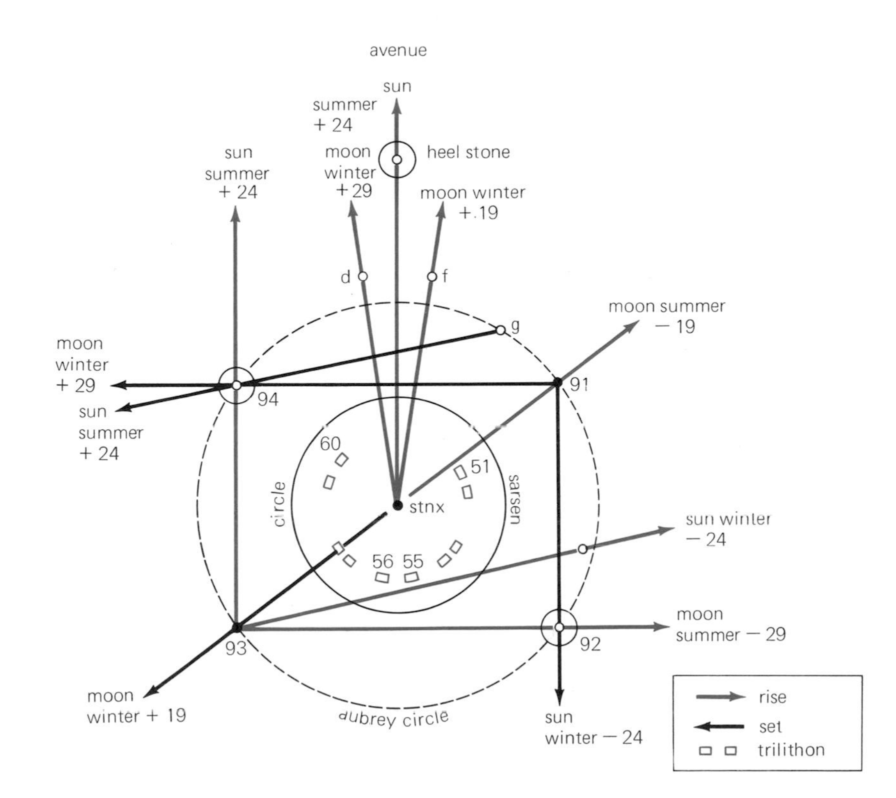


**FIGURE 2.26**
The ground plan of Stonehenge as annotated by G. S. Hawkins. Numbers with + and − symbols are the declinations of the sun and moon in different seasons. The other numbers and letters identify key stones or holes. Stnx is the intersection of the diagonals of the station stones, 91, 92, 93, and 94.

**FIGURE 2.27**
Ra, the Egyptian sun-god, travels along the celestial river to the upper left in this early Egyptian view of the Universe. (Yerkes Observatory photograph)

Around the base of the mountains was the great celestial river whose principal branch was the Nile. The sun god, Ra, was carried in the solar boat westward across the sky once each day. At night, Ra continued his journey in the celestial river along the horizon, going from west to north to east by the next morning (Figure 2.27). The ancient Hindus believed the universe was encircled by a giant cobra. The earth was supported at the north, south, east, and west points by four elephants, who in turn stood on the back of a giant tortoise floating on a sea of milk (Figure 2.28).

The Homeric poems present a good picture of early Greek cosmology. The earth is a flat, circular disk surrounded by the river Okeanos. The heavens hang like a huge bell over all of this. Between setting and rising, it is assumed that the heavenly bodies do not go under the earth. Homer mentions Okeanos as the origin of everything. This theme of water being the origin of everything is also seen in the Babylonian and Egyptian world views.

## The Shape of the Earth

The Babylonians assigned no particular shape to the earth. By the time we get to the Greeks, the earth is described as flat and circular. But then mysticism and legend yielded to rational explanations of our world. The Greeks began to study celestial events in a scientific way. The first step toward a new world view is a spherical earth. Pythagoras (c. 530 B.C.) reasoned that the earth was a globe. By the time of Plato (427–347 B.C.), the spherical view was generally accepted. Three reasons were cited most often to justify it. (1) Ships' masts disappeared last on the horizon. (2) Traveling north and south on the surface of the earth caused the positions of the constellations to shift slightly south or north. (3) The shadow of the earth during an eclipse of the moon is always round regardless of the season.

After accepting the idea of a spherical earth, the next great challenge is to measure its size. The best size measurement of antiquity was made by Eratosthenes about 230 B.C. Using a purely geometric

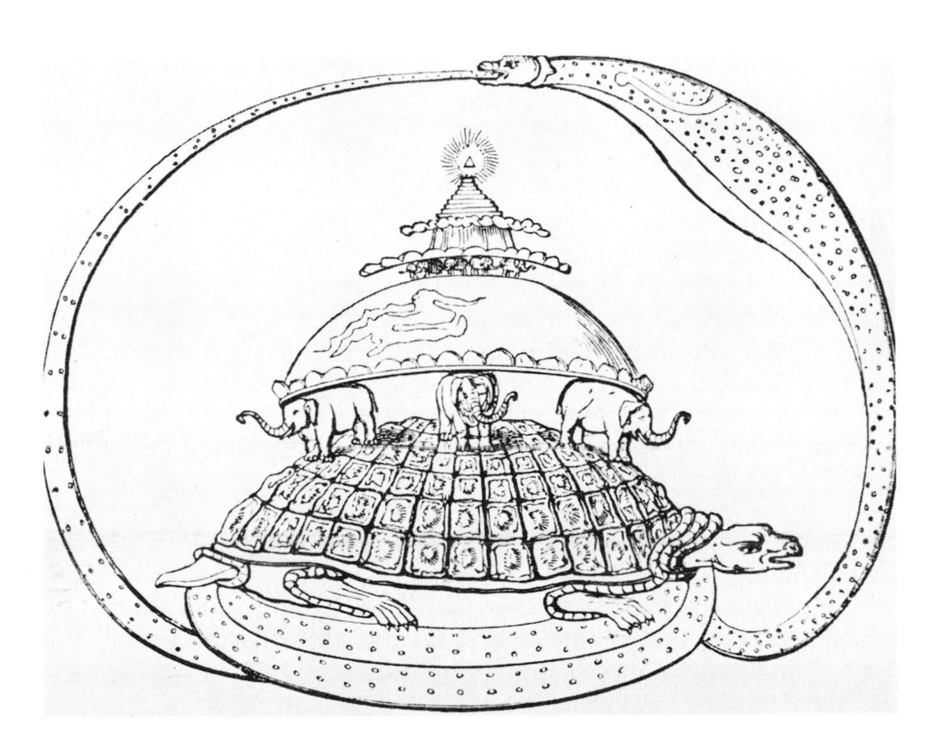

**FIGURE 2.28**
An early Hindu Universe was enclosed by a cobra. The terrapin formed a base for the elephants which supported the earth at the north, south, east, and west points. (The Bettmann Archive, Inc.)

method, he calculated the diameter of the earth to be 12,600 kilometers (km), only 100 km shorter than the modern calculation.

Today, space-age photographs show us that the earth is indeed spherical. More precise measurements indicate that the earth is slightly flattened at the poles. The distance from pole to pole is 12,700 km, compared to the equatorial diameter of 12,756 km. This shape is called an **oblate spheroid.** At the next level of sophistication, satellite information tells us the earth is actually slightly pear-shaped.

## More on Planets

Philolaus, a disciple of Pythagoras (c. 582–500 B.C.), took the next step. He speculated that the earth rotated and that the motions of the stars were only apparent. Though the idea of the rotation of the earth surfaced several times in ancient Greece, it was never proved and the arguments against it prevailed. If the earth were spinning, would not objects thrown into the air be left behind? A central and immobile earth was at the heart of the Greek world view. Greek scholars concentrated on describing the complex planet motions around the earth. The bold assumption they made was that planets must move in perfect circles or combinations of circles. Without this assumption, their cosmology would have been chaos. Even today, in our sophisticated world view, basic assumptions not unlike those of the Greeks are necessary as we will see in Chapter 13.

The Greek mathematician Eudoxus devised a complex system of earth-centered spheres to describe planet motions. Each planet had a nest of four spheres coupled together in such a way so as to produce its correct motion. The spheres were not thought to exist physically, but they provided a mathematical framework for a model of planet motions. Three centuries after Eudoxus came the most famous development in Greek cosmology, Ptolemy's epicycles. Earlier in this chapter we described this system of circles on circles. The assumption of circular motion was still held to be valid. Though Ptolemy's system did not describe reality, it did an excellent job of describing planet positions and accounting for their retrograde motions.

Ptolemy himself would probably have been surprised to learn that his cosmology was regarded as the final word for 1,400 years. Those fourteen centuries spanned the rise of Christianity, the fall of the Roman Empire, the Middle Ages, and the Reformation. During this time, science was at a standstill. There were no significant advances in astronomy; there were even some setbacks. The Bible was interpreted literally and scientists with different views were punished as heretics. But finally, the early Greek ideas began to filter back into western Europe. One of these was Aristarchus' idea that the earth was moving around the sun. The **heliocentric theory** that surfaced about the mid-16th century is that of the Polish mathematician and astronomer Nicholas Copernicus. Copernicus' sun-centered view of planet motions caused a major revolution in thought. He still employed circular orbits and epicycles, but his central sun simplified the world view. Actual proof that the earth moved around the sun, however, would not come for two hundred years.

During those two centuries, the assumption of circular motion was discarded, Kepler developed his laws of planetary motion, and the gravitational astronomy discussed in the next chapter was born.

## SUMMARY

To the unaided eye, the sky looks the same today as it did thousands of years ago. The constellations are arbitrary groupings of stars that happen to lie in about the same direction in space. Many of the individual star names and the legends about the constellations originated with ancient cultures. Another way to name stars is to use the system called Bayer's letters. To describe the brightnesses of stars, we use a system of magnitudes first developed by Hipparchus.

The concept of a celestial sphere infinitely far away and completely surrounding the earth is a convenient way to locate the positions of objects in the sky. Ancient observers thought that the celestial sphere moved around the earth. A few fanciful world views of ancient cultures were described to show that they considered the earth to be flat, stationary, and centrally located. Ptolemy's earth-centered view of planet motions prevailed for many centuries. Ancient cosmology was mostly involved with describing and predicting the motions of the sun, moon, and planets. The days of the week are named after these objects.

Copernicus introduced the heliocentric theory of planet motions that revolutionized the world view. As soon as the earth is regarded as a planet that moves, explanations for sky observations become more simple. The rotation of the earth causes the sun and stars to appear to move. The daily paths of the sun and stars depend on our location on earth. Except for circumpolar stars, the stars appear to rise in the east and set in the west each day. The revolution of the earth around the sun causes different constellations to appear in different seasons. Each year the sun traces out a path called the ecliptic in front of the zodiacal constellations. The looping or retrograde motions of the planets is due to the combined effect of the earth's orbital motion and the planets' orbital motions. The planets travel in elliptical orbits around the sun.

The 23 1/2° tip of the earth's axis is responsible for the changing seasons on earth. There are three basic types of calendars: solar, lunar and luni-solar. Today we use the Gregorian calendar, a solar type. Several times each year, straight-line configurations of the moon, earth, and sun will produce solar and lunar eclipses.

### *FALLACIES AND FANTASIES*

*The Big Dipper is an official constellation.*
*The earth is closest to the sun in our summer.*
*The year 1900 was a leap year.*
*Planetary orbits are circles.*

## REVIEW QUESTIONS

_____ 1. How did ancient observers tell the difference between a star and a planet?

_____ 2. Compare the magnitudes of $\beta$ Aurigae and $\epsilon$ Aurigae. Which star would look brightest? How do you know?

_____ 3. Which star is brighter, a third-magnitude star or a fourth-magnitude star?

_____ 4. Sketch the celestial sphere and the apparent daily paths of the stars as viewed from your location on the earth.

_____ 5. (a) What connection, if any, is there between astronomy and astrology? (b) Describe the difference between the signs of the zodiac and the constellations of the zodiac.

_____ 6. Outline the rules used to keep the Gregorian calendar in step with the seasons.

_____ 7. When is the earth closest to the sun?

_____ 8. The earth always casts a shadow into space. Describe the shape of the earth's shadow.

_____ 9. Imagine looking straight down on the north pole of the earth. Which way would the earth be rotating, clockwise or counterclockwise?

_____ 10. Star Maps 2 and 4 do not seem to have as many stars as Maps 3 and 5. How might you explain this?

# 3 Discoveries: From Celestial Motions to Distances

## BETWEEN THE HELIOCENTRIC THEORY AND ITS PROOF

### Kepler's Laws

Between the presentation of the heliocentric theory and its proof were two hundred fruitful years of astronomical research. There were continual refinements in telescopes and clocks and bright new discoveries about the stars and planets. The scholars whose lives and works filled this period are remembered as those who laid the foundations of modern astronomy, mathematics, and mechanics. Fifty years after Copernicus published his treatise on the heliocentric theory, *De Revolutionibus,* Kepler (1571–1630) was hard at work describing the orbits of the planets.

Kepler's three laws of planetary motions helped to promote acceptance of the heliocentric theory. These laws were described in Chapter 2; here we summarize them in a more formal style.

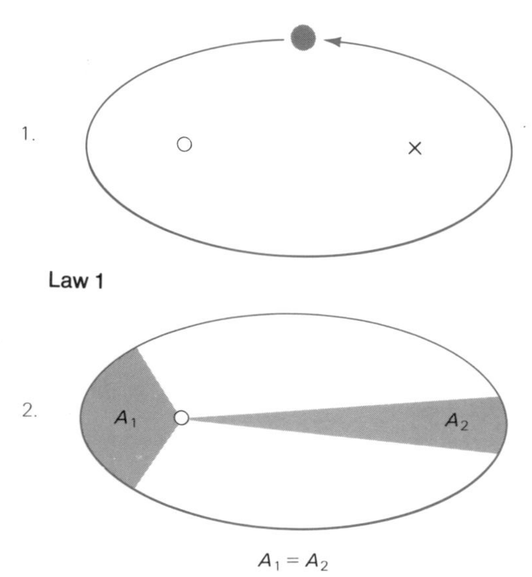


1. *Law of orbits.* The orbits of the planets are ellipses with the sun at one focus.
2. *Law of equal areas.* Equal areas are swept out by the radius vector of an orbit in equal intervals of time. The radius vector is the imaginary line connecting the planet and the sun.
3. *Harmonic law.* The cube of the planet's distance from the sun divided by the square of its orbital period is a constant.

Kepler derived these laws solely from observations of the positions of the planets. Laws based on observations are called empirical laws. Most of the planet observations Kepler used had been made by Tycho Brahe. It was after some difficulties with Tycho's heirs that Kepler managed to obtain his magnificent observations. Then Kepler set to work to see if he could determine the courses of the planets from the observations. For Kepler's purposes, Tycho's observations covered an especially favorable period in the orbits of the planets.

Kepler's investigation of the shape of the planet orbits is a good illustration of scientific procedure. He had been introduced to the Copernican theory in school and was a confirmed believer in its correctness. Assuming circular orbits, Kepler noted that Mars, being more distant from the sun than the earth, revolved more slowly around the sun than the earth did. Thus, he knew that if the sun, earth, and Mars were lined up in space, it would take several years for that alignment to occur again. Such an alignment is called an **opposition** because Mars is exactly opposite the sun as viewed from the earth. By observation, it was determined that oppositions of Mars occur every 780 days (Figure

3.1). Kepler also knew the mathematical relationship between the earth's year, the 780-day observed opposition interval, and the true year of Mar's orbit. After measuring the necessary angles, he calculated that the true year of Mars is 687 days.

Thus, every 687 days Mars is in the same position in its orbit. This key fact helped Kepler in his solution of the shape of the planet orbits. Kepler searched Tycho's data for pairs of observations of Mars separated by 687 days. Every time he found a pair, he reasoned that the different directions to Mars must be due to the differing positions of the earth. Thus, Kepler could make a geometrical model of the situation (Figure 3.2). Using plane geometry, he found the distance to Mars in terms of the radius of the earth's orbit. (We call the radius of the earth's orbit one **astronomical unit,** or one AU.) Kepler repeated this geometry over and over again with different pairs of observations for Mars in different parts of its orbit. Finally, he had determined the distance and position of Mars enough times to sketch out the true orbit. What he saw was definitely not a circle. By trial and error, Kepler found that an ellipse would fit all the positions.

Being a good scientist, Kepler checked this result against observations for another planet. In this case, he chose Tycho's extensive observations of the slower moving Saturn and showed that Saturn's orbit was also an ellipse. Kepler had made a major breakthrough. He could now explain the way the planets moved, although he was unable to explain why they moved as they did. Why would not be understood for another sixty years. In the meantime, Galileo Galilei (1564–1642) was making scientific discoveries in Italy.

## Galileo's Contributions

Galileo also became interested in how objects move—not just a planet, but any object at all. He was aware, as was Copernicus, that there was a force involved. He could not generalize the problem, however, so he performed many experiments on rolling and falling bodies. Among other things, Galileo found that the distance falling bodies covered was proportional to the square of the time that they fell. For example, if a ball fell 5 meters (m) in one second, it would fall four times as far in two seconds (Figure 3.3). This distance-time relationship is almost certainly the key that led Isaac Newton to his laws of mechanics.

Galileo was a very practical person. He had studied medicine and was a physician at the age of seventeen. At about this time, while

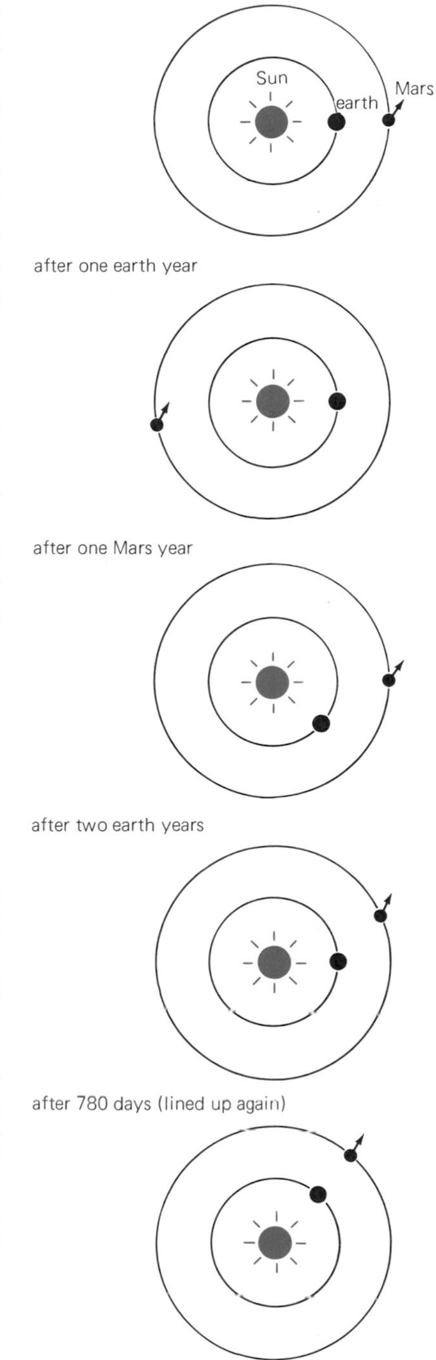


**FIGURE 3.1** Configuration of earth and Mars as one earth year passes, one Mars year passes, two earth years pass, and so on. In this way we can determine the true orbital period of Mars.

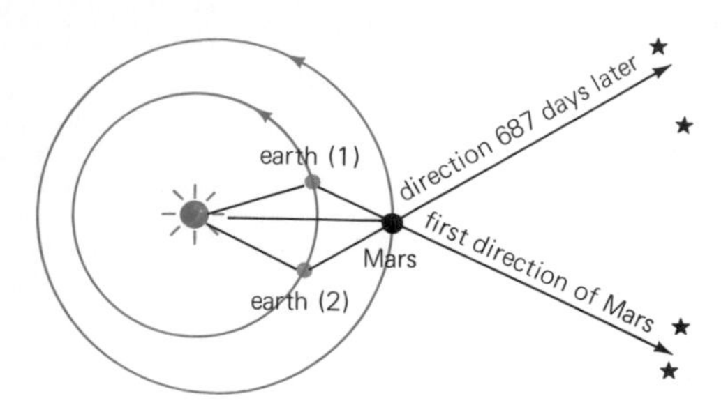


**FIGURE 3.2**
Kepler's method of determining the orbit of Mars. Pairs of apparent places on Mars separated by its period of 687 days gave the planet's distance from the sun at many points around the orbit.

listening to a long and probably tedious sermon, he noticed that according to his pulse, the swinging chandelier above the pulpit had a constant period. That is, the time required for the pendulum to swing from one extreme to the other and back again is always the same. He checked this observation with other pendulums and found that the period of a pendulum depends upon its length (actually the square root of its length). Using this information, Galileo constructed a pendulum with a specific period. He could then check a patient's pulse against this constant period. A pulse faster than the standard pendulum probably indicated illness. A pulse equal to or slightly slower than the standard would indicate that the person was healthy.

In addition to laying the foundation for the theory behind the planet motions, Galileo was also collecting observational support for the Copernican system. In 1610, Galileo became the first person to look at the heavens with a telescope. Since he was so experienced at experimenting and keeping records, he made an excellent observer. His observations of the sun, moon, planets, and the Milky Way were the beginning of a new era of far-reaching discoveries. Here we will mention some of the observations that helped to defeat the earth-centered world view.

When Galileo observed Venus through his telescope, he saw that it displayed all the phases that our moon does—from crescent to full to new and so on (Figure 3.4). This observation discredited Ptolemy's system, in which there could be no full phase of Venus. Galileo's discovery of four bright moons orbiting Jupiter indicated that not everything orbited the earth, as some people believed. (Today, we know there are at least 14 moons orbiting Jupiter.) The observations of Jupiter's moons also dispelled the notion that our moon would be left behind if the earth orbited the sun (Figure 3.5). It is interesting to recall that Galileo lived at a time when discoveries such as his were looked upon with great disfavor by the Church. The publications of Copernicus, Kepler, and Galileo were on the Church's Index of Prohibited Books until 1835.

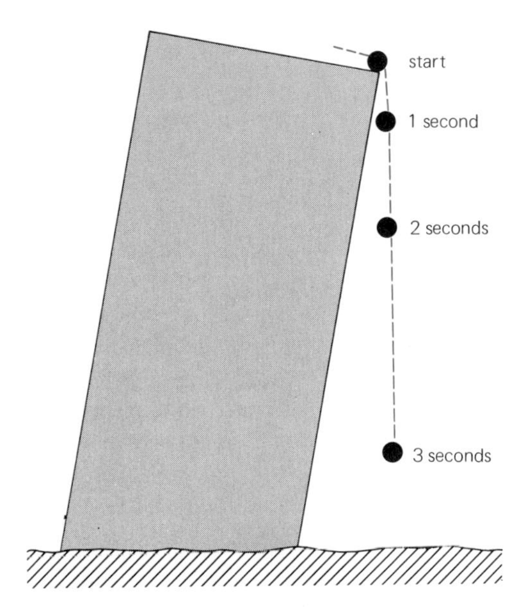


**FIGURE 3.3**
A ball falling off a building accelerates rapidly, as Galileo observed.

## Newton and Gravitational Astronomy

Isaac Newton (1642–1727) was an avid reader, and we surmise from how he presented his research that he was fully aware of Copernicus' treatise and Galileo's papers. Indeed, it is obvious that those works led him to his three laws of motion and to his law of gravitation. During Newton's lifetime, astronomy went from merely observing the courses of the planets to inferring the gravitational forces of the sun and the planets that controlled their motions. For the first time, Newton could explain why the planets moved the way they did. He concluded that the same pull of gravity holding us on the earth extends to the moon and holds it in orbit around the earth. Likewise, the pull of the sun's gravity holds all the planets in orbit around the sun.

Newton's **law of gravitation** states that the force of attraction between any two bodies was proportional to the product of their masses divided by the square of the distance between the two bodies. To understand the consequences of this law, we can perform a hypothetical experiment using the earth and sun as the two bodies. We can place the earth at various distances from the sun and determine the

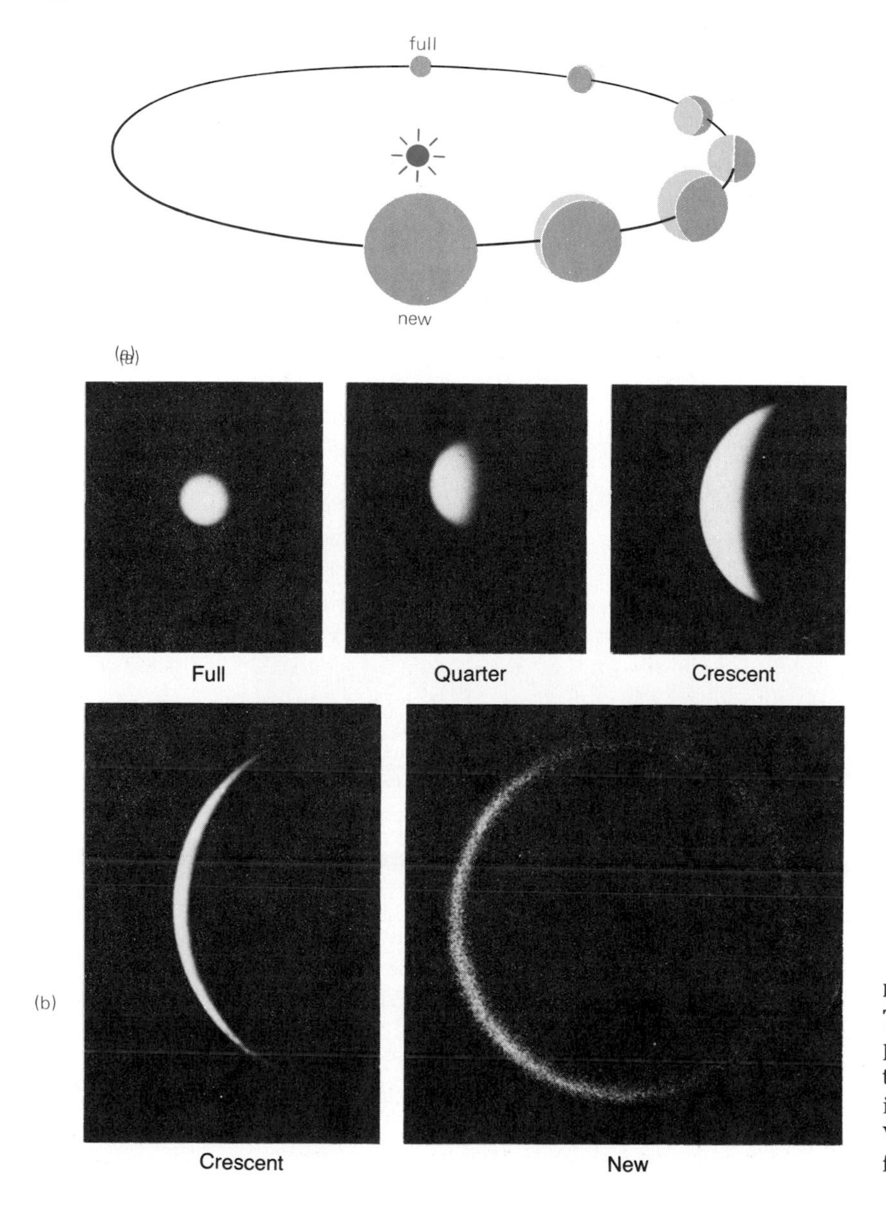


**FIGURE 3.4**
The phases of Venus in its orbit (a). The full phase proves that Venus is heliocentric and that the Ptolemaic view of Venus is incorrect. Notice how the apparent size of Venus changes as it is nearer and farther from us (b).

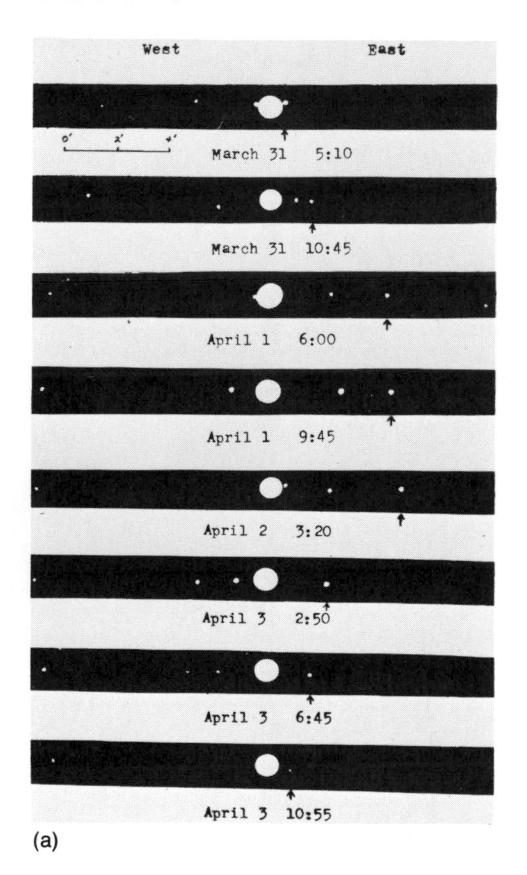

(a)

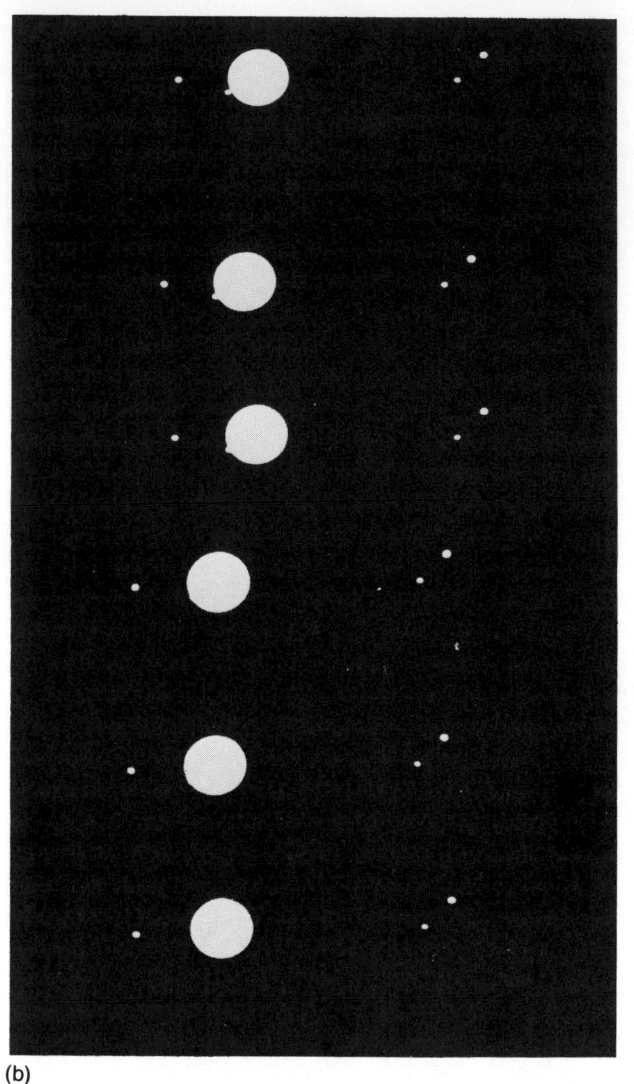
(b)

**FIGURE 3.5**
Jupiter's Galilean satellites, Io, Europa, Ganymede, and Callisto, move around the planet. In (a), Ganymede is indicated by the arrows. Io's motion is shown in (b) as it disappears behind Jupiter. These photographs were taken only 30 minutes apart. (Photograph (a) courtesy of R. Culver)

force of attraction or the effect of gravity. The only variable we are concerned with in this example is the distance of the earth from the sun. If the earth is at a distance of 2 AU, the force of attraction is one-fourth of the force at 1 AU. At 3 AU, the force is only one-ninth as much. At 10 AU, the force is one one-hundredth as much, and so on. Evidently, the farther a body is from the sun, the easier it would be to escape the gravitational attraction.

Since the law of gravitation explained the *why* of planet motions, Newton began with the law and reproduced Kepler's three empirical laws of planetary motion. When he did this, he found that the harmonic law was not quite correct. The harmonic law as corrected by Newton is that the ratio of the cube of the distance of a planet from the sun to the square of its period equals the sum of the masses of the sun and planet. Mass is a fundamental property of any object. It is a measure of the amount of material the object contains. Since the sum of the sun's mass plus any planet differs very little from the mass of the sun, we can see why Kepler's simple statement had served well enough for many years.

Newton's three laws of motion describe what happens to objects when forces are applied. The most famous law is the third one which says that for every action there is an equal and opposite reaction. Newton's laws, coupled with his law of gravity, give us a very powerful set of tools. Every object that passes near another object is attracted by that object, and vice versa. Thus, the earth affects Mars' orbital motion and Mars affects the earth's orbital motion more when they are near each other than when they are farther apart. By these laws, the path of a spaceship from the earth to the moon and back can be worked out with absolute confidence. The small changes in the orbit of one body imposed by another body are called **perturbations** and must be allowed for in computing the orbits of the planets, the orbit of a spacecraft, or the journey of a spacecraft to a distant planet such as the journey of Pioneer 10 to Jupiter and beyond.

## PROOF OF THE EARTH'S MOTION AROUND THE SUN

### The Idea behind a Stellar Parallax

After Newton's remarkable contributions, it was clear that a deterministic philosophy (one thing leads to another) ruled the Universe: the Universe was Copernican and Newtonian mechanics governed its motions. But there was still no proof that the world view was Copernican. Now it was up to observers, in the tradition of Tycho, to verify or refute the Copernican world view.

If the earth is really moving around the sun, then nearby stars ought to appear in slightly different positions as we observe them first from one side of our orbit and six months later from the other side (Figure 3.6). One-half the amount of the angular shift, indicated by the symbol $\pi$ on the figure, is called the **parallax** of the star. The nearer the star, the greater its parallax would be. A similar parallactic shift of position can be demonstrated by looking at any object from two dif-

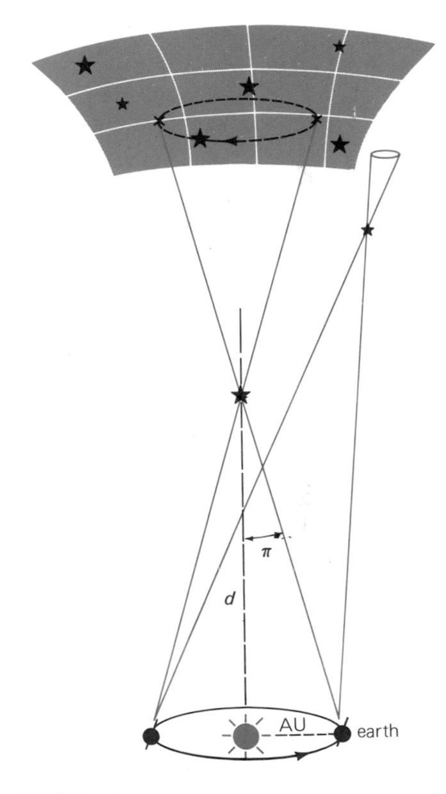


**FIGURE 3.6**
The parallax of a star. The nearer star appears to oscillate annually relative to more remote stars because of the earth's revolution.

ferent directions. Even the small separation between your eyes can produce the effect. Hold a pencil in front of you and look at it first with one eye and then the other. It will appear to jump back and forth compared with the rest of the background. The closer you hold the pencil, the greater the shift will be. Thus, the amount of the shift is a measure of distance.

Some simple calculations applied to the parallax of a star will give the distance to the star. To be able to find the distances to the stars was another desirable result of discovering a stellar parallax. The biggest observational challenge for Copernicans was to see a stellar parallax and prove the earth's revolution without question.

Tycho, with instruments capable of measuring a shift as small as 1 minute (min) of arc,* rejected the Copernican theory because he could not see any periodic motion whatsoever. If Tycho could not see a parallax shift, that would mean the stars were over 3,400 times farther away from the earth than the sun. Tycho just could not imagine that the Universe extended to such a distance. Almost two thousand years earlier, Aristotle ventured the same opinion for the same reason.

## Proper Motions of Stars

Other observers persevered in the search for parallax, and even though they failed they happened upon some remarkable discoveries. In 1718, Edmund Halley was observing certain bright stars for parallax. As in the case of other observers, Halley concentrated on the bright stars. He simply assumed that if all stars had about the same absolute brightness, then the bright stars must be nearby. This seemed reasonable enough as a start. Instead of discovering parallax, Halley found that the stars he was observing were moving in random directions with respect to one another. He had discovered the individual motions of stars across the sky. These are called the **proper motions** of stars. Proper motions are small but many are much larger than the largest parallax. A motion of several seconds of arc per year would be considered very large. The proper motions of stars will eventually change the familiar appearance of the constellations. However, that will take tens of thousands of years (Figure 3.7).

Actually, the stars do not move just in the plane of the sky. Stars can and do move in random directions. Part of this motion, called the **space motion** of a star, can be toward or away from us and part can be across the sky. The component along the line of sight toward or away from us is called the star's **radial velocity.** The other component is its proper motion (or **tangential velocity**), as mentioned above.

*There are 60 min. of arc in 1° and 360° in a circle; 1 min. of arc is a tiny angle that represents 1/21600 of a circle.

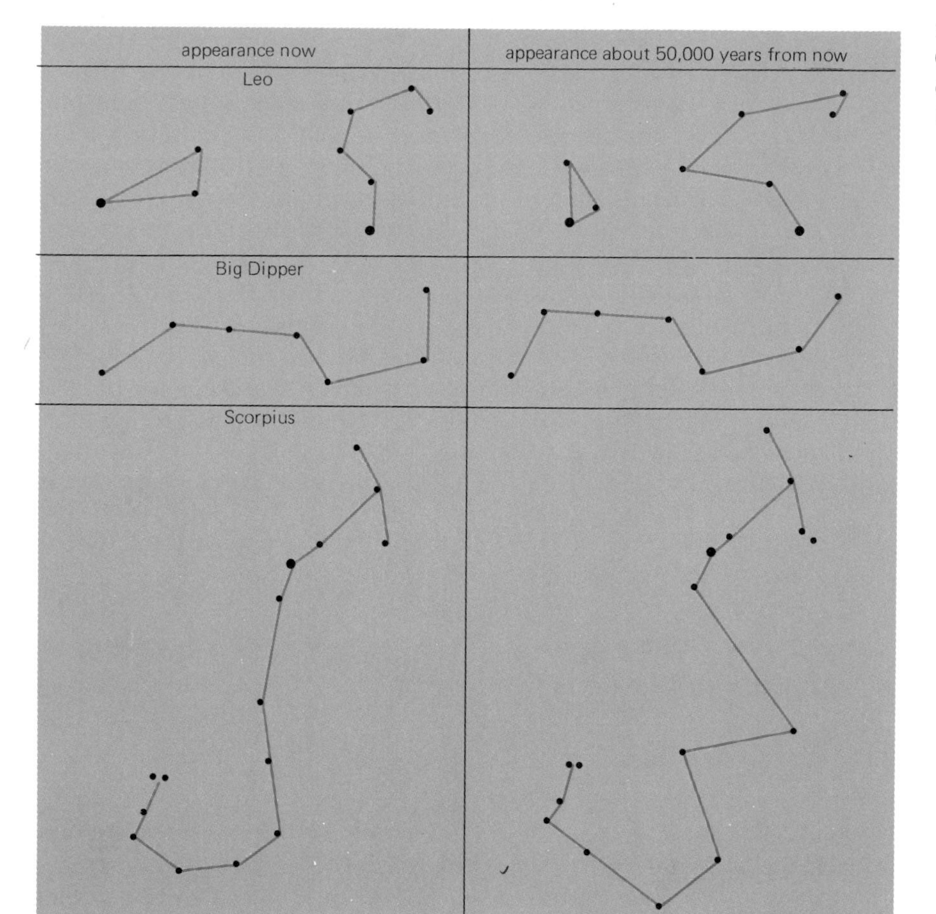


**FIGURE 3.7**
Changes in the shapes of familiar constellations due to proper motions over a 50,000-year interval.

## A Serendipitous Proof of Revolution

A short time later, in 1727, James Bradley was trying to measure the parallax of $\alpha$ Virginis (Spica) when he observed a curious effect that had profound implications. Bradley was observing $\alpha$ Virginis in a **meridian telescope.** This kind of telescope is fixed to look only along the **celestial meridian,** an imaginary line in the sky running from north to south and through the point directly overhead called the **zenith** (Figure 3.8). Once each day, as the earth rotates, every star seems to drift across the meridian. Accurate times of star crossings can be recorded with a meridian telescope. Bradley found that what should have been the midnight crossing of $\alpha$ Virginis occurred about 1.3 seconds (sec) of time earlier than expected. He confirmed this same ef-

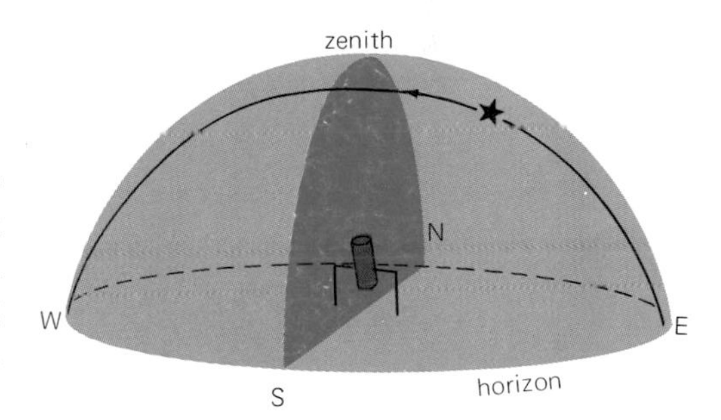


**FIGURE 3.8**
A meridian telescope is mounted in such a way that it moves up and down along the meridian and is used to measure when stars cross the meridian.

fect for other stars. After carefully considering all sources of error and other possible causes to explain this effect, Bradley concluded that he was observing the **aberration of starlight.** Aberration results from a combination of a finite speed of light and the revolution of the earth.

Light traveling with a finite speed and entering the front of a telescope takes some small amount of time to reach the viewing end. If the earth moves during that time, the starlight coming through the telescope will be offset just slightly (Figure 3.9). Bradley could see and measure this offset in his meridian telescope. Since the finite speed of light had already been demonstrated by Ole Roemer in 1677, Bradley correctly concluded that the earth must be moving around the sun. The aberration of starlight can amount to a variation in a star's position of about 20 seconds (sec) of arc. Observing that effect was indisputable proof that the earth moved. The sought-for proof, stellar parallax, was still in the future.

(a)

(b)

**FIGURE 3.9**
Stellar aberration. Raindrops (a) appear to fall toward a runner due to their finite velocity and that of the runner. So it is with telescopes and starlight. The telescope needs to be tilted slightly in order for the light to fall onto the eyepiece (b).

## MORE DISCOVERIES ON THE WAY TO FINDING A STELLAR PARALLAX

### The Sun's Motion

Now that the motion of the earth around the sun was proved, there must be a parallactic effect. The search for stellar parallax went on.

Based on his own study of the proper motions of only thirteen stars, William Herschel (1738–1822) made a remarkable discovery in 1783. He found that the stars seem to be moving systematically out of the constellation Hercules, past the sun, and toward the constellation Columba (Figure 3.10). This movement of stars seemed unlikely to Herschel; rather than all the stars moving past the sun, it seemed more likely that the sun was doing the moving. This is analogous to a car driving along a highway. To a person in the car, the trees and bushes and other scenery appear to be moving past the car. However, it is much easier to assume that the car is doing the moving. Herschel had discovered the sun's motion in space. Compared to the average motions of all the stars, the sun is moving toward Hercules. The point toward which the sun is moving is called the apex of the sun's way, or more simply, the **solar apex.** It is marked in Figure 3.11. The sun and the entire solar system are moving toward the solar apex at about 20 km/sec. The direction from which the sun is moving is called the **antapex** and is also shown in Figure 3.11.

## Binary Stars

In the late 1700s, Herschel was attempting to observe a parallax for the bright star Castor in the constellation Gemini. He was using a meridian telescope and recording the meridian crossing time. Looking at Castor, he observed two stars very close together in the sky. Both stars could be seen in the telescope at once. Without changing any settings for the telescope, Herschel timed the meridian crossings of each star. Since the stars were so close, differential measures of the crossing times were very accurate (Figure 3.12). After repeated observations on several nights each year for twenty-four years, it was clear to Herschel that these stars were orbiting around each other. Other observers had noticed the same effect, but Herschel was the first to discover that the stars were a physical pair. He reported this in 1803. Newton's law of gravitation could be used to explain their orbits around each other. We call these physical pairs of stars **binary stars.**

Herschel's discovery marked the first time that Newton's law of gravitation was observed to be valid for objects beyond the sun and planets. Now it could truly be called the universal law of gravitation. The gravitational analysis of binary star systems is a powerful tool for studying distant stars. One of the most important items that can be determined is the mass of the stars. As we will see, the mass of a star determines its basic character. Gravitational analysis is extremely useful, since it turns out that approximately half of all stars are in either binary or multiple systems. Today we know that Castor, the first known binary, is really a system of six stars. All six stars are gravitationally bound to one another in a complex pattern of orbits.

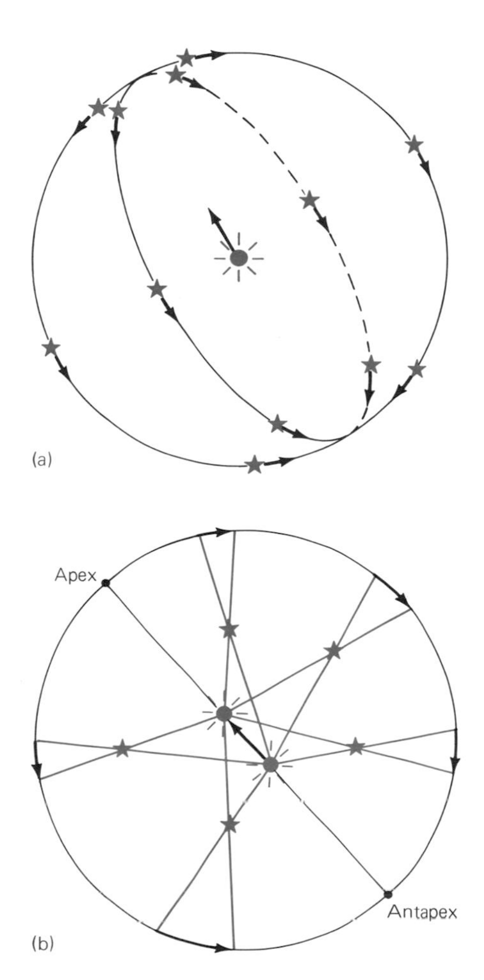


**FIGURE 3.10**
Stars appear to drift across the sky out of one region and into the opposite region (a) due to the sun's motion. The actual effect as projected on the celestial sphere is shown in (b).

## The Titius-Bode Relation

By the late 18th century, the world view had expanded considerably. Lord Rosse was sketching spiral nebulae as viewed through his great telescopes (Figure 3.13), opticians were making better and better telescopes, and a dentist by the name of Titius hit upon a strange number combination that gave the spacings of the planets. If we follow Titius and write down the numbers 0, 3, 6, 12, 24, 48, and 96, where every number after three is just the number in front doubled, add four to each number, and divide by ten, we obtain numbers that compare closely to the real distances of the first six planets (Table 3.1).

An astronomer named Bode noted the Titius relationship and promoted it. Thus we call this unexplained and curious sequence of numbers the **Titius-Bode relation.** The relation reasonably stated distances of all known planets of the time except for a gap at 2.8 AU.

**Table 3.1**

| Titus-Bode Distance | Planet | Real Distance in AU |
|---|---|---|
| ( 0 + 4)/10 = 0.4 | Mercury | 0.39 |
| ( 3 + 4)/10 = 0.7 | Venus | 0.72 |
| ( 6 + 4)/10 = 1.0 | Earth | 1.00 |
| (12 + 4)/10 = 1.6 | Mars | 1.52 |
| (24 + 4)/10 = 2.8 | ? | ? |
| (48 + 4)/10 = 5.2 | Jupiter | 5.20 |
| (96 + 4)/10 = 10.0 | Saturn | 9.54 |

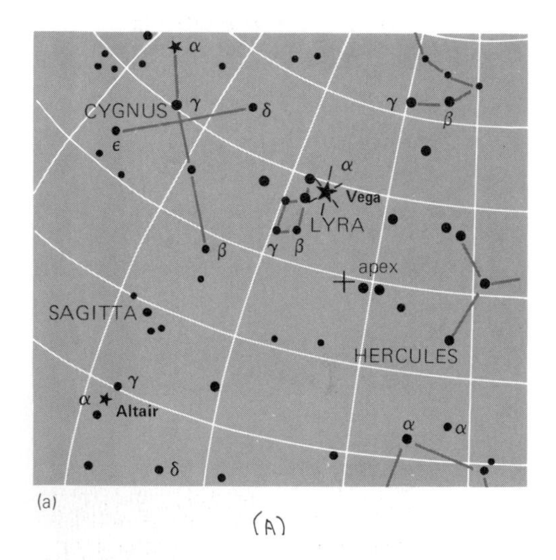


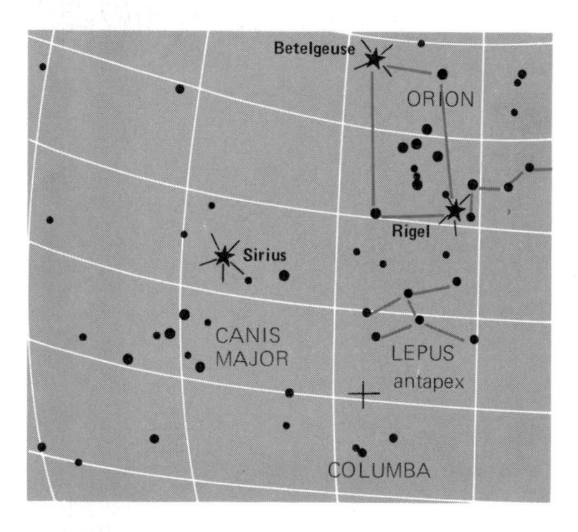


**FIGURE 3.11**
The location of the apex (a) and antapex (b) on the sky.

Surely there must be a planet at that distance. Astronomers began searching for a planet between Mars and Jupiter.

If we extend the Titus-Bode relation one more step, the next planet ought to be at a distance of 19.6 AU. In 1781, Herschel discovered Uranus at a distance of 19.18 AU. The relation still seemed to work. The search for a planet at 2.8 AU continued. Finally, on New Year's Day 1801, Piazzi found a small one named Ceres. Ceres is only 780 km in diameter. Soon after, other small planets were found. Today there are thousands of small planets known to be orbiting the sun at distances averaging about 2.8 AU. Some are no bigger than a kilometer across. We call them minor planets, or **asteroids.**

The world was clearly deterministic. Reinforcing this deterministic view was the fact that the new planet, Uranus, deviated from its predicted positions. Such deviations might be due to gravitational forces of a more distant planet. The nature of the deviations would indicate where to look for the planet in the sky. Sure enough, in 1846 Galle found the new planet, later named Neptune, within one degree of its predicted position. Little attention was paid to the fact that it did not fit the Titius-Bode relation. Neptune's distance from the sun was only about 30 AU, whereas the Titius-Bode distance would be 38.8 AU.

In time, Neptune also deviated from its predicted positions. Was there a planet beyond Neptune? After a long search, Pluto was discovered in 1930 at the predicted position by Clyde Tombaugh working at the Lowell Observatory in Arizona (Figure 3.14). Later, it was shown that the calculations used were wrong, so the discovery of Pluto was a fortunate accident. Observational surveys completed after the discovery of Pluto indicate that there probably is not another planet beyond Pluto. If a planet is ever detected beyond Pluto, it will be either quite small or at least five times farther from the sun than Pluto.

By the time of the discovery of Neptune, observations showed that Mercury's orbit was being distrubed more than gravitational theory would allow (Figure 3.15). Further observations confirmed this. Such a disturbance might be due to some inner planet closer to the sun than Mercury. Astronomers began to search for a planet they called Vulcan. They carried out the searches during eclipses of the sun. The reasoning behind this was simple: If Vulcan existed, it would be difficult to see in daylight, even more difficult than Mercury, which itself is seldom seen. Since Vulcan would never be far from the sun's position in the sky, it would rise or set so close to sunrise or sunset that it would not be observable by twilight. Therefore, the only time to look for Vulcan is when the sun is covered by the moon during a solar eclipse. But repeated searches were futile. Vulcan does not exist. The real reason for the disturbance of Mercury's orbit was explained about a hundred years later with the introduction of the theory of relativity by Albert Einstein.

# OBSERVING A STELLAR PARALLAX AND WHY IT TOOK SO LONG

## The First One

A stellar parallax was finally announced by F. G. W. Struve in 1837. Applying the same reasoning as Halley, i.e., that the bright stars are nearby, Struve was observing bright stars for parallax. He found a measurable parallax for the brilliant star Vega in the constellation of Lyra. Vega is located near the Milky Way in the sky. When viewed in a telescope, many faint background stars are also visible for comparison. By timing the meridian crossings of Vega throughout the observing season and comparing them to the crossings for the background stars, Struve was able to detect a slight shift in Vega's position (Figure 3.16). When Vega was viewed from one side of the earth's orbit and then six months later from the other side, it appeared to shift 0.25 sec of arc. The quoted parallax is, by convention, one-half of that shift, or 0.125 sec of arc. Sometimes it is simply written 0″.125.

No wonder Tycho had problems. The parallax shift of Vega is more than two hundred times smaller than Tycho could have hoped to measure. Vega's parallax represents a distance of 26.5 lt-yr. This means Vega is about 2 million times farther away than the distance from the earth to the sun. The background stars would have to be even more distant.

Although such vast distances were incomprehensible to Tycho, they were fully anticipated by a contemporary of his, Thomas Digges. Digges had simply extended Copernicus' celestial sphere. Instead of having all the stars on the surface of the sphere, Digges' view was of an infinite three-dimensional volume uniformly filled with stars. The idea was a major advance in cosmology, but went relatively unnoticed for more than a century.

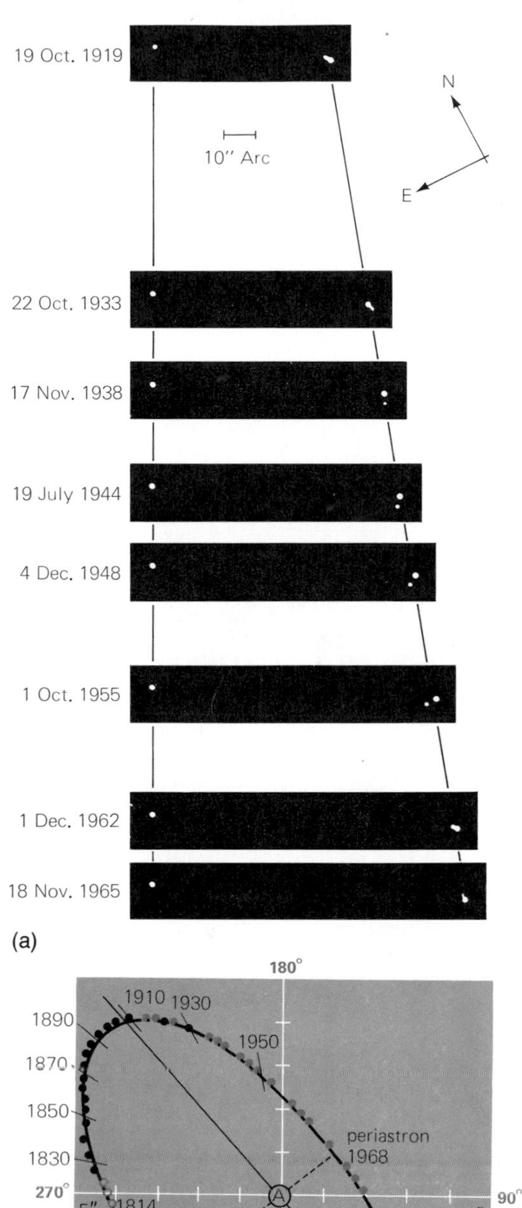


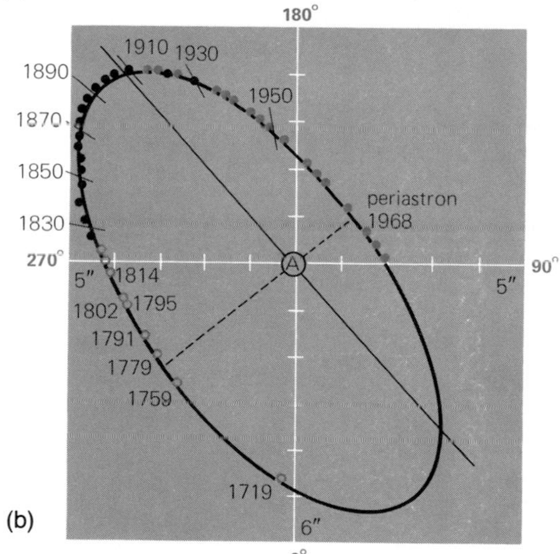


**FIGURE 3.12**
(a) Composite photographs of the binary star Kruger 60 on the right referenced to an optical companion star on the left. Note the orbital motion and the effect of the stars' proper motion. (Leander McCormick Observatory and Sproul Observatory photographs)
(b) The apparent orbit of Castor with the proper motion removed. Note Herschel's observations from 1759 to 1802. (Courtesy of K. Aa. Strand)

## Others

One year after Struve's announcement of the parallax of Vega, F. W. Bessel announced the parallax of a pair of stars called 61 Cygni. 61 Cygni has a parallax of 0.″3 and is closer to the earth than Vega. Even though Vega is farther away from the earth, it is brighter than 61 Cygni. In the following year, T. Henderson announced the parallax of Alpha Centauri. Alpha Centauri has a parallax of 0.″76 and is the closest star to the sun (actually Alpha Centauri is a triple-star system). Even though it is much closer than Vega, it is only slightly brighter than Vega. Here is the first evidence that all stars are not of the same brightness. Most of the brightest stars in the sky show no parallactic shift at all. Since these stars do not appear bright because of their closeness to the earth, they must be intrinsically bright stars.

In practice, measuring stellar parallaxes takes many nights of observing over a period of many years. Great patience is required. In the seventy years before star positions were recorded on photographs, only 55 parallaxes were measured. In the seventy years since the introduction of photography, parallaxes have been measured for over 7,000 stars.

## The Role of Telescopes and Clocks

We might quite naturally wonder why it took until 1837 to measure the first stellar parallax when the prediction was made in 1543. The answer lies in the development of instruments and the nature of measurements. Each new discovery made by astronomers searching for stellar parallaxes involved measuring smaller and smaller angles.

Prior to the introduction of photography, many observations depended upon fixed instruments and accurate timekeeping. Tycho, with his open sights (sextants and crosstaffs), could measure relative angular positions to about 1 min of arc (Figure 3.17). A better method for measuring positions was to time various celestial objects as they crossed the meridian and then to calculate their separations. The application of the meridian telescope combined with excellent timekeeping allowed those measurements. Figure 3.8 shows a meridian telescope.

Accurate clocks became possible with Galileo's discovery of the constant period of the pendulum and C. Huygens' development in 1656 of an escapement mechanism to keep the pendulum swinging. Clocks of Huygens' design could keep time with a random error of only 1 sec per day. This accuracy enabled Ole Roemer to time eclipses of the moons of Jupiter and with some clever calculations determine the speed of light.

A major advancement in timekeeping occurred in 1715, when G. Graham developed a device called the deadbeat escapement. This device reduced the random daily error of the pendulum clock to well under one-tenth of a second per day (Figure 3.18). With a clock incor-

**FIGURE 3.13**
Lord Rosse's sketch of spiral nebula. Compare with Figure 1.6

**FIGURE 3.14**
A small section of photographic plates used to discover Pluto. (Lowell Observatory photograph)

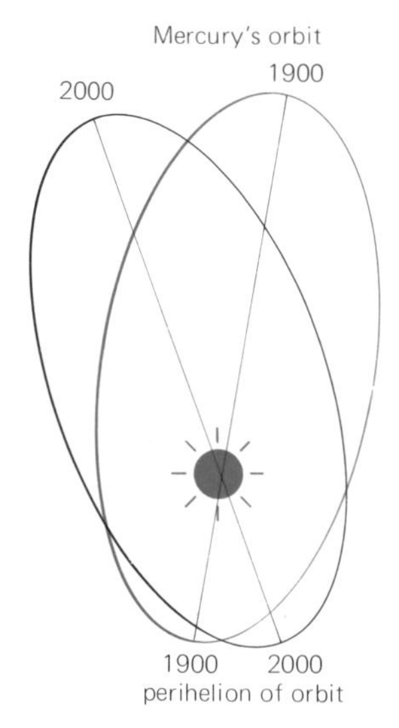


**FIGURE 3.15**
The distrubance of the orbit of Mercury. The change is too large to be fully explained by Newton's Law of gravitation.

porating this development, Bradley could easily note the aberration in Spica's position. He detected a systematic shift in Spica's position of 1.3 seconds of time over a period of 3 months.

The clocks developed by Graham were accurate enough to allow astronomers to measure parallax. The real problem now was to record the passage of stars across the meridian with sufficient accuracy. This was accomplished by having the clock turn a recording drum or cylinder. Each time a star crossed the meridian, the observer, who was looking through the telescope, would press a handle that would mark the drum. At the end of the night, the observer could measure the time between marks on the drum and hence reduce the observations to meaningful numbers.

An interesting human problem was uncovered because of this

technique. Two observers recording on the same drum seldom recorded the same event at the same time. That is, any given observer has a certain response time and also tends to anticipate the moment of crossing slightly differently from another observer. This effect is called the **observer's personal equation.** For a given observer, the personal equation is essentially constant; it might change only very slowly over decades. Thus, when measurements made by a single observer were analyzed, the minute differences of time required to detect stellar parallax could be seen. The maximum parallactic shift of Vega's position corresponds to only two one-hundredths of a second of time. Even the parallax shift of Alpha Centauri, the nearest star system, corresponds to only one-tenth of a second of time. Stellar parallaxes are so small that it is not hard to understand why their discovery took so long.

Astronomical observations continued to demand improvements in timekeeping. By the early part of the twentieth century, pendulum clocks with random errors of only two-thousandths of a second per day were in use. In the 1940s, quartz crystal clocks were introduced having errors of only a few ten-thousandths of a second per day. Since the late 1960s, atomic clocks having errors of less than one-billionth of a second per day are in common use (Figure 3.19). Timekeeping technology has progressed so far that now we can buy wristwatches that have about the same accuracies as the clocks used to determine the first stellar parallax.

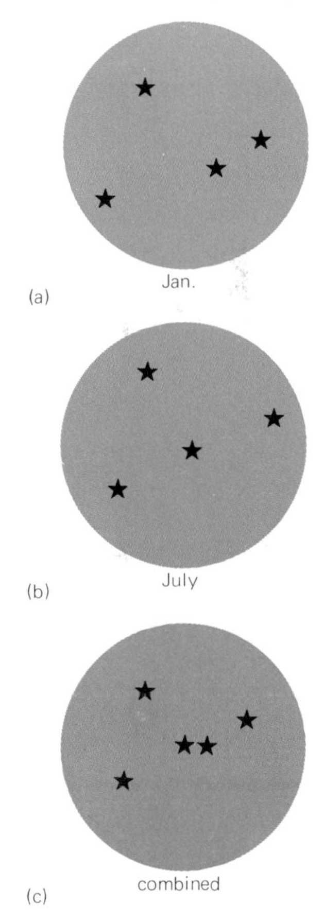


**FIGURE 3.16**
Two observations made in January (a), and July (b) are combined (c) to show the parallactic effect.

# THE TRUE BRIGHTNESS OF STARS

## Absolute Magnitude

Observing a stellar parallax does prove the earth's revolution around the sun, but the greatest value of a stellar parallax measure is for calculating the distance to the star. Parallax is the only direct method astronomers have for determining star distances.

When astronomers adjust the apparent brightness of stars for the distances to the stars, they can compare the true brightness of the stars. To make the comparisons easier, they put all stars at a standard distance of 32.6 **light-years** (32.6 lt-yr).* A star's magnitude at a distance of 32.6 lt-yr is called its **absolute magnitude** (M).

*A light year is the distance light travels in one year. It is approximately 9,500,000,000,000, km.

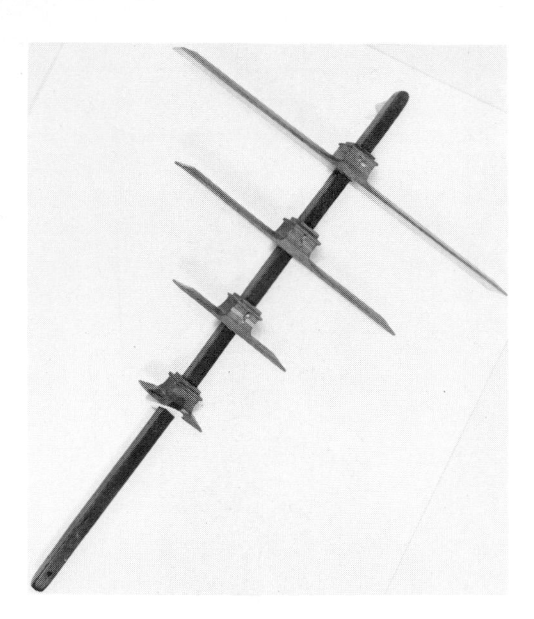

FIGURE 3.17
An old 16th-century cross-staff such as those used by Tycho Brahe. This instrument was used to measure angles between stars and other celestial objects. (Adler Planetarium photograph)

## Brightness and Magnitude

Magnitude as a measure of a star's brightness is not a straightforward relationship. Our eyes respond to brightness in a unique way. The psychologist Fechner found that if a person said one candle at a given distance had a certain brightness, he or she would say that ten candles clumped together were much less than ten times as bright. It would take one hundred candles to even get the person to say they were five times brighter than the single candle. What this means is that when we casually look into the sky and see a star that appears five times brighter than another one, it really is not just five times brighter. We say the star is 5 magnitudes greater, but it is actually one hundred times brighter than the other star.

In 1856, following a suggestion by John Herschel and anticipating Fechner's study, Pogson proposed that the magnitude scale be formalized. He suggested that a difference of 5 magnitudes correspond to a factor of exactly 100 in brightness. In comparing any two stars, if we see a difference of 1 magnitude, the brighter star is really 2.512 times as bright. If the difference is 2 magnitudes, the brighter star is really 6.31 times as bright, and so on.

The magnitudes of stars as they appear to our eyes are called **apparent magnitudes** (m) (Figure 3.20). To set all the stars on the same apparent magnitude scale, some arbitrary star had to be selected to have a known apparent magnitude. Vega ($\alpha$ Lyrae) was assigned an apparent magnitude of 0.0, and other stars were adjusted accordingly. Sirius is brighter than Vega and had to be given an apparent magnitude of −1.6. The sun has an apparent magnitude of −26. This formalization of the magnitude scale conforms very nicely with Hipparchus' original one (Chapter 2). The star maps in Chapter 2 indicate the apparent magnitudes of the stars.

## Brightness and Distance

To change an observed apparent magnitude (m) to an absolute magnitude (M) means hypothetically moving the star to the standard distance of 32.6 lt-yr from the sun. The relationship between the brightness of an object and its distance is fairly simple. As we move objects away, their brightness goes down by the square of the distance we move them. Suppose a star is 16.3 lt-yr away and we want to adjust its brightness to the standard distance of 32.6 lt-yr. We would be moving the star two times farther away, so its brightness would go down by four times (2 squared). Likewise, bringing a star closer would increase the brightness by the square of the distance. Once the brightness adjustment for distance is made, astronomers sometimes want to convert the brightness to a star magnitude. That calculation is done using the formalized brightness-magnitude scale explained above.

Until the discovery of a stellar parallax, most astronomers thought that all stars had the same absolute magnitudes. Differences in apparent magnitudes were thought to be differences in distances alone. Now we know that there are real intrinsic differences in star brightnesses. To compare and understand these differences, astronomers continue to observe parallaxes and apparent magnitudes, to calculate distances, and to compute absolute magnitudes of stars.

**FIGURE 3.18**
A pendulum clock. The perfection of the pendulum clock was a major early advance in timekeeping. Pendulum clocks have been replaced by electronic clocks at observatories. (Leander McCormick Observatory photograph)

## SUMMARY

Many of the astronomical discoveries of the 18th and 19th centuries involved detecting various motions of the earth, sun, and stars. We discovered stellar aberration and parallax. Both these effects prove that the earth is revolving around the sun. Observations of the proper motions of stars told us that all stars are moving through space. The sun is moving toward the direction of the constellation of Hercules. As soon as a stellar parallax was discovered, we determined the actual distances to stars. The Titius-Bode relation prompted a search for a missing planet between Mars and Jupiter. Instead of a single planet, thousands of asteroids were discovered. Beyond Jupiter, Uranus was discovered by accident, Neptune was predicted and discovered, and finally, in the 20th century, Pluto was discovered.

Two things triggered this great rush of discoveries. One was rapidly developing technology. The telescope was first used for astronomical purposes by Galileo in 1610. From then on, telescopes were continually improved and enlarged to enable us to see more and more detail in the heavens. Paralleling that was the development of accurate clocks, which also began with Galileo and his work on the principles of the pendulum. The second important ingredient was the development of theoretical studies in mathematics and physics. Why the planets moved as they did was discovered by Newton. Kepler's laws of planetary motions could be derived by Newton using his theory of gravitation. Gravitation is the universal force governing all celestial motions.

**FIGURE 3.19**
A cesium beam atomic clock. With an accuracy of one-millionth of a second over a period of 3 weeks, this clock is a recent major advance in timekeeping. (Official U.S. Naval Observatory photograph)

As soon as the distances to stars were measured, it became apparent that all stars do not have the same brightness. Absolute magnitude is a measure of the true brightness of stars. The magnitude scale was formalized so that a difference of 5 magnitudes represents a factor of 100 in brightness. There is a direct relationship between the brightness of a star and its distance. Apparent magnitudes are affected by distances.

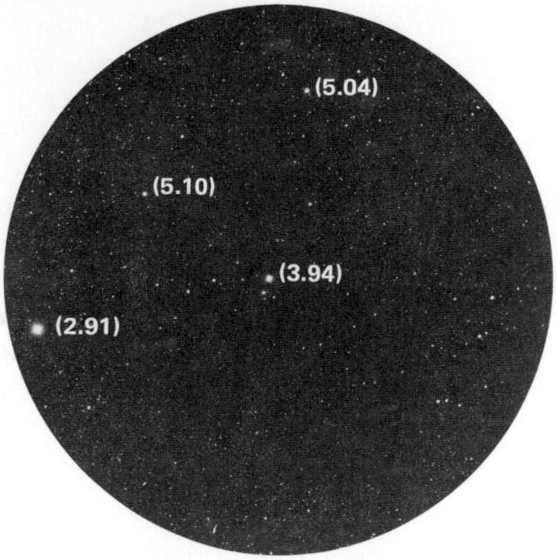

**FIGURE 3.20**
A star field centered on omicron ο Persei. Star images on photographs are larger for brighter stars. (The size of the image has nothing to do with the size of the star, since all stars are so distant that they are essentially points of light.) The brightest appearing stars are not necessarily nearest. The numbers are apparent magnitudes of the stars. (Lick Observatory photograph by G. Herbig)

### *FALLACIES AND FANTASIES*

*The stars do not move.*
*The speed of light is infinite.*
*All stars are the same brightness.*

## REVIEW QUESTIONS

______ 1. Does a planet have a greater orbital speed at aphelion or perihelion? Why?

______ 2. If the earth is spinning, why don't we fall off? Why don't we feel the spinning?

______ 3. Distinguish between the earth's rotation and revolution.

______ 4. Define one astronomical unit (AU). Compare it to one light year (lt-yr).

______ 5. What two parameters determine the amount of gravitational attraction between any two objects? Name two objects that show evidence of a strong gravitational attraction. Name two objects where the gravitational attraction is not noticeable.

______ 6. (a) Which is harder to determine, absolute magnitude or apparent magnitude? Why? (b) A star has an absolute magnitude of $+4$ and an apparent magnitude of $+4$. What can you conclude about the distance to the star?

______ 7. Name a specific scientific contribution of Kepler, Galileo, Newton.

______ 8. Describe the observations that told us the sun is moving through space.

______ 9. Describe the first evidence that the force of gravity is universal, operating outside the solar system.

______ 10. Suppose Mars was our home instead of the earth. Would an observation of stellar parallax from Mars prove it orbited the sun? Would it have been easier to discover a stellar parallax from Mars? Explain briefly.

# 4 Radiation: From Starlight to Understanding Stars



## EXAMINING LIGHT

### Light from Space

Astronomers are unique among scientists. For them, there is only one laboratory. It is the Universe, the same Universe that we all see at night. Astronomers cannot touch the objects in their laboratory. They cannot experiment by moving things around. They cannot dissect a star or stir up the great clouds of gas. Astronomers can only work with the one thing that reaches us, light or radiation.

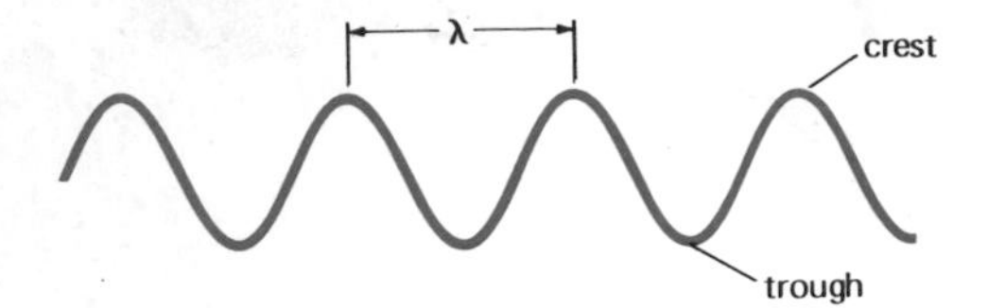


**FIGURE 4.1**
Wavelength of a wave is the distance from crest to crest or trough to trough. It is indicated by the Greek letter λ.

Radiation is continually sent out in all directions from stars and other objects in the Universe. It travels through the vacuum of space at the speed of light, about 300,000 km/sec. When the radiation enters our eyes, some of it produces a chemical change in our retinas. Our brains read this change as the sensation of light.

## Light Waves

Is light a wave or a stream of individual particles? For many years there was a lot of uncertainly about the answer to that question. As it turned out, two views about light are correct. For certain purposes, light must be represented as a stream of particles. Each particle is called a photon and has a certain amount of energy. At other times, light must be thought of as a wave. Light waves spread out in all directions from a star. The waves can be compared to ripples spreading over the surface of a pond when a stone is dropped into the water. The **wavelength** of light is the distance from crest to crest of successive waves (Figure 4.1).

All light in space, no matter what the wavelength, travels at the speed of light. Each second the light moves about 300,000 km. Suppose a star is radiating very long wavelengths of 1000 km each. By the end of one second, that star would have sent out 300 waves. The number of waves sent out from a star every second is called the **frequency** of the light. At times, frequency is quoted instead of wavelength. Everyone has probably heard a local radio station sign off by saying, "This is station WLS broadcasting with a frequency of 890 kilohertz." A **kilohertz** (kHz) is a thousand cycles per second (1000 cps) or a thousand waves per second. So the station is sending out 890,000 waves per second during its broadcast. The direct relationship between wavelength and frequency is as follows: The more waves sent out per second, or the shorter the wavelength, the higher the frequency (Figure 4.2).

Astronomers use both the particle and the wave theories of light. First, consider light as a wave. Visible light, it turns out, is only a small part of the radiation emitted by stars. Our eyes only respond to wavelengths between .000038 cm (centimeters) (violet light) and .000075 cm (red light).* Anything longer or shorter is not perceived by our eyes. The stars send out many wavelengths of light at the same time. Some waves are many meters long. These are called **radio waves.** The shortest waves are called **gamma rays.** The entire range of possible wavelengths is shown in Figure 4.3. We call the entire range of radiation the **electromagnetic spectrum.**

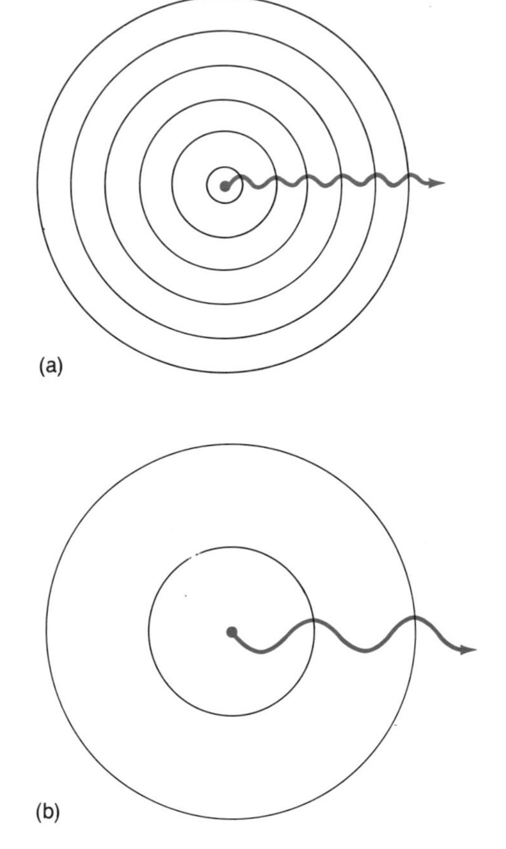


**FIGURE 4.2**
Short wavelengths (a) correspond to high frequency; long wavelengths (b) correspond to low frequency.

*Very large or very small numbers will sometimes be written in scientific notation (powers of 10). In Figure 4.3 the wavelength of violet light, 0.000038 cm, is expressed as $3.8 \times 10^{-5}$ cm. A review of scientific notation is found in the Appendix.

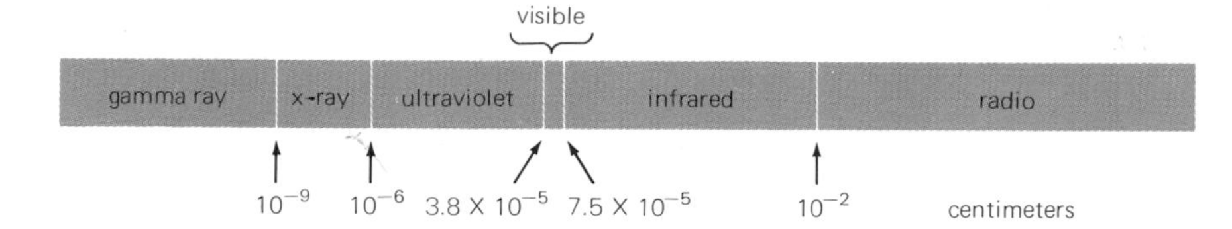


**FIGURE 4.3**
A schematic representation of the electromagnetic spectrum. The boundaries between regions are defined by convention.

Not all the radiation from the stars reaches the surface of the earth. In fact, most of it does not. Only visible light and radio waves clearly penetrate the earth's atmosphere. At those wavelengths, we have so-called windows to see the Universe (Figure 4.4). Since much of the radiation that we cannot see is harmful to life, we are fortunate to have such a protective atmosphere. The ozone in the earth's atmosphere is especially helpful in preventing the sun's dangerous ultraviolet radiation from reaching us. Since the advent of the space age, satellites orbiting above the atmosphere have been collecting some of the more interesting radiation from space. No matter what the wavelength or whether or not it penetrates the earth's atmosphere, some common properties are shared by all radiation. We will discuss these in the following sections.

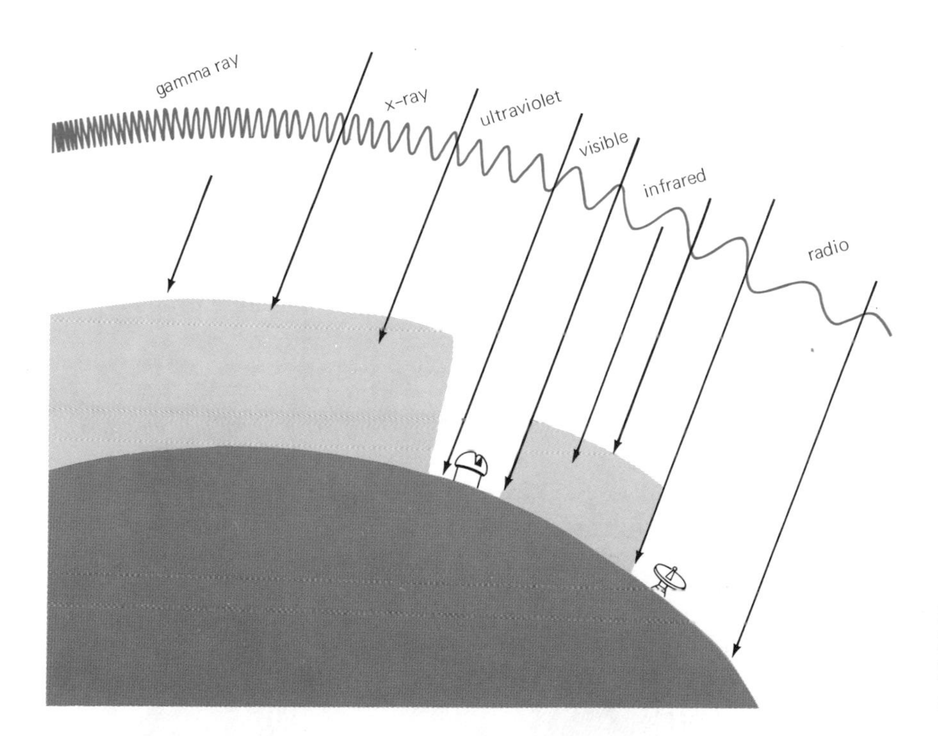


**FIGURE 4.4**
Radiation as it impinges on the atmosphere. Not all wavelengths reach the surface of the earth.

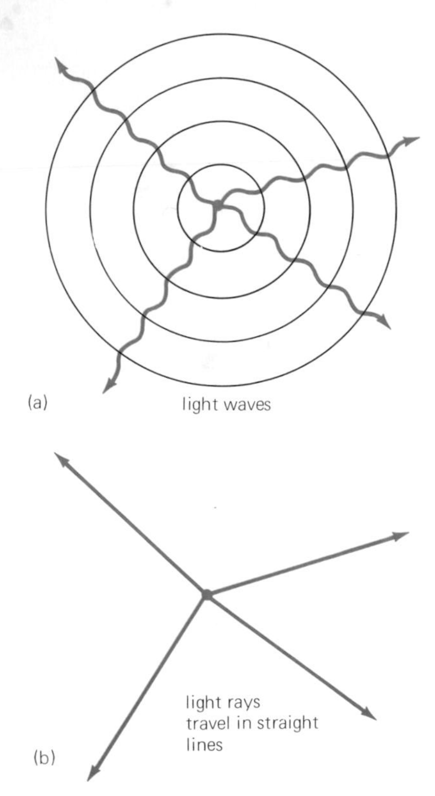


**FIGURE 4.5**
The propogation of light represented as waves (a) and rays (b)

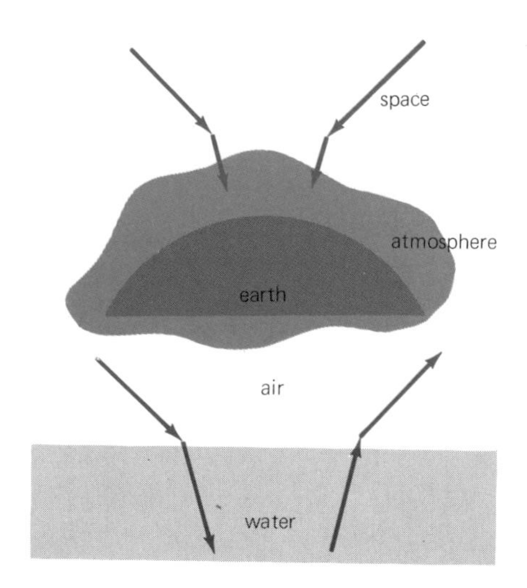


**FIGURE 4.6**
Light is refracted or bent when it passes from a vacuum into air (a) or air into water (b). It is also bent when it goes the other way.

## Refraction of Light

In general, radiation travels in straight lines. The direction the waves are moving away from a source could be indicated by a series of straight lines that are sometimes called light rays (Figure 4.5). These light rays keep moving in straight lines unless they encounter a new medium. In space, light travels at the speed of light. Any light that enters the earth's atmosphere is slowed down slightly, and the light ray is actually bent. Light in the air that enters a pool of water is slowed a little more, and the rays are bent more (Figure 4.6). The bending of light rays from one medium to another is called **refraction.**

The bending of light entering water gives rise to many familiar illusions. Put a pencil into a glass of water so that part of it is above the water. If we look at the half-submerged pencil, it appears broken right in the place where it enters the water. Our brains know from experience that light travels in straight lines. When light is actually bent slightly, such as going from air to water or vice versa, our eyes do not adjust for the bending. They see the pencil where it would be if the light were coming straight back to us. Our eyes are fooled and the pencil looks broken (Figure 4.7).

When we see the sun on the horizon, it is not really there. Sunlight coming through the atmosphere is always bent so that we see the sun a little higher in the sky than it actually is (Figure 4.8). At sunset and sunrise, we see the sun on the horizon, when in fact it is below the horizon. This refraction affect gives us several extra minutes of sunlight every day.

Light is also bent as it passes through glass. Refraction of light is the fundamental principle behind the functioning of eyeglasses and refracting telescopes. When our eyes do not focus light properly onto our retinas, we wear glasses. The lenses of glasses are specially shaped to adjust the bending of light and concentrate it on our retinas (Figure 4.9). Refracting telescopes generally have a set of lenses through which light passes and is brought to a focus (Figure 4.10).

## Reflection of Light

Light rays can also be reflected. A light ray striking a flat mirror will bounce off in a straight line and at a similar angle to its striking angle. Figure 4.11 shows several light rays striking a mirror at different angles. The angle of reflection is always the same as the striking angle.

The reflection principle of light is the basis for reflecting telescopes. The mirrors in reflecting telescopes are not flat. They are specially curved so that all the incoming light is reflected to a point called the focus (Figure 4.12).

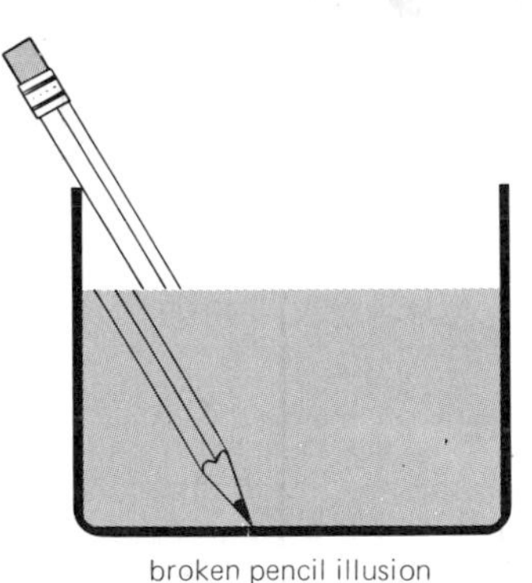


**FIGURE 4.7**
Refraction causes many illusions. The apparently broken pencil in a water glass is a common example.

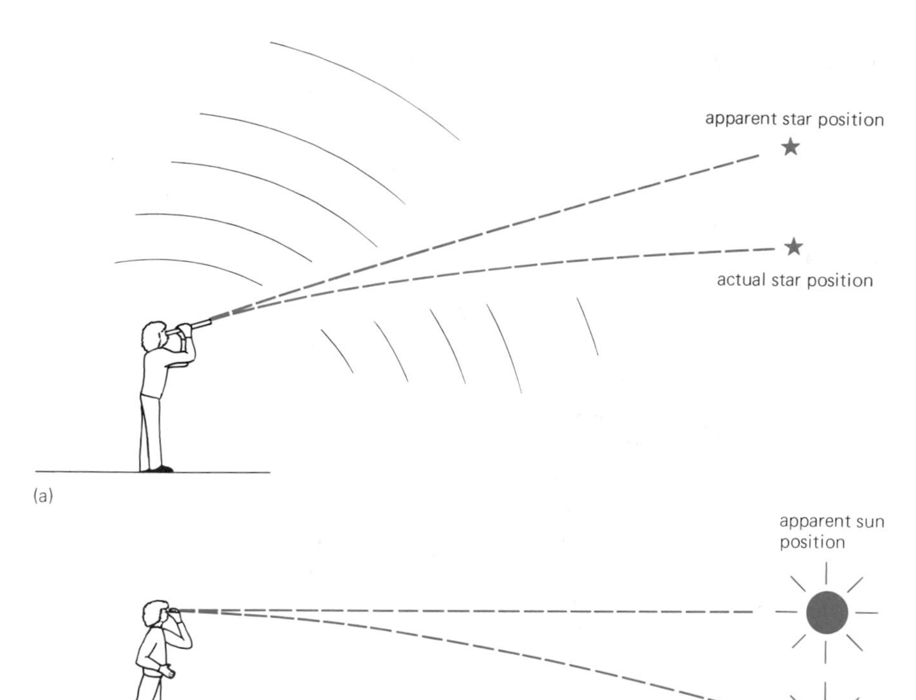


**FIGURE 4.8**
The positions of celestial objects are different depending upon the objects' distances from the zenith. The atmosphere refracts starlight and causes a star to appear higher in the sky than it actually is (a). The same is true for the sun on the horizon (b).

## Dispersion of Light

Light can also be dispersed. White light is made up of all different colors. When white light is passed through a prism, each color is bent by a slightly different amount (Figure 4.13). The prism disperses the colors into a rainbow from violet to red. This rainbow of light is called a

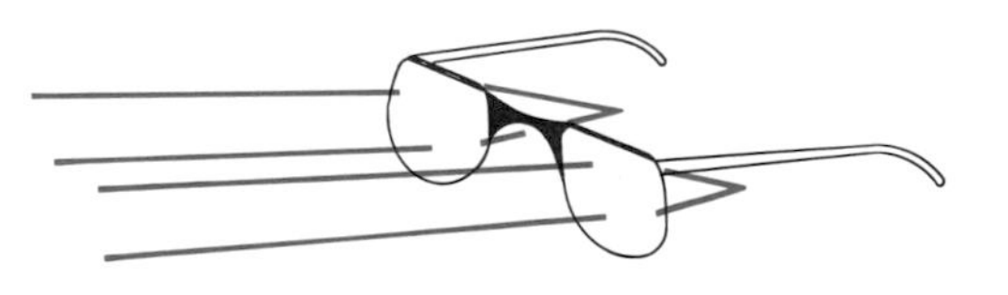

**FIGURE 4.9**
Spectacles are used to bend light to correct a person's vision.

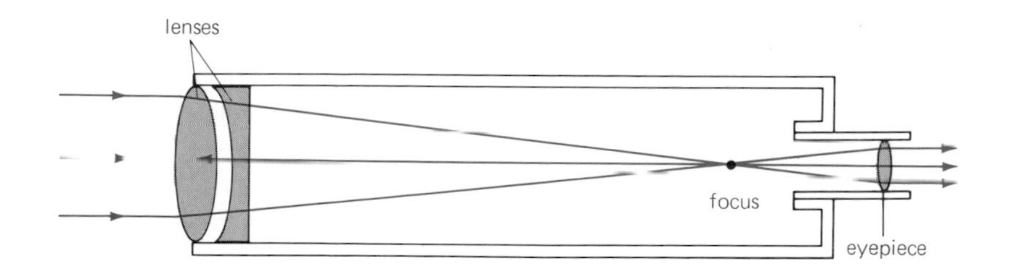


**FIGURE 4.10**
The principle of a refracting telescope. Light is collected and then refracted so that it comes together at the focus. An eyepiece is used to view the image formed.

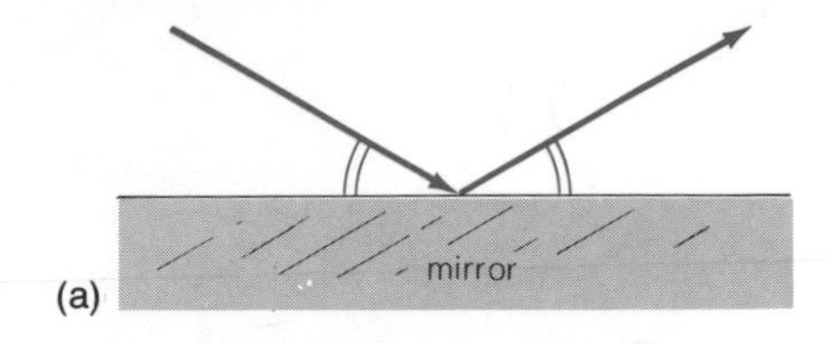


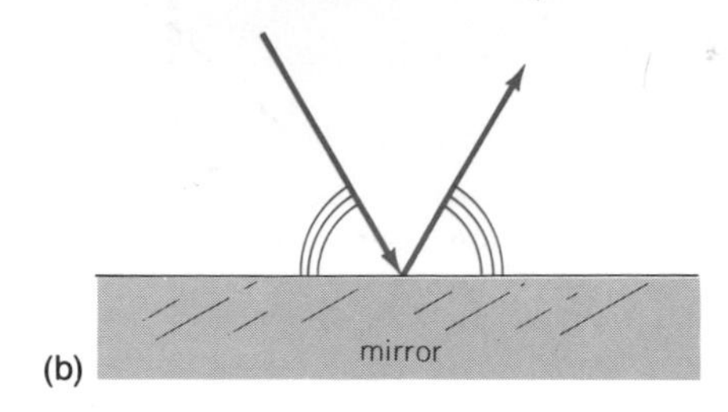


**FIGURE 4.11**
Light is reflected from a mirror at the same angle that it falls upon the mirror. If it strikes the mirror at a shallow angle (a) it is reflected at a shallow angle. If it strikes at a large angle (b), it is reflected at a large angle.

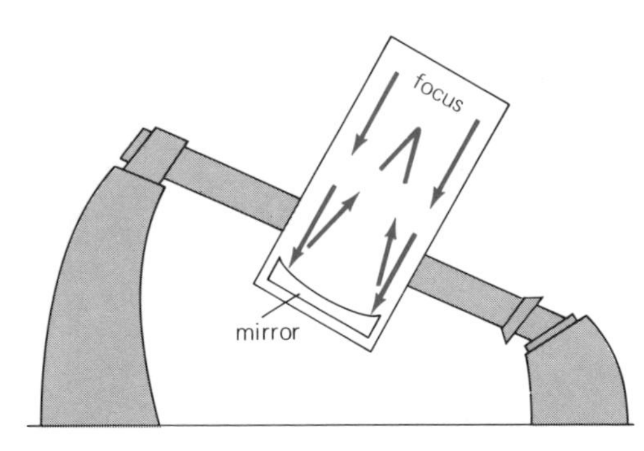


**FIGURE 4.12**
A reflecting telescope makes use of the principle of reflection. Light falling on the mirror is reflected to the focus of the mirror.

**spectrum** (the plural of spectrum is spectra). Light can be dispersed over any wavelength range, not just the visible one. Information contained in the spectrum of a star is discussed later in this chapter.

## Doppler Shift of Light

A **Doppler shift** of light occurs when the source of the light is moving either toward you or way from you. If you have heard a train whistle change its pitch as it passes by you, then you have experienced the sound wave version of the Doppler shift. While the train is approaching, its whistle pitch is raised. As soon as the train passes, its whistle pitch drops. The exact amount of change depends on the speed and direction of the train. Of course, light does not "whistle" in the same sense as sound, but the characteristic frequency and hence wavelength of the light changes when the source and/or receiver move with respect to each other.

Suppose the light source in Figure 4.14 is moving from left to right. It is moving directly away from observer A and directly toward observer B. As the source moves, it continually sends out light waves. The light waves in the direction of motion get crowded together. Observer B, watching the source approaching, would see shorter wavelengths or bluer light than the original. Waves in the opposite direction appear stretched out and longer or redder to observer A. The greater the speed of the source, the greater the Doppler shift of the wavelengths. Even though the source is moving, observer C sees the true wavelength.

When we observe distant galaxies, grand assemblies of stars, gas, and dust, their light always appears redder than it should. If this is interpreted as a Doppler shift of light, it means that the distant galaxies are moving away from us. Their light is red-shifted. Measuring the amount of the shift allows us to calculate the line of sight velocity, or **radial velocity,** of these objects. We will return to the Doppler shift when we get to the realm of galaxies and the expanding Universe (Chapter 12).

There is really only one thing that astronomers continuously do with starlight. They collect it. On every clear night, hundreds of astronomers at observatories all over the world are collecting starlight. They collect it through telescopes using the refraction and reflection properties of light. The larger the telescope, the more light they can collect. While the light is collected, astronomers can analyze the sum total of the light, or at the other extreme they can filter out all but a single wavelength of light. The maximum amount of information about stars can be determined by looking at the light received from them at every possible wavelength.

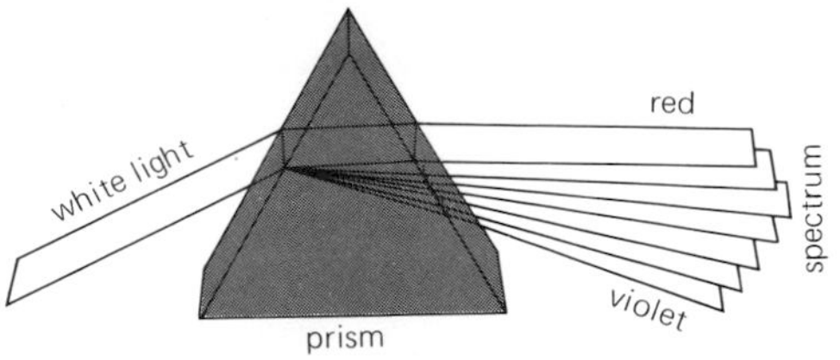

**FIGURE 4.13**
White light falling on a prism is dispersed into its various colors. This principle is used in making a simple spectroscope.

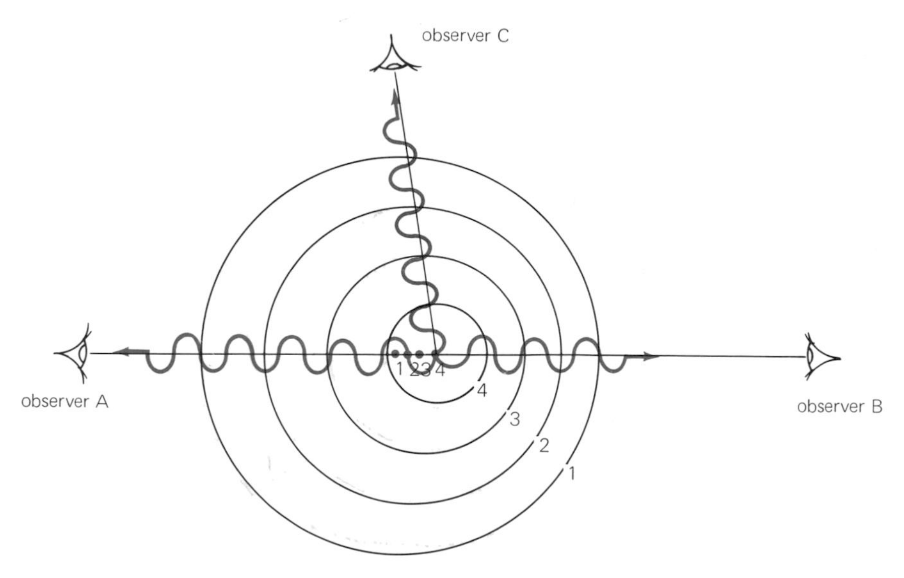


**FIGURE 4.14** The Doppler effect. The radiation from a moving source is shifted to shorter wavelengths in the direction that the source is moving and longer wavelengths in the direction away from which it is moving. Note that the true wavelength is seen by observer C.

## READING THE INFORMATION IN STARLIGHT

### The Whole Spectrum

Any star radiates energy at all the different wavelengths of the electromagnetic spectrum. The light at all wavelengths is not equally intense. Figure 4.15 shows the strength or intensity of the sun's radiation over many wavelengths. The sun's strongest wavelength is about 0.00005 cm. That happens to be right in the middle of the visible light

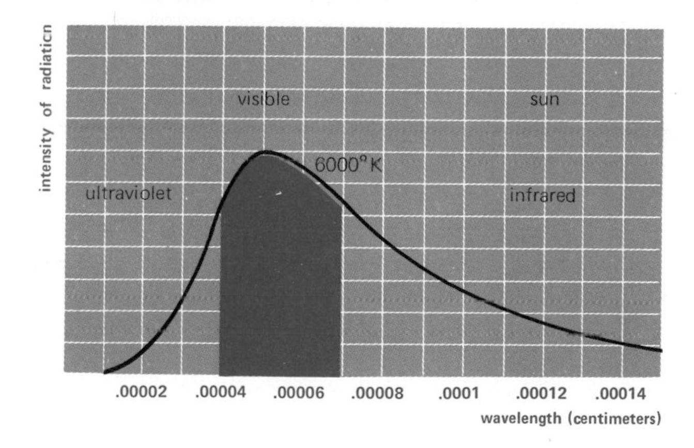


**FIGURE 4.15**
The radiation curve of the sun. Note that the maximum intensity is at visible wavelengths. The visible radiation is indicated by shading.

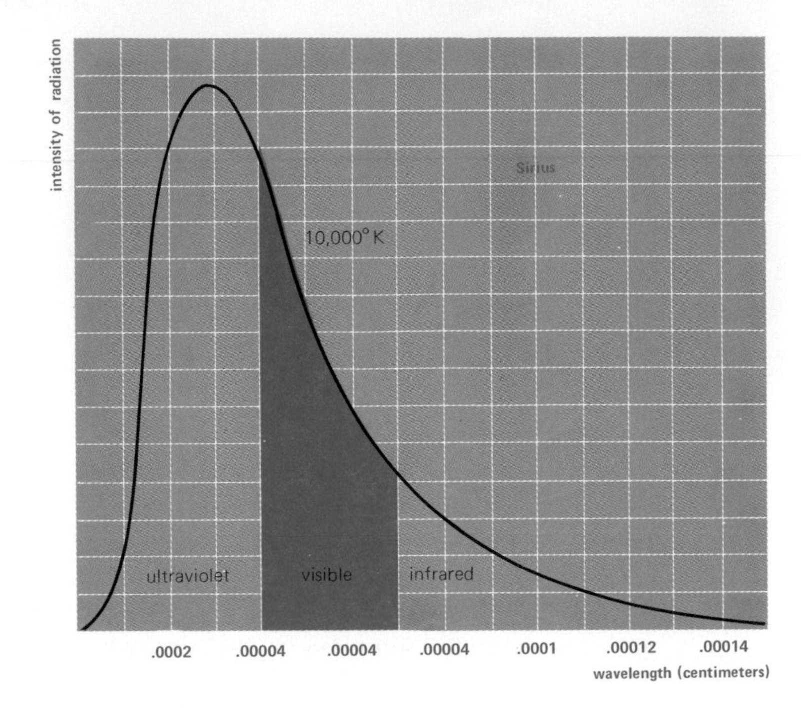


**FIGURE 4.16**
The radiation curve of Sirius. Note that the maximum intensity is at ultraviolet wavelengths.

range, where our eyes are most sensitive. That our eyes are so sensitive to the sun's strongest radiation is no accident. We do not receive the sun's weaker ultraviolet and infrared radiations because they do not penetrate the earth's atmosphere. If we extend the chart to long wavelengths, we would find that the sun radiates some energy at radio wavelengths.

Figure 4.16 shows the intensity of radiation for a star hotter than the sun, like Sirius. For Sirius, the strongest wavelength is in the ultraviolet region. When we look at Sirius in the summer sky, all we see is the visible light intensity. Figure 4.17 shows that there is not much visible radiation coming from the star called Ross 614. This star is cooler than the sun and difficult to see in a small telescope. Most of its radiation, and its brightest radiation, is in the infrared region.

The single most important factor in determining which wavelength is brightest is the temperature of the star.* The cooler the star, the longer the wavelength for the maximum strength. Since the wavelength of light determines the color of light, we can replace the term wavelength with color. Then we can state the relationship again. The

*Temperatures of stars are given in degrees Kelvin. To change Kelvin to Celsius, subtract 273°. A comparison of the major temperature scales is found in the Appendix.

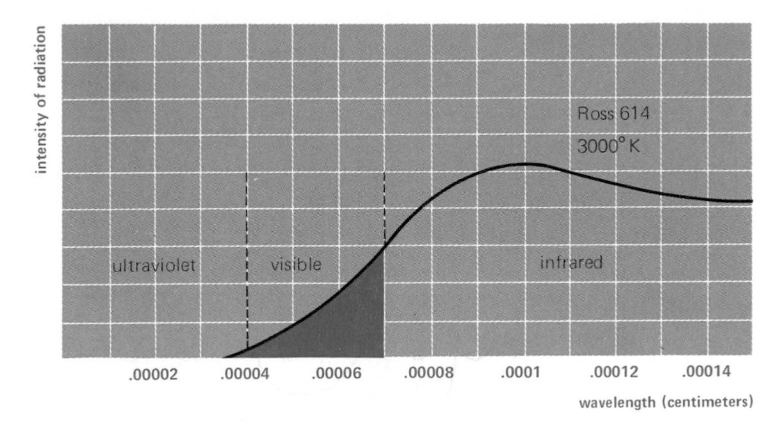


**FIGURE 4.17**
The radiation curve of Ross 614. Note that the maximum intensity is at infrared wavelengths.

single most important factor in determining which color is brigthest is the temperature of the star. The cooler the star, the redder the color.

The relationship between color and temperature can easily be demonstrated with a simple household dimmer switch. When we turn the lights up full, their color will be a bright blue-white and the lights will be at their hottest. As we dim the lights, their color will change until they appear quite dull and noticeably reddish. At this time the lights are cooler than when they are in the full up position. Obeying the same relationship, stars with hotter surface temperatures are blue-white, while those with cooler surface temperatures are red.

Astronomers use sophisticated electronic equipment attached to telescopes to measure the strength of a star's radiation at a selected wavelength. Usually, astronomers measure the radiation coming from two or three wavelengths. Comparing the strengths gives a measure of the star's color and hence its temperature. Even with your eye, the different colors of some stars can be detected. Certain red stars are especially easy to spot. Aldebaran and Betelgeuse are the reddest stars we see in the winter sky. Arcturus is an orange star visible in the spring evening sky. In the summer sky, the star Antares shows its redness.

## Three Types of Spectra

To observe a spectrum we usually need more than a simple prism; we need an instrument called a spectroscope. In what follows we will use the term prism to mean a spectroscope. Three kinds of spectra can be produced in a laboratory:

1. Passing the light from a glowing gas at low pressure through a prism produces a bright-line or **emission-line spectrum** (Figure 4.18). This spectrum shows only a series of bright lines. The bright

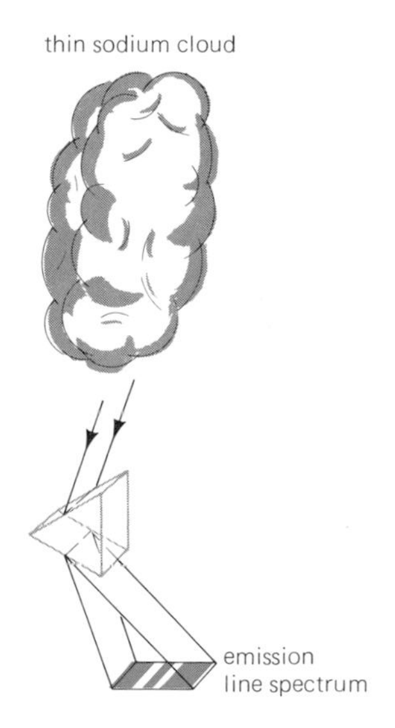


**FIGURE 4.18**
A thin glowing cloud emits radiation characteristic of its atoms. The radiation is at discrete wavelengths (frequencies) and an emission line spectrum results.

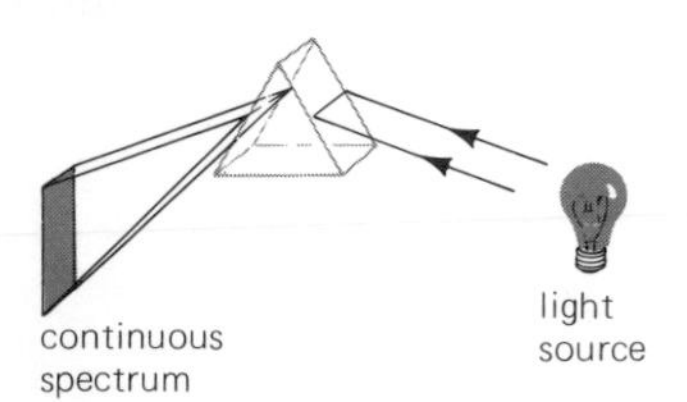


**FIGURE 4.19**
An incandescent filament is an example of a glowing solid (liquid or dense gas) and emits a continuous spectrum.

lines are radiations of the gas at specific wavelengths. If you look through a prism at a mercury vapor or sodium vapor street lamp, you will see bright lines of emission.

2. Passing light from a luminous solid, liquid, or gas under high pressure through a prism produces a **continuous spectrum.** No lines appear in this spectrum (Figure 4.19). Radiation is received continuously at all wavelengths. All the radiation blurs together and forms a continuous band of emission. If you look through a prism at a regular household light bulb, you will see a continuous spectrum.
3. Passing light that would normally produce a continuous spectrum through a gas that would normally produce a bright-line spectrum and then through a prism produces a dark-line or **absorption-line spectrum.** An absorption spectrum has a continuous spectrum in the background with dark lines on top of that (Figure 4.20). A gas can absorb radiation at the same wavelengths that it emits radiation.

If you pass sunlight through a prism, you will produce a nice rainbow of colors, a continuous spectrum. From that observation, we can learn that the sun is a luminous gas under high pressure. If you spread the sunlight even more by stretching the spectrum out, dark lines will appear. These are absorption lines (Figure 4.21). The dark lines in the solar spectrum were first discovered by Fraunhofer at the beginning of the 19th century. In general, the spectra of stars are dark-line spectra. This fact tells us that stars have cooler gases under low pressure surrounding the hot, high-pressure gases that produce the continuous part of the spectrum. These cooler gases make up the atmosphere of a star. A layered view of a star helps us to explain the dark absorption lines, but in reality, the structure of a star is much more complex as we will see in Chapters 10 and 11.

Each dark line in a star's spectrum can be identified with a

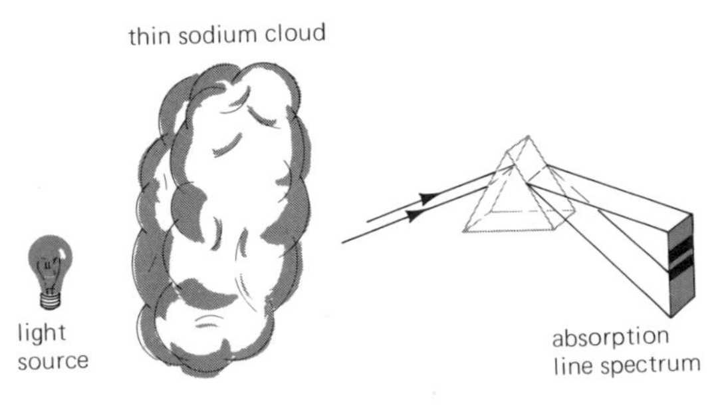


**FIGURE 4.20**
Light from a continuous source passing through a low-density cool cloud will have removed from it light characteristic of the atoms making up the cloud. The absence of light leaves an absorption line spectrum.

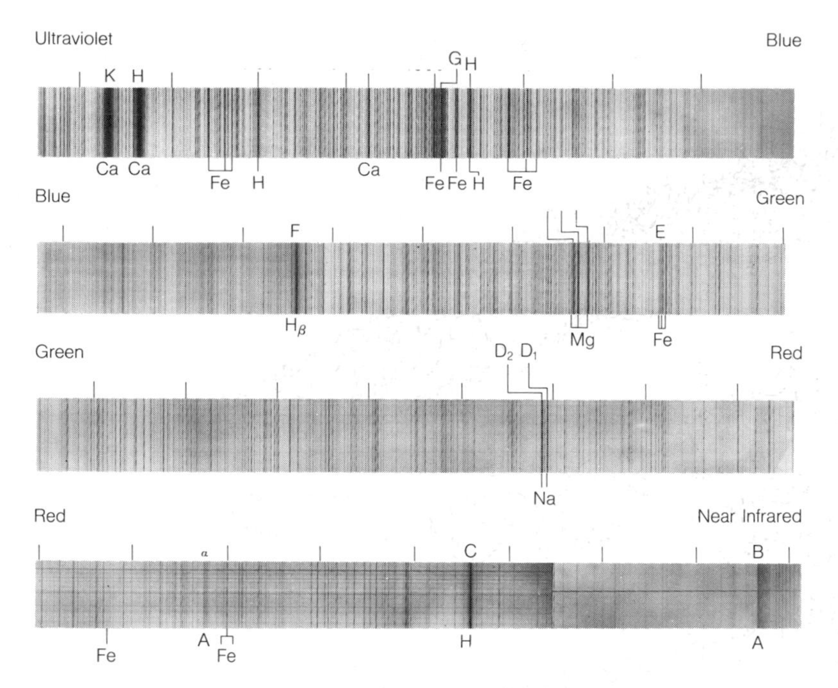


**FIGURE 4.21**
The spectrum of the sun. The spectrum is an absorption line spectrum because the cool, low-density atmosphere of the sun lies above hotter gases and absorbs selected outgoing radiation. The Fraunhofer lines are marked B through K, above the spectrum. Lines of specific elements are identified below the spectrum.

particular wavelength and with a specific chemical element. In order to understand how the spectral lines are produced in stars, we must examine the atoms that make up the star.

## The Fingerprints of Atoms

Atoms are the building blocks of all material. They are composed essentially of electrons, protons, and neutrons. The electron is the lightest of these constituents. It carries a unit negative charge of electricity. The proton is 1,836 times as massive as the electron and carries a unit positive charge. The neutron has about the same mass as the proton and is electrically neutral. Table 4.1 (p. 76) lists the elements from hydrogen through iron along with how many electrons, protons, and neutrons make up the most common form of the element.

The nucleus is the heart of the atom. It consists of protons and neutrons. In the normal atom, the nucleus is surrounded by negatively charged electrons equal in number to the protons, so that the atom as a whole is electrically neutral. Hydrogen is the simplest atom. It has one proton and no neutrons in its nucleus. Circling around the nucleus is a single electron (Figure 4.22). Helium has two protons and two neutrons in its nucleus, which is surrounded by two electrons. Lithium has three

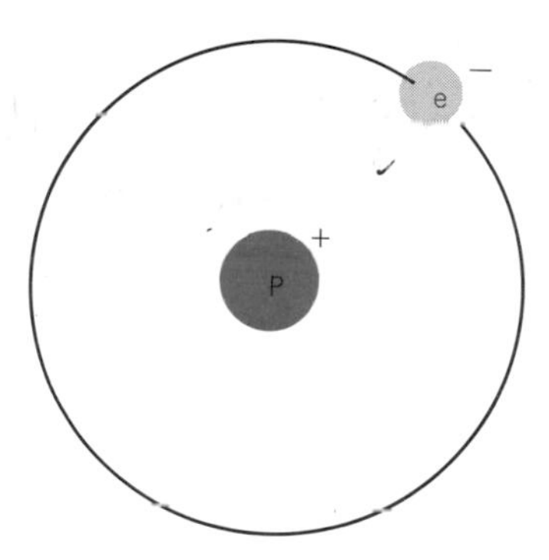


**FIGURE 4.22**
A simple schematic of the hydrogen atom.

**TABLE 4.1 The Chemical Elements***

| Element | Symbol | Atomic Number | Protons Number | Neutrons Number | Electrons Number | Atomic Weight |
|---|---|---|---|---|---|---|
| Hydrogen | H | 1 | 1 | 0 | 1 | 1.0080 |
| Helium | He | 2 | 2 | 2 | 2 | 4.0026 |
| Lithium | Li | 3 | 3 | 4 | 3 | 6.941 |
| Beryllium | Be | 4 | 4 | 5 | 4 | 9.012 |
| Boron | B | 5 | 5 | 6 | 5 | 10.811 |
| Carbon | C | 6 | 6 | 6 | 6 | 12.011 |
| Nitrogen | N | 7 | 7 | 7 | 7 | 14.007 |
| Oxygen | O | 8 | 8 | 8 | 8 | 15.999 |
| Fluorine | F | 9 | 9 | 10 | 9 | 18.998 |
| Neon | Ne | 10 | 10 | 10 | 10 | 20.179 |
| Sodium | Na | 11 | 11 | 12 | 11 | 22.990 |
| Magnesium | Mg | 12 | 12 | 12 | 12 | 24.305 |
| Aluminum | Al | 13 | 13 | 14 | 13 | 26.98 |
| Silicon | Si | 14 | 14 | 14 | 14 | 28.09 |
| Phosphorus | P | 15 | 15 | 16 | 15 | 30.97 |
| Sulfur | S | 16 | 16 | 16 | 16 | 32.06 |
| Chlorine | Cl | 17 | 17 | 18 | 17 | 35.45 |
| Argon | A | 18 | 20 | 20 | 18 | 39.95 |
| Potassium | K | 19 | 19 | 20 | 19 | 39.10 |
| Calcium | Ca | 20 | 20 | 20 | 20 | 40.08 |
| Scandium | Sc | 21 | 21 | 24 | 21 | 44.96 |
| Titanium | Ti | 22 | 22 | 26 | 22 | 47.90 |
| Vanadium | V | 23 | 23 | 28 | 23 | 50.94 |
| Chromium | Cr | 24 | 24 | 28 | 24 | 52.00 |
| Manganese | Mn | 25 | 25 | 30 | 25 | 54.94 |
| Iron | Fe | 26 | 26 | 30 | 26 | 55.85 |

*Most common isotope is used in this table.

protons and four neutrons in its nucleus, which is surrounded by three electrons. Each successive atom has a more complex nucleus and added electrons. According to a special set of rules, atoms absorb and emit radiation. The rules are easier to describe if we now think of radiation or light as a stream of photons (bundles of energy). It will also simplify things if we begin with hydrogen, the simplest and most abundant element in the Universe.

We have already described the normal hydrogen atom. It has one proton in its nucleus and one electron orbiting the nucleus. Unlike the earth's orbit around the sun, which is always along the same path, there are numerous possibilities for the orbit of the electron around the nucleus. At the turn of the 20th century, Niels Bohr showed that there were a whole series of possible orbits for the electron. Figure 4.23 shows a representation of the hydrogen atom with circles indicating possible orbits of the electron. These orbits are also called **shells** of the

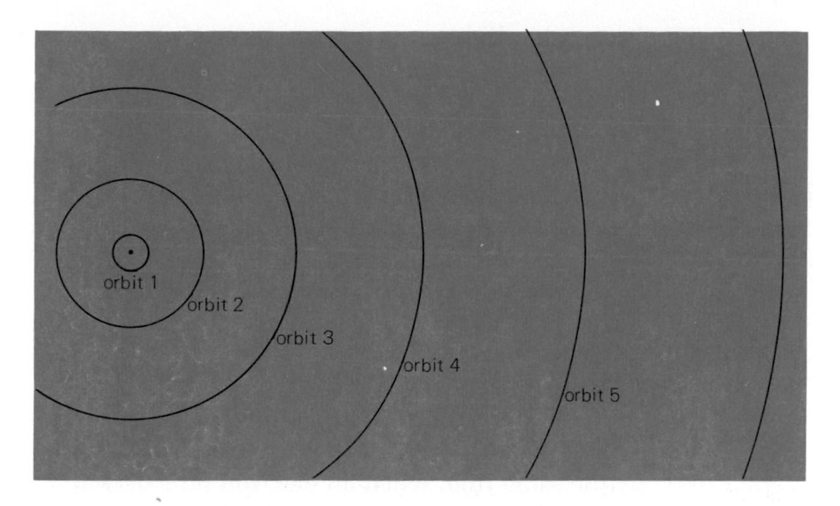


**FIGURE 4.23**
A simple schematic of the hydrogen atom showing five of the many allowed orbits.

atom. This particular way of looking at the atom has come to be known as the **Bohr model** of the atom.

When the electron is in the closest orbit to the nucleus, we say the atom is in the **ground state.** The ground state is the lowest energy state. When the electron is found in any other orbit, we say that the hydrogen atom is in an **excited state.** One way to excite an atom is by using photons. If a photon of just the right energy comes along, it can be absorbed by the atom. The extra energy delivered by the photon gives the electron a boost out to one of the larger orbits (Figures 4.24).

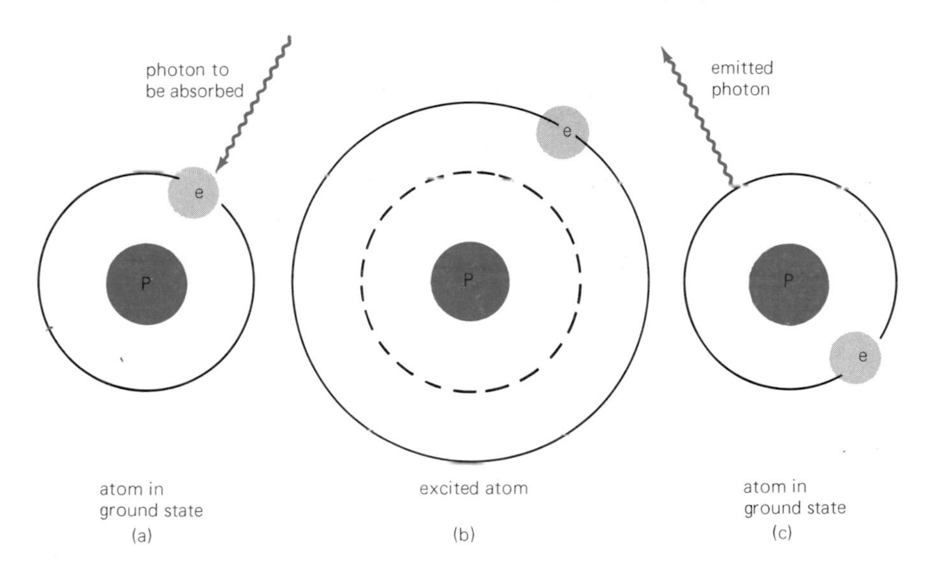


**FIGURE 4.24**
When a photon has the proper energy and is absorbed by an atom, the atom goes to an excited state. The atom immediately radiates a new photon of the same energy and deexcites. The new photon can go off in any direction.

Usually, the electron hardly gets out to an excited orbit when it spontaneously falls back to the ground state and the atom emits the same energy photon that it absorbed only a moment ago.

The absorption and emission of photons by the hydrogen atom is a never-ending process. The hydrogen atoms in stars are continually bombarded by photons. Sometimes a photon of just the right energy will boost the electron many orbits at once. A very high-energy photon may completely free the electron from the atom. An atom that has lost an electron is said to be **ionized,** and the atom is called an ion. An ion of helium would have two neutrons, two protons, and only one (or even no) electron. The special part of the Bohr model of the atom is that only whole jumps between orbits are possible. An atom cannot absorb a photon that would boost its electron 2½ orbits. It can only absorb photons that would raise it exactly one, two, three or more whole orbits. By the same rule, when an electron falls back it must fall an exact number of orbits. This restriction to whole orbits means that absorptions and emissions will only occur at certain energies.

Whenever the electron changes its orbit, a spectral line is produced. If the electron moves to a larger orbit, the atom takes in energy or absorbs radiation. If the electron falls to a lower orbit, the atom releases energy or emits radiation. Every hydrogen atom whose electron jumps from orbit 3 to orbit 2 is emitting the same amount of energy and producing the same spectral line. Other electrons going from orbit 5 to orbit 3 are producing a spectral line at a different wavelength. When billions of hydrogen atoms are absorbing and emitting radiation the way they are in stars, many lines are produced. The series of lines for hydrogen is unique to that element, a true fingerprint of the atom.

More complex atoms than hydrogen have more electrons and a different shell structure for the possible orbits. With more electrons, more jumps between orbits are possible. Since every jump produces a spectral line, there are many more lines in the spectra of complex atoms (Figure 4.25). The placement of the spectral lines for any element can be determined experimentally in laboratories. In reality, the spectrum of a star contains the lines of many elements together. Some of the lines crowd together and even overlap. It takes a good sleuth to separate the lines and positively identify elements in the star. Over sixty elements

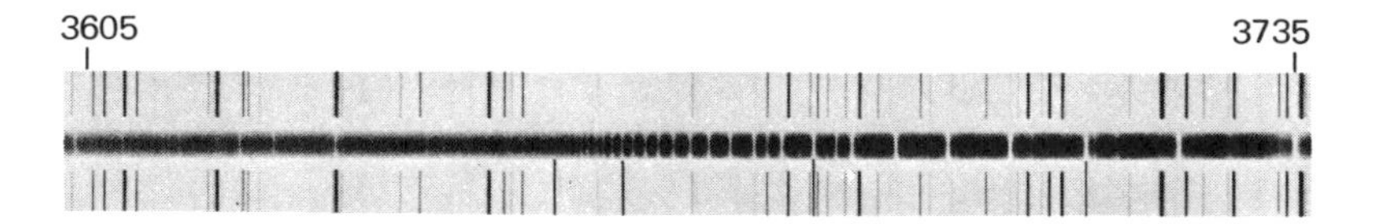


**FIGURE 4.25**
The absorption line spectrum of hydrogen flanked top and bottom by the emission line spectrum of iron and neon. The iron and neon spectrum is more complicated than that of hydrogen because many more electron energy levels are involved. This is a negative print so dark lines are actually bright emission lines and white areas and lines are dark absorption lines. Region of the spectrum is indicated at the top. (Hale Observatories photograph)

have been observed in the spectrum of the sun. The lines of helium were identified in the sun's spectrum before that element was observed on earth, hence the name helium after the Greek word for sun, *helios*.

In addition to identifying elements, the strength of the spectral lines tells astronomers something about the temperature of the gas producing the lines. Consider the hydrogen atom again. A whole set of spectral lines is produced by the electron jumping from the ground state to any larger orbit. Another set of lines is produced by the electron jumping from the second orbit to any larger orbit. Still another set of lines is produced by the electron jumping from the third orbit to any larger orbit and so on. Electrons can change orbits when the atoms bump into each other. As hydrogen gas is heated up, more atoms are found in excited states because the bumping or collision is more vigorous and more frequent. Suppose most of the hydrogen electrons are excited to the second orbit. Then the set of lines produced when these electrons change orbits will be stronger than any other lines. By comparing the strengths of different sets of spectral lines, astronomers can determine how excited the gas is, and that leads them directly to the temperature of the gas.

There is one last point to be made about stellar spectra. In almost all cases, the lines in a stellar spectrum are dark absorption lines. This may seem odd, since absorption and emission always go on together. We might think they would cancel each other out. The key to this puzzle is the direction of the event. When an atom absorbs energy, it generally soaks up a photon that would have come straight through to us. When the atom emits that same photon, it can send the photon off in any direction. Some photons may come back in our direction, but most atoms will emit the photons in other directions. Because we see all the absorption and only some of the emissions, the spectral lines are dark.

## Spectral Classification of Stars

When we look at the spectra of many stars, it becomes obvious that some of them are strikingly similar. As early as 1860, an attempt was made to classify the stars according to their spectra. An Italian astronomer, Father Secchi, grouped the spectra of about 4,000 stars into four distinct types. By the late 19th century, spectral classification of stars was proceeding rapidly. Astronomers had discovered how to record the spectra of many stars at once. They learned to place a thin wedge of glass called an objective prism in front of the telescope. The objective prism dispersed the light entering the telescope and made the image of each star into a spectrum of light. A single photograph taken through the telescope might show dozens of stellar spectra (Figure 4.26).

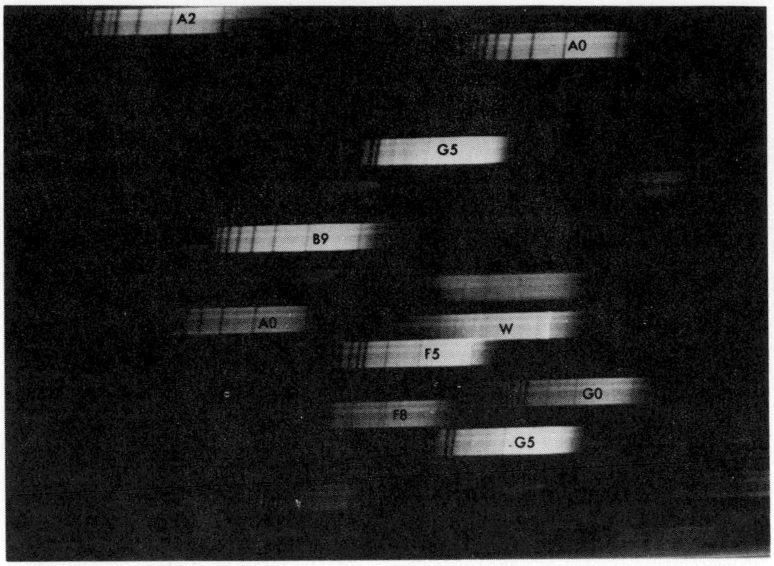

**FIGURE 4.26**
An objective prism plate showing the spectra of eleven stars. Note the various absorption line spectra and the one emission line spectrum marked W. The letters represent spectral types discussed later in this chapter.

By 1890, over 10,000 stars had been photographed and their spectra classified at the Harvard College Observatory. These classifications were published in *The Draper Catalogue of Stellar Spectra*. The classification types were similar to Father Secchi's, only there were more. Sixteen separate star types were designated by the letters A through Q (except J, since it can be mistaken for I).

The first classifications of stars were all done without an understanding of the Bohr model of the atom and without knowing what produced the lines in the spectra. In the early part of the 20th century, all that changed. The lines were explained. It was also recognized that all stars were quite similar in makeup. They were composed mainly of hydrogen and helium with a fraction of other elements. The differences in star spectra were due primarily to differences in the temperatures of the stars. A set of hydrogen lines produced when an electron jumped from orbit 2 to any other orbit was strongest in the A stars. For hotter stars, the hydrogen atoms were more excited and even ionized so the visible lines weakened.

In order to make the star classification show a smoothly varying temperature sequence, the original types were changed slightly. Several types were combined and some types omitted. Those that were left were arranged in order of decreasing temperature. This left the following spectral types: O B A F G K M (R N S). The last three types (R, N, S), along with type W, were added later. They represent relatively rare stars. Each major type is divided into ten subdivisions except for O stars which start at O5 and go to O9. For example, there are A0, A1, and so on to A9 stars. The sun is a G2 star. The O stars are the hottest, with surface temperatures of about 35000°K. G stars like the sun have surface temperatures of about 6000°K. At the cool end of the sequence are the M stars with surface temperatures of about 2000°K.

For years, astronomy students have used the mnemonic "O Be A Fine Girl Kiss Me, Right Now Sweetheart" to remember the spectral types in order. More recently, other students have come up with variations such as

> "O Bring A Fully Grown Kangaroo, My Recipe Needs Some"
> "Oh Brutal and Fearsome Gorilla, Kill My Roommate Next Saturday"
> "Oven-Baked Ants, Fried Gently, Kept Moist, Retain Natural Succulence"
> "Out Beyond Andromeda, Fiery Gases Kindle Many Red New Stars"*

With the new spectral types and an understanding of what produced spectral lines in stars, spectral classification became less arbitrary. Specific lines in the spectra of stars were compared in strength. Just how strong or dark one line was, compared to another line, determined the spectral type of the star. In the early part of the 20th century, A. J. Cannon, working at Harvard, single-handedly classified nearly a quarter of a million stars. The *Draper Catalogue*, extended and revised, was published in 1924. It gives the approximate positions, magnitudes, and spectral types of 225,300 stars. Some of the characteristics of the individual spectral types are as follows (see also Figure 4.27):

**Type O.** Lines of ionized helium, oxygen, and nitrogen are prominent in the spectra of these very hot stars, along with lines of hydrogen.
**Type B.** Lines of neutral helium are most intense at B2 and then fade, until at B9 they have practically vanished. Hydrogen lines increase in strength through the subdivisions. Examples are Spica and Rigel.
**Type A.** Hydrogen lines attain their greatest strength at A2 and then decline through the remainder of the sequence. Examples are Sirius and Vega. Thus far, the stars are blue.
**Type F.** Lines of metals† are increasing in strength, notably the

*From *Mercury* magazine, from students in Owen Gingerich's classes.

†Since most of the Universe is made up of mostly hydrogen and helium, astronomers tend to refer to all of the other elements collectively as metals.

**TABLE 4.2 Stellar Types and Surface Temperatures**

| *Spectrum* | *Temperature (°K)* |
|---|---|
| MAIN SEQUENCE | |
| O5 | 35,000 |
| B0 | 21,000 |
| B5 | 13,600 |
| A0 | 9,700 |
| A5 | 8,100 |
| F0 | 7,200 |
| F5 | 6,500 |
| G0 | 6,000 |
| G5 | 5,400 |
| K0 | 4,700 |
| K5 | 4,000 |
| M0 | 3,300 |
| M5 | 2,600 |

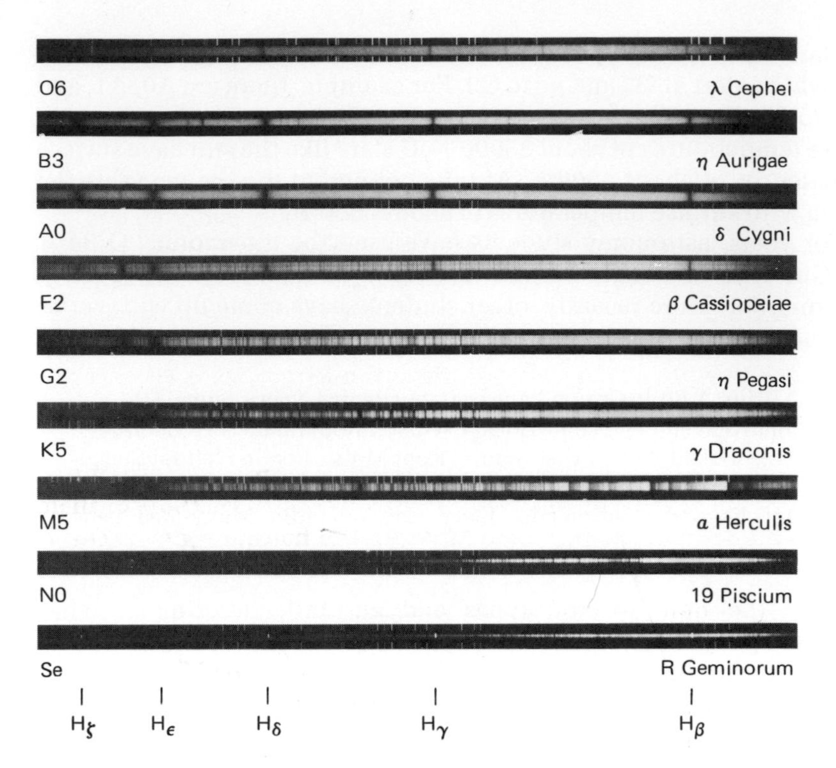


**FIGURE 4.27**
Typical stellar spectra. Note that the hydrogen lines (Hβ, Hδ, etc.) are strongest in the spectrum of δ Cygni (AO). Many metal lines appear in the G and K stars, and molecular bands can be seen in the M, N, and S spectra.

Fraunhofer H and K of ionized calcium. These are yellowish stars. Examples are Canopus and Procyon.

**Type G.** Metallic lines are numerous and conspicuous in the spectra of these yellow stars. The sun and Capella belong to this type.

**Type K.** Lines of metals surpass the hydrogen lines in strength. Bands (crowded lines) of cyanogen and other molecules are becoming conspicuous. These cooler stars are reddish. Examples are Arcturus and Aldebaran.

**Type M.** Bands of titanium oxide become stronger up to their maximum at M7. Vanadium oxide bands strengthen in the still cooler divisions of these red stars. Examples are Betelgeuse and Antares.

Four additional and less populous types branch off near the ends of the sequence. Type W near the blue end comprises the Wolf-Rayet stars, which have broad lines in their spectra. Near the red end, types R and N show molecular bands of carbon and carbon compounds, and type S has conspicuous bands of zirconium oxide.

# ORGANIZING THE STARS

## The H-R Diagram

Millions of stars have been observed by astronomers, but billions more have not been observed. Although observing every star is impossible, there is a way to organize what is known to learn something about all stars. Making a graph using absolute magnitude (Chapter 3) and spectral type brings an immediate order out of seeming chaos.

The horizontal line of Figure 4.28 is labeled with spectral types in order of temperature. The blue-white, high-temperature stars are to the left. The dull red, cool-temperature stars are to the right. The vertical line represents absolute magnitudes or luminosities of stars. The brightest stars are at the top and the faintest stars are at the bottom. When stars are organized this way, it turns out that over 90% of them lie along a band from the top left corner to the bottom right corner. A few stars are in the top right corner above the band, and a few others are below the band to the left.

The principal band of stars is called the **main sequence** (Figure 4.29). The positions of stars on this graph have nothing to do with their real space positions. The graph merely shows a relationship between

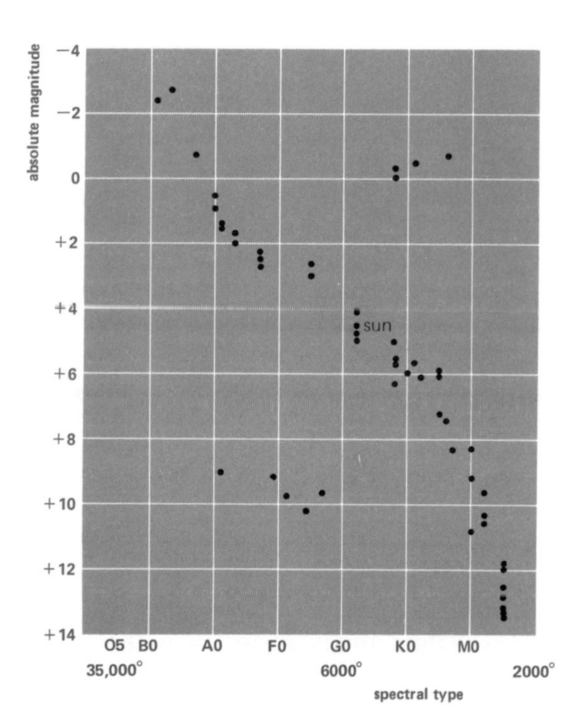


**FIGURE 4.28**
An H-R diagram for selected stars. The horizontal scale is given in spectral types with a few temperatures beneath. The vertical scale is given in absolute magnitudes.

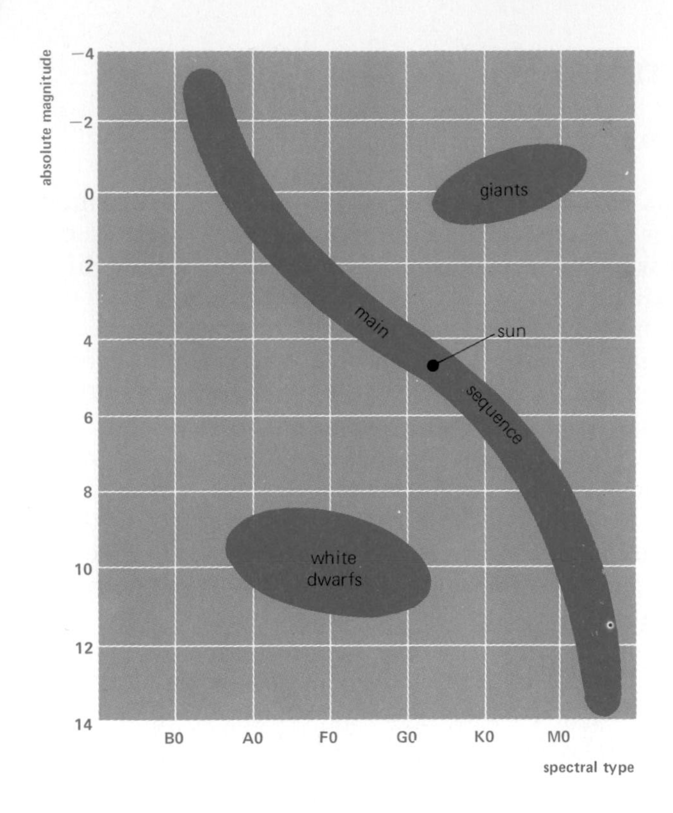


**FIGURE 4.29**
A schematic H-R diagram showing the location of the main sequence, red giants, and white dwards.

two characteristics of a star, spectral type and absolute magnitude. It is a little like charting the relationship between height and weight for adults. The vast majority of people will fall within a certain range or band on the chart (Figure 4.30) ranging from the taller and heavier down to the smaller and lighter. They are not physically on the band. In the same way, most stars fall within a certain range, the main sequence. The sun is a main sequence star. Its position is marked in Figure 4.29. Compared to other main sequence stars, the sun is average in both spectral type and absolute magnitude. For main sequence stars, the relationship between spectral type and absolute magnitude may be stated several ways. Basically, the hotter and bluer the star, the more luminous it is, or the cooler and redder the star, the fainter it is.

Stars that are in the two areas of the chart off the main sequence are more exotic and perhaps more interesting. The stars located below the main sequence are called **white dwarfs**—white because they have fairly high temperatures, and dwarfs because they are not very bright or luminous. The white dwarf stars are also small stars. Some of them are only a few times larger than the earth, yet they contain the same

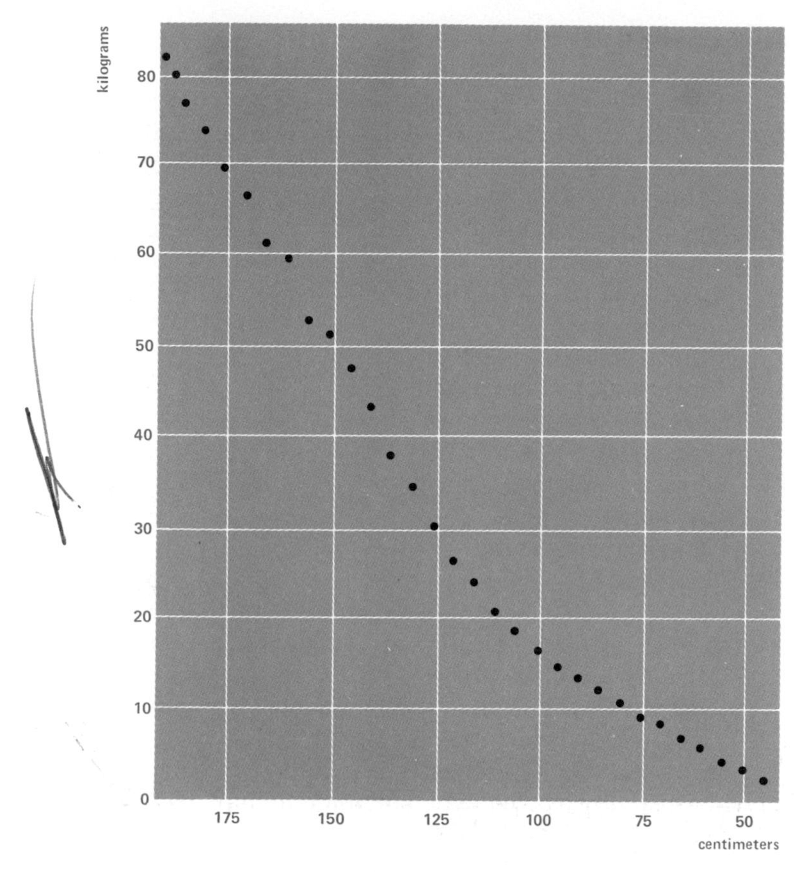


**FIGURE 4.30**
A graph of weight versus height for average U. S. males. This is analogous to the stellar main sequence.

amount of material as an average-size star such as the sun. If you can imagine squeezing the sun down to earth size, then you can realize how dense white dwarfs are. A teaspoon of matter from a white dwarf star would weigh over 1200 kg (kilograms) if brought to earth.

In the opposite corner of the chart are the **red giant** stars. They are red in color and extremely bright or luminous. Red giants are giants in size as well as brightness. The brightest red stars, like Betelgeuse in Orion's shoulder, outshine the sun by a hundred thousand times and are called supergiants. In the story of the birth and death of stars (Chapters 7 and 11), red giants and white dwarfs play important roles.

Studies of the relationship between spectral type and absolute magnitude were first completed about 1911 by two astronomers, Henry Russell and Ejner Hertzsprung. In their honor, this kind of chart is called a Hertzsprung-Russell Diagram or, simply, an H-R diagram.

## Red Giant or White Dwarf

It is obvious from looking at the H-R diagram that there are at least two possible kinds of G stars. A G star could be a main sequence star or a red giant. To distinguish between these possibilities, **luminosity classes** of stars are indicated following the spectral type of the star. The luminosity classes were first described and assigned by Morgan and Keenan. Their system has six main divisions, beginning with the most luminous stars.

I Supergiants
Ia Most luminous supergiants
Ib Less luminous supergiants
II Bright giants
III Normal giants
IV Subgiants
V Main sequence stars
VI Subdwarfs

Figure 4.31 shows where the luminosity classes fall on the H-R diagram. Some examples of the use of luminosity classes are these: Deneb, A2 Ia, means Deneb is an A star and a most luminous supergiant; sun, G2 V, means G-type main sequence star.

Many stars do not fit neatly into the spectral types and luminosity classes just described. Most conspicuous are the different types of variable stars discussed in Chapter 11.

## Star Limits

How big? How heavy? How bright? How hot? Are there any star limits? The answer to the last question is yes. The standard H-R diagram gives the limits on brightness or luminosity and temperature. Figure 4.32 shows an H-R diagram with temperature and luminosity expressed in terms of the sun's temperature and luminosity. It shows that temperatures of stars can be about four times greater or four times less than the sun's. The luminosity of stars has a much larger range. Stars can be as much as one hundred thousand times brighter or fainter than the sun.

Sizes and masses of stars are not as easy to determine as luminosity and temperature. One long and complicated method for determining sizes and masses involves applying Kepler's Harmonic Law to certain double stars. Here we will only show some results.

In Figure 4.33, the circles represent the actual sizes of stars relative to the sun's radius of 696,000 km. The sizes of stars on the main se-

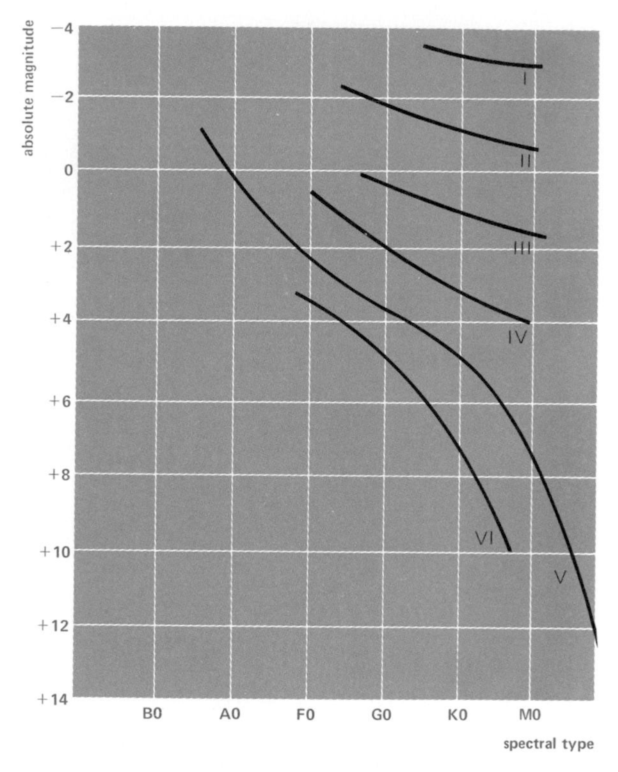


**FIGURE 4.31**
A schematic H-R diagram showing the locations of the principal luminosity classes: supergiants (I), bright giants (II), giants (III), subgiants (IV), main sequence or dwarfs (V), and subdwarfs (VI).

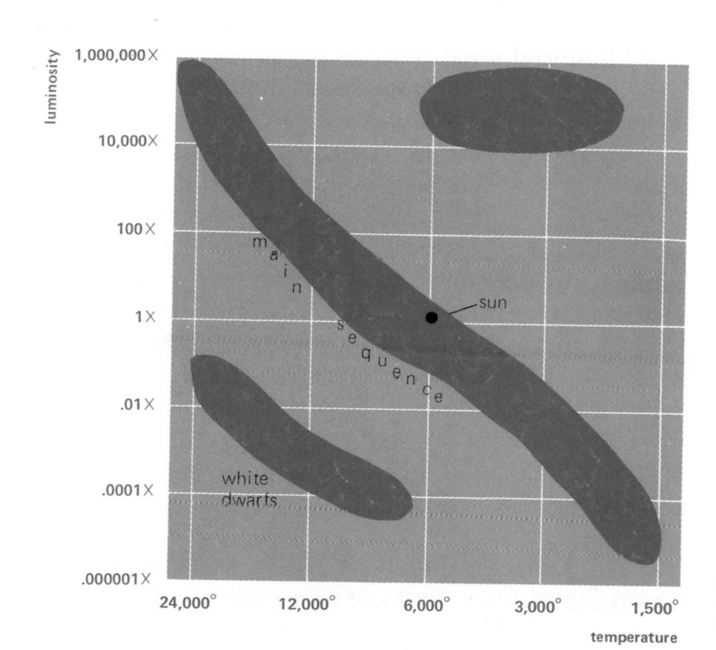


**FIGURE 4.32**
The luminosity-temperature diagram. This is a modified H-R diagram where the horizontal scale is given in temperature and the vertical scale is in terms of the brightness of the sun.

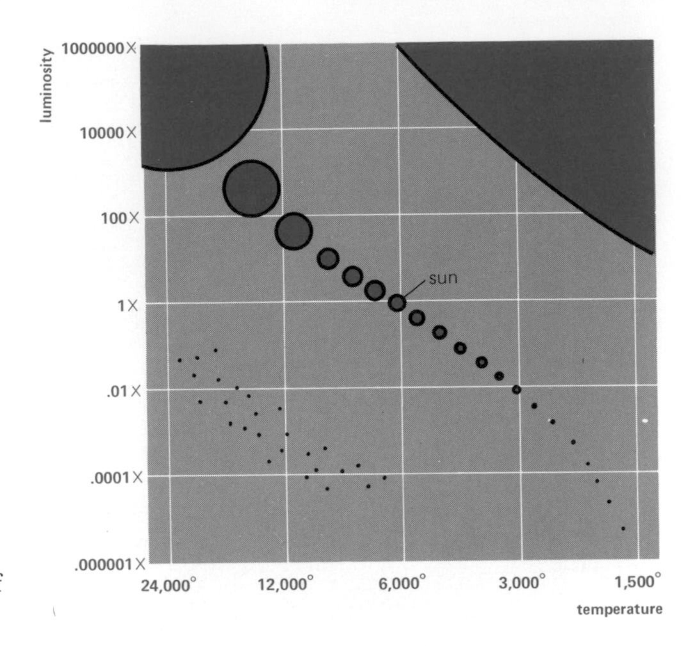


**FIGURE 4.33**
A luminosity-temperature diagram showing schematically the relative sizes of stars.

quence vary between ten times more and ten times less than the sun's size. The white dwarfs are fifty times smaller than the sun and not much bigger than the earth. The red giants and supergiants can be up to eight hundred times the size of the sun. A star that big, placed where the sun is, would swallow up comfortably the sun and the orbits of Mercury, Venus, earth, and even Mars.

Finally we look at the masses of the stars. In Figure 4.34, the numbers along the main sequence represent the masses of stars relative to the sun's mass of $2 \times 10^{33}$ g (grams). Just as there are limits to the luminosities, temperatures, and sizes of the stars, the masses of stars range to a limit of about forty times more or twelve times less than the sun's mass. More significant is the fact that the main sequence is arranged in order of increasing mass. The least massive stars are cool, small, and dim. The most massive stars are hot, big, and bright. It is the mass alone, as a single physical attribute, that determines the luminosity, temperature, and size of any star on the main sequence. This simple relationship does not hold for stars that are not on the main sequence. In the story of the life cycles of stars, mass will be a critical quantity.

We could spend a lifetime viewing stars and not see them all. Thanks to the orderliness of nature, that is not necessary. The stars fall

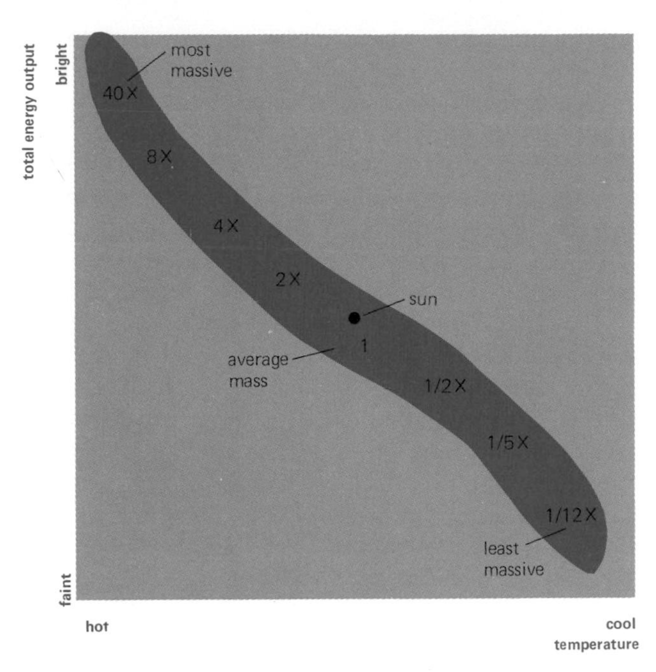


**FIGURE 4.34**
A very schematic luminosity-temperature diagram showing the masses of main sequence stars in terms of the sun's mass.

within certain limits of size, mass, temperature, and brightness. These limits show up quickly after observing only a small sample of the multitude of stars.

## SUMMARY

Analyzing radiation is the only way for astronomers to study the stars. Stars send out radiation in all regions of the electromagnetic spectrum. Only visible light and certain radio waves reach the surface of the earth. Gamma rays, x rays, ultraviolet, and infrared radiation can be studied from above the earth's atmosphere in orbiting satellites. All radiation has several common properties: It is bent or refracted when it passes from one medium to another, it can be reflected from a surface, and it can be dispersed into a spectrum. Astronomers make use of all these properties when collecting starlight. If a light source is moving toward or away from us, the wavelength of the light will be Doppler shifted.

There are three basic types of spectra: emission line, continuous, and absorption line. Most stars display absorption line spectra. The

Bohr model of the atom can be used to explain the different lines and patterns of lines found in spectra. Every element produces a set of unique lines. The composition of all stars is essentially the same: mostly hydrogen and helium with a small fraction of other elements. The primary differences in stellar spectra are caused by a single factor—the temperature of the star. Stars have been classified in order of decreasing temperature into the following spectral types: O B A F G K M (RNS). Luminosity classes, indicated with Roman numerals from I to VI, show differences in the absolute magnitudes of the stars.

An H-R diagram shows the relationship between the spectral type and the absolute magnitude of stars. When organized in this way, most of the stars fall into a band called the main sequence. There are some very bright red giant stars, and some very faint white dwarf stars. The sun is an average star in absolute magnitude or luminosity, size, temperature, and mass. Other stars vary within certain limits around the sun's values.

### FALLACIES AND FANTASIES

*The sun is the largest star.*
*Red stars are hot.*
*All stars are the same brightness.*

## REVIEW QUESTIONS

______ 1. Describe the wavelength of light and the frequency of light.

______ 2. Over the visible light range, are the long wavelengths red or blue?

______ 3. Suppose the light from a star shows a Doppler redshift. What does that mean about the light? about the star?

______ 4. Describe two general properties of light that are used to build telescopes.

______ 5. If we built an observatory on top of Mt. Everest, could we observe x rays? Explain.

______ 6. Using the Bohr model of the hydrogen atom, answer the following questions: (a) What happens to the position of the electron when it absorbs energy? (b) What happens to the position of the electron when it loses energy? (c) Which effect produces a dark line in stellar spectra? (d) Since the hydrogen atom has only one electron, why do we observe, in stellar spectra, whole series of dark hydrogen lines rather than just one line? (e) What one factor most influences which lines will appear in a star's spectrum? (f) What is the difference between an excited hydrogen atom and an ionized hydrogen atom?

______ 7. Answer the following questions: (a) Beginning with the hottest spectral type, list the main Harvard spectral types in order of temperature. (b) Which type is reddest? bluest? (c) List the luminosity classes for stars, indicating their common names as well as numbers. (d) What is the sun's spectral type and luminosity class?

______ 8. What kinds of objects produce continuous spectra?

______ 9. Why are most stellar spectra absorption line spectra?

______ 10. On the following diagram, draw a freehand sketch of the main sequence. Indicate several giant stars with open circles, and white dwarfs with solid dots.

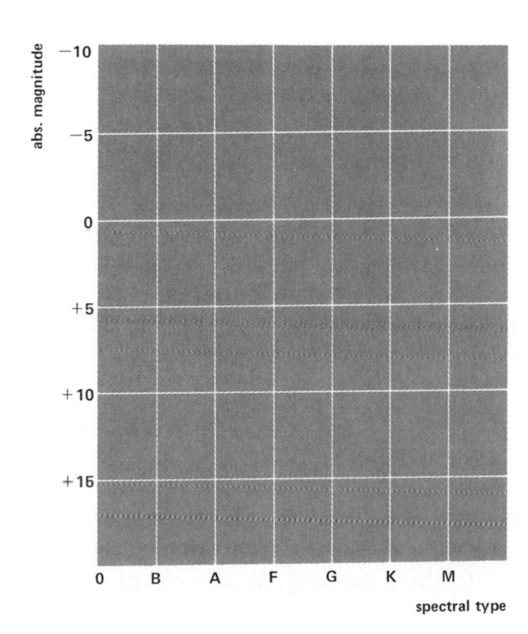

# 5 Star Clusters: From Their Types to Their Locations



## BACKGROUND

### Some Building Blocks of Galaxies

There is a certain orderliness about the arrangements of stars in space. A clustering of stars seems to be the rule rather than the exception. The **galaxies**, each with billions of stars, are the largest star clusters. They have been called the building blocks of the Universe, and studying them leads us to a better knowledge of the Universe (Chapter 12 is devoted entirely to the galaxies). In this chapter, we focus on smaller clusters of stars within the galaxies. These smaller star clusters are some of the building blocks of galaxies, and studying them gives us information about the size and shape of our Milky Way.

**Star clusters** are gravitationally related groups of stars. We generally think of all stars in a cluster as forming at about the same time from the same galactic cloud. The stars in a cluster are closer together than are other stars. The average space between stars is about 4 lt-yr, but in a cluster the stars may be only 2 lt-yr apart, or even less. Cluster stars move together through space. After many, many years, disruptive forces may dilute a cluster until it is no longer recognizable as such. Despite these disruptive forces, most clusters keep their identity for a very long time (Figure 5.1).

Even though the stars of a cluster have a common origin and are born about the same time, we see many different sizes and types of stars in a cluster. The reason for this is that the most massive stars to form go through every stage of their life cycle faster than less massive stars. Though all of the stars will be the same age in number of years, at any given time, there will be stars in every stage of development. Studying the stars in a cluster gives us important information on the life cycle of stars, a topic treated in detail in Chapter 11.

**FIGURE 5.1**
The beautiful open cluster M67 located in Cancer. This is visible with binoculars on a dark clear night.

## Naming and Identifying Clusters

Star clusters visible to the unaided eye have acquired proper names and places in our sky lore and legend. The Seven Sisters, or Pleiades, is probably the most famous bright cluster. It is in the constellation of Taurus (Figure 5.2) and is prominent on clear winter evenings. Since the pattern of bright stars in the Pleiades looks like a dipper, many people mistake this cluster for the Little Dipper. Today only six of the cluster stars are easily visible to the eye. Several cultures have legends explaining the apparent disappearance of one of the Seven Sisters. It is unlikely that a seventh bright naked-eye star was ever visible. With binoculars, several dozen Pleiades stars can be seen;

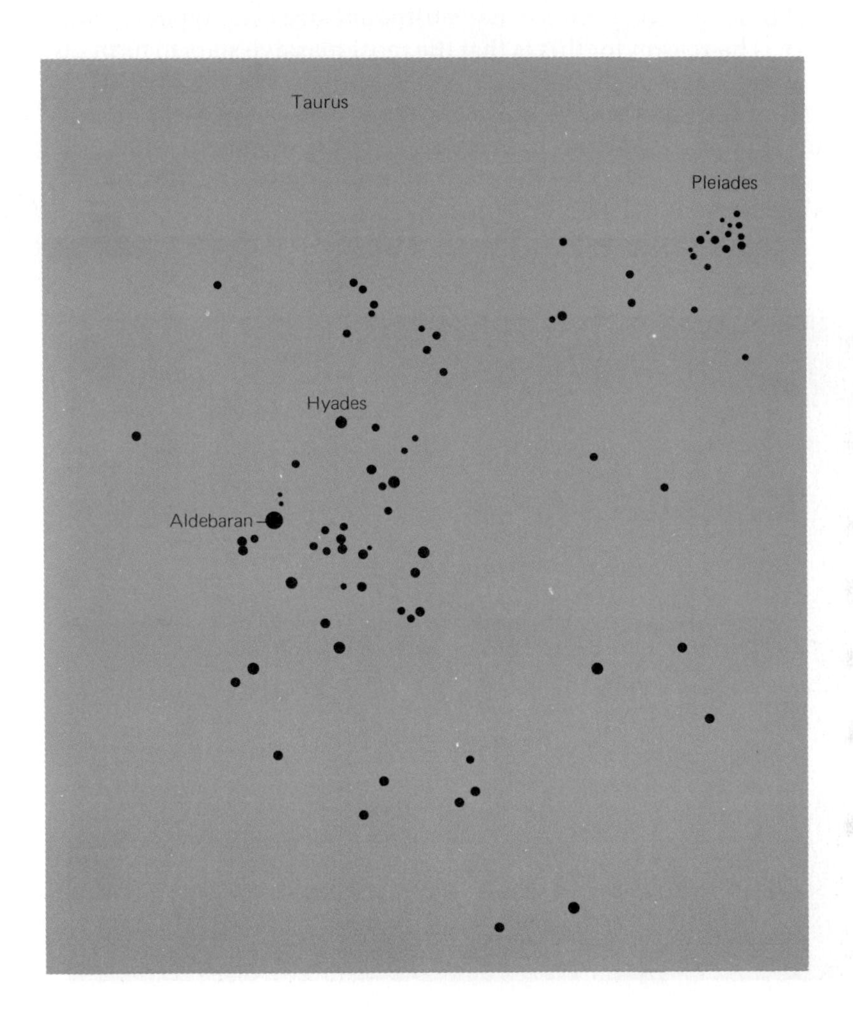


**FIGURE 5.2**
The region of Taurus showing the main portion of the Hyades cluster and the Pleiades cluster.

with a telescope, over a hundred stars become visible. A time-exposure photograph of the Pleiades shows that the brighter stars are surrounded by gas and dust which reflect the star light (Figure 5.3). The Pleiades, about 410 lt-yr away, is one of the few clusters close enough so that we can see its individual stars with our eyes.

Another close cluster is the Hyades in Taurus. This cluster is the V-shaped group of stars near the bright star Aldebaran. The Hyades cluster has a total of about 400 stars at a distance of 133 lt-yr from us. Aldebaran is in front of the Hyades and is not a member of the cluster.

How do we identify which stars are actually members of a cluster? How are they distinguished from other stars in the same general direction in space? The key to identifying cluster members lies in observing their proper motions. For our purposes, we may simply assume that all the stars in the cluster are at the same distance from us and are moving together in space. These conditions give each star in the cluster approximately the same proper motion. A noncluster star will have a proper motion of a different amount and probably also in a different direction. In Figure 5.4, the proper motion of the star Aldebaran is shown compared to stars in the Hyades cluster. Using common proper mo-

(a)

(b)

**FIGURE 5.3**
Two pictures of the Pleiades cluster. (a) A short exposure with a small telescope. (b) A long exposure at a large telescope shows gas and dust. (Kitt Peak National Observatory Photograph)

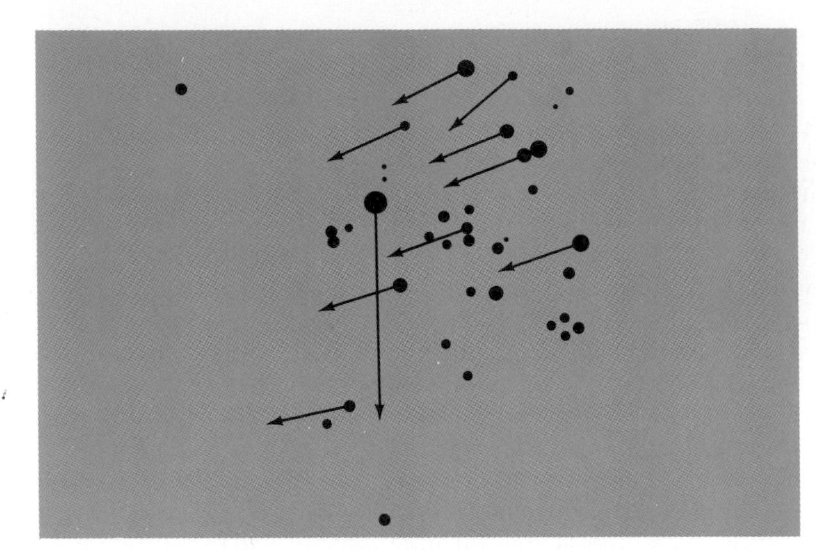

FIGURE 5.4
The proper motions of Hyades cluster stars are all of the same size and in the same direction. The bright star Aldebaran which appears to be in the Hyades cluster clearly is not a member as its proper motion is larger and in a different direction.

tions to identify stars of a cluster is the same technique used to determine which stars are gravitationally related binary stars. When a cluster is so far away that motions are no longer detectable, membership is assigned simply by a star's proximity to the cluster.

The closest star cluster to the sun is the Ursa Major cluster. Five of the brightest Big Dipper stars are members of this cluster (all but the end stars $\alpha$ and $\eta$). Fainter stars in the constellation Ursa Major are also part of this cluster (Figure 5.5). The center of the cluster is 72 lt-yr away from us. Since we are so close to this cluster, its stars seem more spread out in space. Actually, the stars in the Ursa Major cluster form a compact group similar to the other clusters.

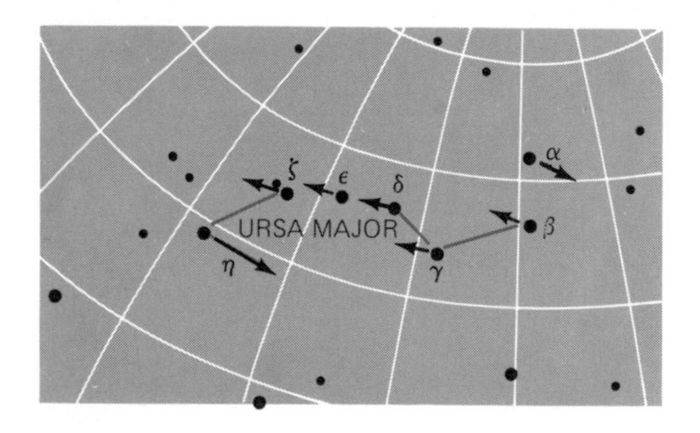


FIGURE 5.5
A portion of the sky showing the proper motions of the stars of the Ursa Major group in the Big Dipper along with $\alpha$ and $\eta$ Ursa Majoris. It is clear the latter stars do not belong to the moving cluster.

Many clusters are just far enough away that their stars blur into a hazy patch of light in the sky. A few of these famous blurs are the Beehive, or Praesepe, cluster in Cancer; h and $\chi$ Persei or the double cluster in Perseus; the Hercules cluster; and the Coma Berenices cluster. There are about a dozen naked-eye star clusters. Five of these are marked on the star maps in Chapter 2.

As soon as you begin looking at the heavens with binoculars or a telescope, many more star clusters become visible. When Charles Messier was searching the sky for comets in the eighteenth century, he kept finding clusters and other hazy patches of light that were definitely not comets. The objects he found never change their positions the way comets do. To aid other observers searching for comets, Messier prepared a list of the positions of 103 of these objects. Messier's objects are indicated by an M followed by the number in his list. For example, the Pleiades cluster is known as M45. Over half of the 103 objects listed by Messier are star clusters. The rest are either nebulae or galaxies.

Since Messier's time, other catalogs of clusters have been compiled. In 1888, Dreyer published the *New General Catalog* (NGC). Then in 1894 and 1908 the *Index Catalogs* (IC), an extension of the NGC, were published. All together, these catalogs list over 13,000 objects. Clusters often have multiple designations. For example, the Praesepe cluster is known by its name as well as its catalog numbers M44 and NGC 2632. The great cluster in Hercules is known as M13 and NGC 6205.

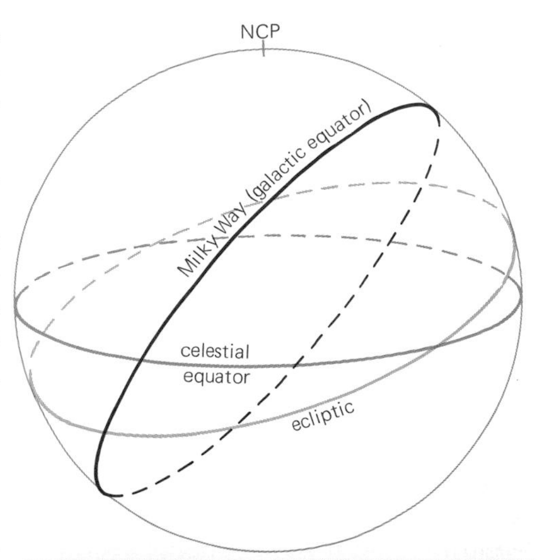


## Types of Clusters

Before we discuss the two types of clusters, we will introduce a new way of looking at the sky. The band of light, called the Milky Way, is the densest part of our galaxy. A line through the center of the Milky Way defines the plane of our galaxy. In this chapter, when we map the sky, we will put the galactic plane through the center of the map like an equator. Any tendency of objects to lie in or close to the galactic plane shows up immediately as we shall see in Figures 5.8 and 5.9.

Most of the star clusters in the galaxy can be divided into two types—**open clusters** and **globular clusters.** The Praesepe cluster is an example of an open cluster (Figure 5.6). The great Hercules cluster, M13, is an example of a globular cluster (Figure 5.7). The Pleiades, Hyades, and Ursa Major clusters mentioned earlier are all open clusters. Just by looking at photographs of the two types of clusters, four differences are obvious.

1. Open clusters have fewer stars than globular clusters. Open clusters generally have from a few dozen to a few hundred stars. Occasionally, a very large open cluster may exceed 1,000 stars. The Praesepe open cluster has about 100 stars in it. Globular clusters have tens of thousands of stars. The total membership of M13, for example, is estimated to be half a million stars.
2. Open cluster stars are not as densely packed as stars in globular clusters. The average spacing between stars in the Praesepe cluster is about 2 lt-yr. Even though it looks as if stars in the center of M13 are touching each other, they are not. The average separation of stars at the center of M13 is about 0.2 lt-yr (over 2 trillion kilometers). If the sun were in the center of a globular cluster, the night sky would have a splendor quite unfamiliar to us. About a hundred times as many stars as we see in our skies would be visible

**FIGURE 5.6**
The fine open cluster Praesepe. This cluster located in Cancer is barely visible to the unaided eye. (Harvard College Observatory photograph)

**FIGURE 5.7**
The globular cluster M13 in Hercules. This cluster looks like a faint smudge of light to the unaided eye. (Official U. S. Navy photograph)

to the unaided eye, and the brightest ones would shine as brightly as the moon does now.

3. Globular clusters have a high degree of spherical symmetry that is not present in open clusters. This globelike appearance gives these clusters their name. Stars of open clusters have a more random distribution.
4. An examination of open clusters shows us that many individual stars in these clusters are younger than those in globular clusters. The ages of clusters can be determined by analyzing the H-R diagram of the cluster. The method used to determine the ages of open clusters is discussed later in this chapter. Open clusters also often contain obvious gaseous and dusty material.

Several more differences between the two types of clusters show up when we look at the numbers of clusters and their positions in the galaxy.

5. Globular clusters are more rare than open clusters. About 120 globular clusters are recognized in our galaxy. Thousands of open clusters have been observed in the galaxy, and thousands more cannot be seen because they are too distant. Since globular clusters

are larger and more luminous than open clusters, we see them at much greater distances. Though some globular clusters may be hidden behind dust clouds in the galaxy, most of them have probably been observed.

6. Most open clusters are found in the plane of the galaxy, (Figure 5.8) but most globular clusters are located generally above and below the plane, where they seem to surround the galaxy (Figure 5.9). The preference for open clusters to be located in the plane of the galaxy has given them a second common name—**galactic clusters.** Maps of the exact positions of star clusters in our galaxy did not exist before the early part of this century. With great patience and inge-

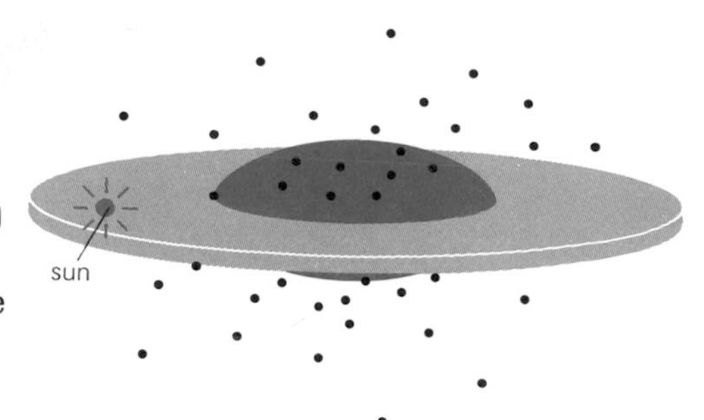


A perspective sketch of the distribution of globular clusters in the Milky Way Galaxy.

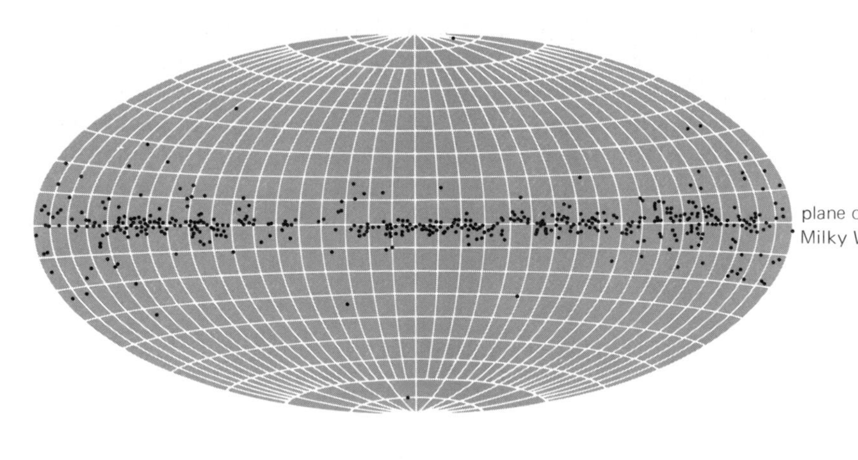


**FIGURE 5.8**
The distribution of galactic or open clusters on a map of the Milky Way. Note how the clusters fall close to the plane of the Galaxy.

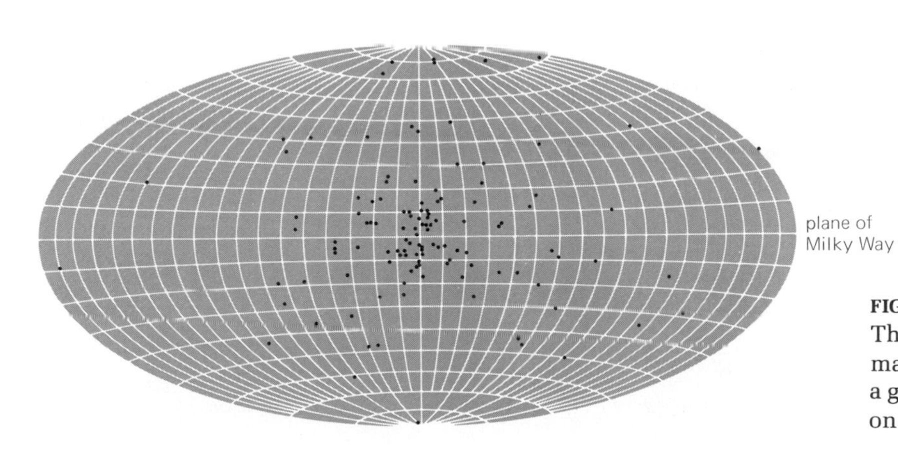


**FIGURE 5.9**
The distribution of globular clusters on a map of the Milky Way. Noto that they cover a great portion of the sky and are centered on the center of the Galaxy.

nuity, astronomers developed methods of determining distances to star clusters. Knowing their distances, we can use star clusters to describe the size and shape of the galaxy.

## DISTANCES TO CLUSTERS

### Parallax

The only direct way to determine the distance to a star cluster is to measure the parallax (Chapter 3) for individual stars in the cluster. This method works well for the nearby clusters, but beyond 200 lt-yr the parallax angle becomes too small to measure accurately. Stellar parallax has been used to find the distances to four or five of the nearest clusters.

Another type of parallax that can be used to find the distance to a few clusters is called **moving-cluster parallax.** With nearby clusters, the proper motions of their stars appear to converge at a single point in space, the way railroad tracks seem to converge in the distance. In Figure 5.10 the arrows represent the proper motions of the Hyades stars. If each of the proper motion arrows were extended, they would appear to meet at the point on the left called the **convergent point.** But we know that the stars of a cluster are really moving about the same direction in space, and that the convergent point is just a matter of perspective.

Here is an area where art can tell us something about nature. The mathematical system of perspective in art was developed by the 15th-

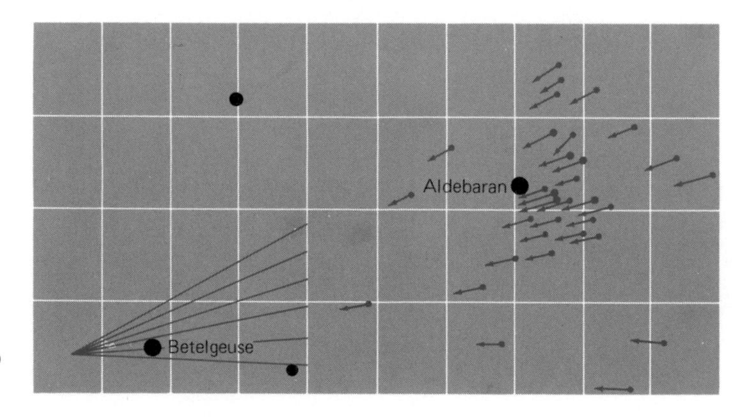


**FIGURE 5.10**
The proper motions of the Hyades stars seem to converge at a point east of Betelgeuse. The length of the arrows shows the proper motions over 50,000 years. Neither Betelgeuse nor Aldebaran belong to the cluster.

century Renaissance painters, at a time when mathematics and art were firmly tied together. Their techniques are still taught in art schools. To make a painting realistic, the parallel lines of a scene are drawn so they all converge at a point called the vanishing point (Figure 5.11). A line drawn from the vanishing point out of the picture to your eye would be parallel to the other lines in the picture.

Turning back to nature and the observations of clusters, we can easily see that the convergent point of the Hyades cluster is merely a vanishing point in mathametical perspective. If we draw a line from our eye to the convergent point, that line would also be parallel to the proper motions of the stars. Once the proper motions and the radial velocities (Doppler shifts) of the stars are measured, there is enough information to calculate the distance to the cluster. Only the nearest clusters clearly show a convergent point in space. For those four or five clusters, the moving-cluster parallax gives a distance that can be used to doublecheck the stellar parallax method.

To go beyond the nearest clusters, astronomers must use indirect methods for determining distances. Most of these are merely clever ways to determine the absolute magnitude of a star or stars in the cluster. Once we have that, the distance may be calculated easily. There is a direct relationship between a star's apparent magnitude (how bright it appears to us), its absolute magnitude (how bright it would appear from a distance of 32.6 lt-yr), and its actual distance from us.

**FIGURE 5.11**
An architectural perspective drawing showing the vanishing point. Straight, parallel lines appear to intersect at a distant point. (The Bettmann Archive, Inc.)

Whenever any two of these three quantities are known, we can solve for the third one. Apparent magnitude is easy to measure observationally with telescopes and instruments. If we can somehow know or determine the absolute magnitude, then we can calculate the third quantity, which is the distance.

The next three methods we describe for finding the distances to clusters are different methods of finding the absolute magnitudes of stars in the cluster.

## Spectral Types and Statistics

The **spectroscopic parallax method** of distance determination relies on a large sample of stars having the same characteristics. These characteristics are then applied to stars of the cluster. If you recall, we can draw some conclusions about stars if we organize them by spectral type and absolute magnitude into an H-R diagram (Figure 5.12). It turns out that most G2 stars, like the sun, have an absolute magnitude of +4.8. Astronomers use a statistic like this to say that all G2 stars will have absolute magnitudes of +4.8. If a G2 star can be identified in a

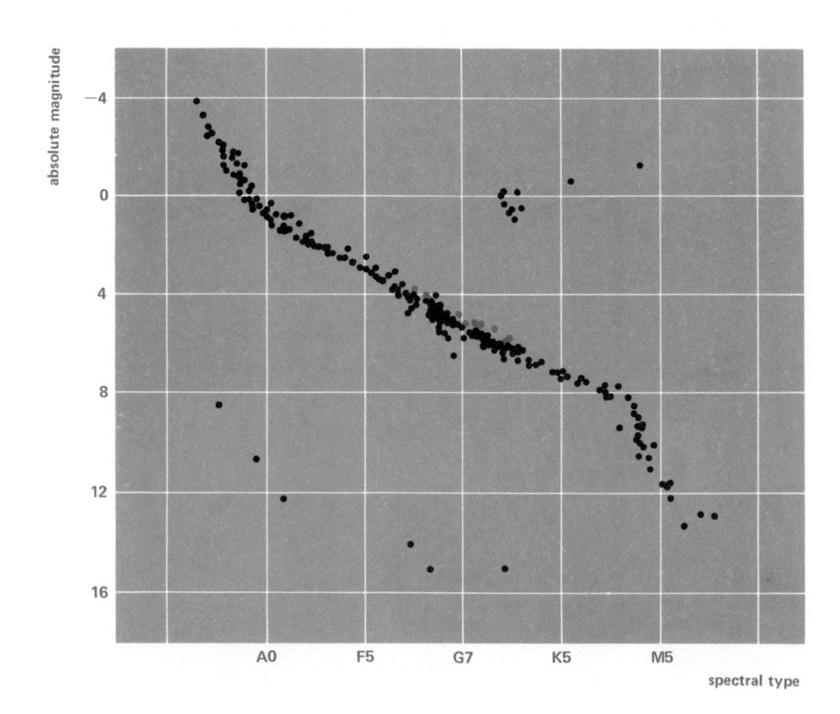


**FIGURE 5.12**
The Hertzsprung-Russell diagram for the nearby stars. The main sequence is well defined.

cluster, its apparent magnitude can be measured and its distance calculated.

But spectroscopic parallax is not quite so simple as this. First, we identify a G2 star—for example, by comparing the spectra of cluster stars to the sun's spectrum. Then, we must make sure it is a main sequence star and not a giant star. There is a subtle difference between the spectra of main sequence stars and giant stars of the same type. The lines in the giant star's spectrum will be noticeably thinner than those of the main sequence star (Figure 5.13). Finally, the apparent magnitude we observe may be affected by gas and dust in space between us and the star. Once the appropriate corrections are made, the distance is computed. This computed distance is only as reliable as the information used for the calculations. To compute the distance to a cluster of stars, we could compute a spectroscopic parallax for every star in the cluster and find the average distance, or we could average *before* we compute the distance by using the main sequence fitting method.

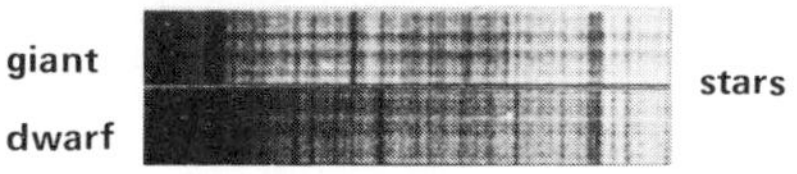

giant
dwarf
stars

**FIGURE 5.13**
Classification spectra of a giant G2 star and a main sequence G2 star. Certain strong, sharp lines in the giant star spectrum are weak and broad in the main sequence star spectrum. (Yerkes Observatory photograph)

## Main Sequence Fitting

If we can observe apparent magnitudes and spectral types for many of the cluster stars, we can organize that information into an H-R diagram. Figure 5.14 shows an H-R diagram for the Praesepe cluster. Since the cluster stars are all at about the same distance, this diagram ought to look just like the absolute magnitude-spectral type diagram except for a shift to adjust the magnitudes for distance. Figure 5.15 shows the

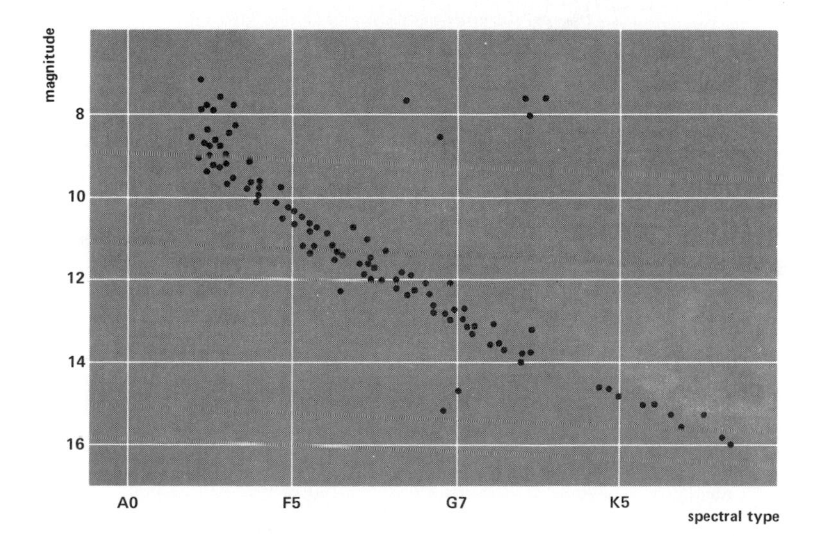


**FIGURE 5.14**
The Hertzsprung-Russell diagram for stars in the Praesepe cluster. The magnitudes used in this diagram are apparent magnitudes.

standard absolute magnitude-spectral type diagram for the Hyades cluster along with the Praesepe cluster apparent magnitude-spectral type diagram. Absolute magnitudes are available for the Hyades stars since we know the distance to the Hyades by other methods. Instead of comparing each star of the Praesepe cluster to a star of the same spectral type in the Hyades and getting its absolute magnitude that way, we compare the entire main sequence of the Praesepe cluster to the standard main sequence of the Hyades. Sometimes this method is called **main sequence fitting.** The main sequence line for the Praesepe cluster is shifted up until it fits over the Hyades main sequence. The amount of shift that is necessary is just the difference between the apparent and absolute magnitudes of the Praesepe stars. This value allows an immediate determination of the distance to the cluster.

The H-R diagram of a cluster also tells us something about the age of the cluster. If many clusters are plotted on top of each other on the

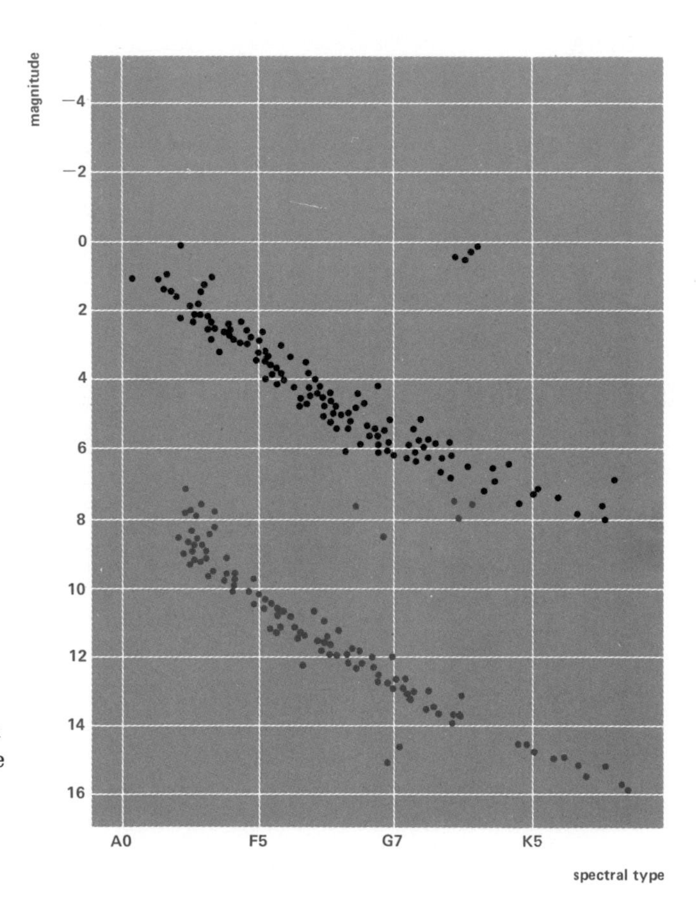


**FIGURE 5.15**
The H-R diagram for the Hyades stars using their absolute magnitudes is plotted in dark color. Praesepe stars are plotted on the same diagram using their apparent magnitudes. The shift between the main sequences is due to the greater distance of the Praesepe cluster.

same chart, the top left side begins to look like a tree with many branches (Figure 5.16). Each cluster tends to bend off to the right at some point along the main sequence. This turnoff point is a measure of the age of the cluster. If you recall from Chapter 4, the most massive, hottest, brightest stars are at the upper lefthand corner of an H-R diagram. These are the first stars to evolve, and they evolve rather quickly. Changes in their spectral types and luminosities cause these stars to move off to the right on the chart. The older the cluster, the more these stars will have aged, or evolved. The turnoff point on the main sequence thus gets lower and lower. The ages of clusters based on the turnoff point are listed on the right side of Figure 5.16. The lower the turnoff point, the older the cluster.

All the clusters shown in Figure 5.16 are open clusters except for M3, which is a globular cluster. The cluster NGC 2362 is the youngest cluster represented. The double cluster, h and $\chi$ Persei, are also in their youth. The Pleiades cluster is middle-aged, and the Hyades and Praesepe clusters are approaching old age. M67, at an age

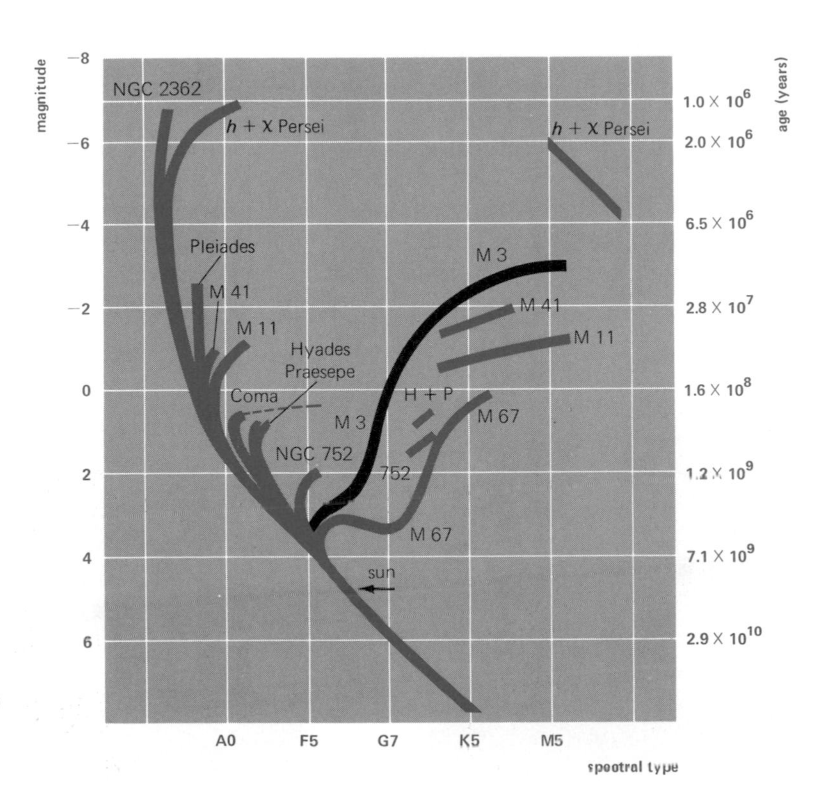


**FIGURE 5.16**
Several clusters fitted to the same main sequence show evolution effects dependent upon age. The point where the main sequence begins to bend to the right gives the cluster's age. Thus, M67 is about seven billion years old.

of 7 billion years, has a longer life than most open clusters. Globular clusters are generally older than open clusters, and the turnoff point for M3 shows its old age.

## Variable Stars

When examined closely, many stars show variations in brightness. Variable stars are discussed in Chapter 11; here we are interested in two types of variables, called cepheid and RR Lyrae stars.

*Cepheid Stars.* Cepheid variables are named after the prototype, δ Cephei. **Cepheids** are single, pulsating stars whose light varies periodically and with extreme regularity and predictability. The regular variations may be as short as a few days or as long as fifty days. The special property of cepheids that makes them useful as distance indicators is that their average absolute magnitudes are directly related to their periods of light variation. For example, a cepheid with a period of 10 days will have an absolute magnitude of -3, and a cepheid with a period of 1 day will have an absolute magnitude of -1.5. Once we identify a cepheid (Figure 5.17) in a cluster and measure its period of variation, we can determine the absolute magnitude from the relationship above, and thus calculate its distance.

This useful relationship between period and luminosity or absolute magnitude was first discovered by Henrietta Leavitt. In 1908 she was observing thirteen cepheid variable stars located in the Small Magellanic Cloud (a small nearby galaxy). The Small Magellanic Cloud is far enough away that she could treat all the stars as though they were at the same distance. She noted that the longer the period of the star's

**FIGURE 5.17**
A cepheid in the globular cluster M14. The arrow points to the variable star near maximum light (left) and minimum light (right). (David Dunlap Observatory photograph, courtesy of Helen Hogg)

variation, the brighter it was on the average. If this relationship were the same for cepheids in the Milky Way galaxy, and if we knew the exact distance to just one of these cepheids, then they would be a marvelous tool for distance determination. Unfortunately, there are no cepheids close enough for a direct parallax determination to be made. But there are some cepheids in clusters of stars. We can use some of the indirect methods to find the distances to the clusters and thus to their cepheids. Once the distance to a cepheid is known, the original period-apparent magnitude relationship can be changed to a period-absolute magnitude relationship, which is what has been done. We still make the assumption that the cepheids in the Milky Way galaxy have the same relationship as those in the Small Magellanic Cloud.

*RR Lyrae Stars.* Following the discovery of cepheids, another type of variable star, named for RR Lyrae, was found to have a direct relationship between period and absolute magnitude. **RR Lyrae** stars all have predictable light variations with periods between one-half and one full day (short compared to the cepheids). The absolute magnitude of all RR Lyrae stars is about +0.7. When an RR Lyrae star is identified in a cluster from its light variation, astronomers immediately know its absolute magnitude and can compute its distance. RR Lyrae stars are commonly found individually and in clusters throughout the galaxy. They are useful distance indicators out to about 300,000 lt-yr. Since cepheids are brighter than RR Lyrae stars, we can use cepheids to determine even greater distances.

## Mapping Cluster Locations

Cepheids can be 100 to 10,000 times as luminous as the sun. About 700 cepheids have been observed in our galaxy. Since they are bright enough to be seen and identified in other galaxies to about 3 million lt-yr away, these bright variable stars have been used to determine distances for the nearest galaxies. We will return to the topic of distances to galaxies in Chapter 12.

Once the distance to star clusters is determined by any of these methods, astronomers can make a three-dimensional space map of the cluster locations. Clusters can then be used to look for the overall structure of the Milky Way galaxy. Are there any symmetries or asymmetries? Are there any obvious patterns shown by the cluster locations? Astronomers are in a forest of stars, looking all around them and trying to describe the forest. How big is it? Are we in the center? Are there any paths or gaps in the arrangement of stars? The first realistic descriptions of the galaxy came at the beginning of this century, when distances to many star clusters were first determined.

## DESCRIBING THE GALAXY USING STAR CLUSTERS

### Open Clusters and the Galactic Plane

We can learn something about the galaxy by looking at the night sky. There are stars in every direction, but they are not arranged randomly. There is a great concentration of stars along the Milky Way, the thickest part of our galaxy. With only these observations, we could conclude that the galaxy is somewhat disk-shaped. As we mentioned earlier in this chapter, open clusters tend to be located in the plane of the galaxy, which is why they are often called galactic clusters.

When we add the distances to these clusters, an interesting picture begins to unfold. The galactic clusters are not randomly sprinkled throughout the central plane of the galaxy. They seem to line up in space. Figure 5.18 shows parts of three lines of galactic clusters. These lines seem to run around the center of the galaxy. This observation is a good indication that our galaxy has spiral arms surrounding a central nucleus. The sun is located in one of these arms. Galactic clusters tell us something about our immediate neighborhood (within 6,000 lt-yr), but you can see in Figure 5.18 that they are not bright enough to tell us the size of the galaxy. That is done by the globular clusters.

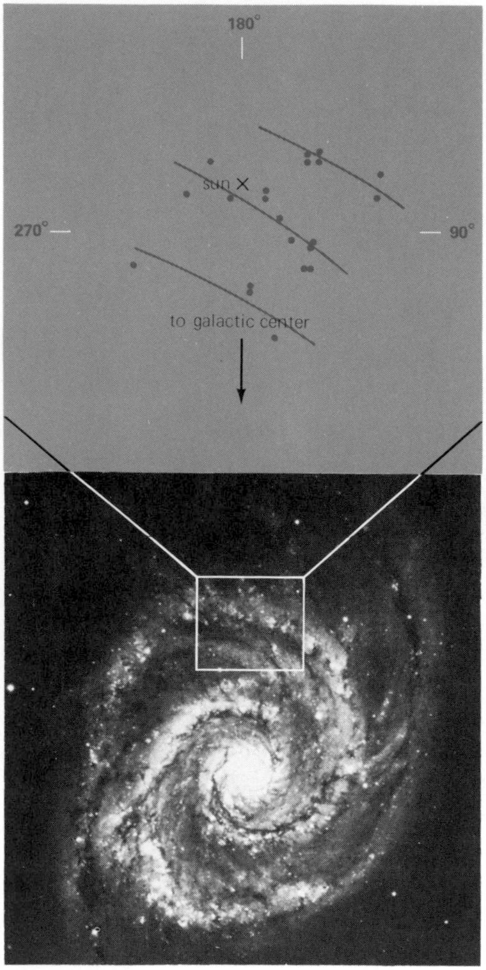

180°
270°
90°
sun
to galactic center

**FIGURE 5.18**
The locations of open clusters are plotted on the plane of the Milky Way. Note how they seem to define three parts of spiral arms. The lower picture shows where this fits into the Milky Way Galaxy. (Kitt Peak National Observatory photograph)

### Globular Clusters and the Galactic Center

Instead of crowding toward the Milky Way as galactic clusters do, globular clusters form a nearly spherical halo around the galaxy. They are somewhat more numerous toward the center of the galaxy (Figure 5.19). All the globular clusters are far away from us; none is within 20, 000 lt-yr, where all the galactic clusters are observed. Because globular clusters are so large and bright, we can see them from one side of the galaxy to the other.

In 1917, Harlow Shapley (1885–1972) took the first important step toward the present understanding of our galaxy by determining the three-dimensional arrangement of the globular clusters in space. He used the RR Lyrae stars in the clusters to determine their distances. Shapley found that the globular clusters occupy a spherical volume of space, which is about 100,000 lt-yr in diameter, surrounding the flat disk of the galaxy proper. The center of the system of globular clusters is about 30,000 lt-yr from the sun in the direction of Sagittarius. Shapley's survey established the dimensions of our galaxy and the eccentric position of the sun in it. As a scientific discovery, the 20th-century revelation that we are not at the center of the galaxy was as important as the earlier discovery that the earth is not the center of the solar system.

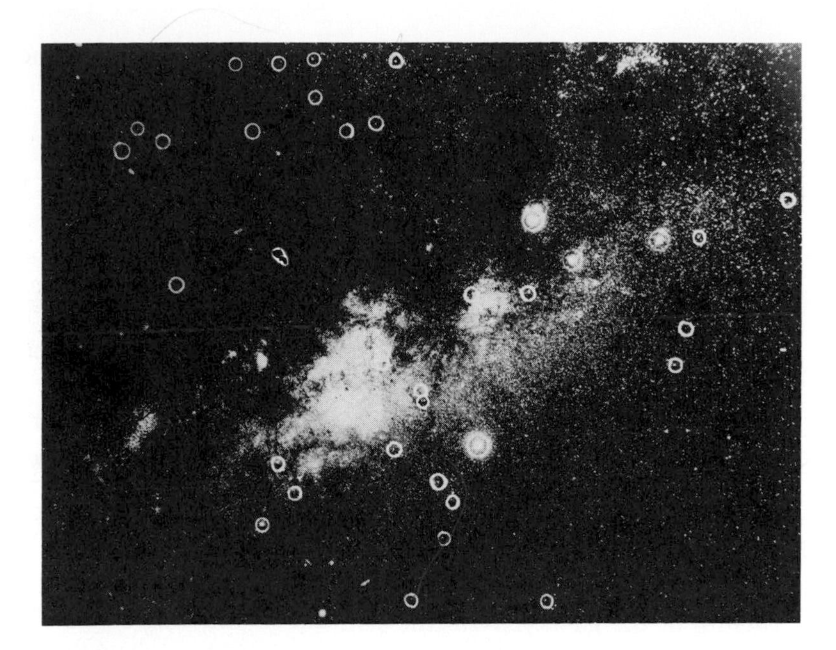

FIGURE 5.19
Distribution of globular clusters in the direction of Sagittarius, toward the center of the Galaxy. The clusters are encircled for easier recognition. Note their absence around the dark clouds, due to obscuration. (Yerkes Observatory photograph)

## Two Stellar Populations

We have been describing the distribution of star clusters in the galaxy. The galactic clusters prefer the disk of the galaxy, and the globular clusters are located around the disk in a spherical halo. The stars that populate the disk and halo of the galaxy can be distinguished in other ways. When we look at a sample of disk stars near the sun, we find that the brightest ones are hot, blue, main sequence stars. There is also a great deal of gas and dust located in the disk of the galaxy. In the halo region, the brightest stars are red giants of spectral type K. There is very little diffuse gas and dust in the halo. The disk stars are also younger than the halo stars.

This differentiation of the stars that populate the disk and halo regions is not unique to our galaxy. A similar differentiation was noted by Walter Baade while he was studying the relatively nearby Andromeda galaxy. The different types of disk and halo star populations showed up on photographs of the Andromeda galaxy taken with the 100-in. (2.5 m) telescope at Mt. Wilson. He named the two arrays of stars populations I and II. Type I population is represented by disk stars similar to those in our neighborhood of space and has an H-R diagram as shown in Figure 5.20. Type II population is represented by the globular clusters in the halo of the galaxy and has an H-R diagram as shown in Figure 5.21. Baade's type I and II populations represent

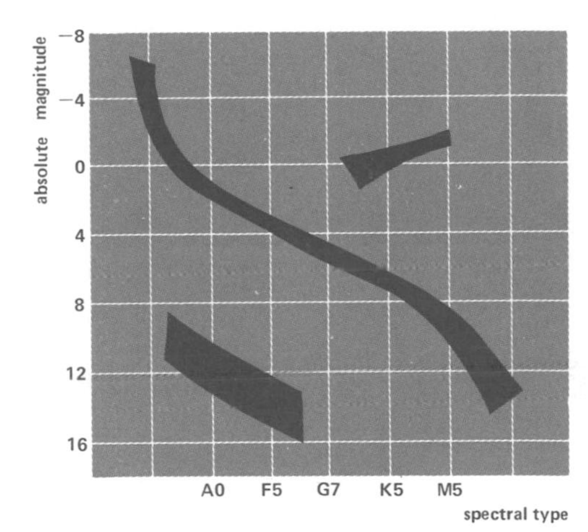


FIGURE 5.20
A schematic H-R diagram for population I stars.

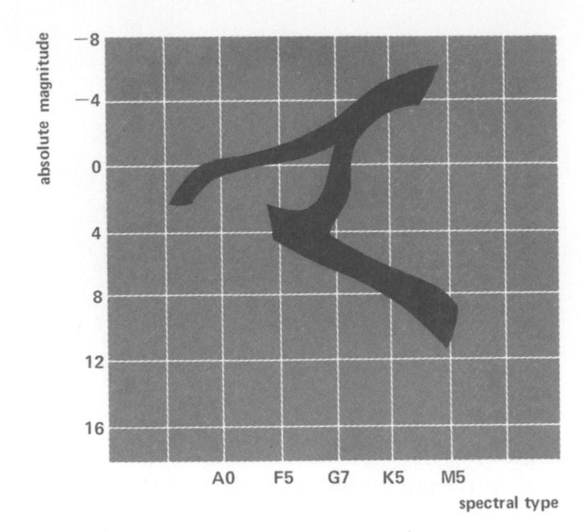


**FIGURE 5.21**
A schematic H-R diagram for population II stars.

extremes of young and old populations of stars. Indeed, it is now accepted that intermediate population types exist as well. The whole set of population types in order of age is listed in Table 5.1

**TABLE 5.1 Population Types**

| *Type* | *Age* | *Galaxy Location* | *Example of Stars in Type* |
|---|---|---|---|
| Extreme population I | Youngest | Disk | Blue supergiants and cepheids |
| Older population I | | Disk | A-stars |
| Disk population | | Disk | RR Lyrae stars |
| Intermediate population II | | Above disk | Long-period variable stars |
| Halo population II | Oldest | Above disk | Globular clusters |

Observations of star clusters helped astronomers describe the gross structure of the galaxy. With the patterns of the clusters established, other experiments and observations were made to give us the detailed picture of the Milky Way described in Chapter 6.

## SUMMARY

There are two basic types of star clusters: open and globular. They differ from each other in number of stars, degree of symmetry, location in the galaxy, and age. Once their exact positions are known, star clusters are used to map the shape and extent of the galaxy.

Parallax (stellar or moving cluster) is the only direct method to determine cluster distances. Measuring the spectral types of cluster stars or discovering cepheid or RR Lyrae variable stars in clusters will give absolute magnitudes for those stars. These can then be used to determine the distance to the cluster. Main sequence fitting is another indirect way to determine a cluster's distance.

The position of star clusters in the galaxy is not random. The globular clusters surround the galaxy in a spherical halo. From their distribution, we determined the location of the galactic center. The open clusters are found near the plane of the galaxy. They actually trace out a portion of the galaxy's spiral arm structure. Open clusters are sometimes called galactic clusters.

In general stars that populate the plane of the galaxy (population I stars) are brighter, hotter, and younger than those which populate the halo region (population II stars).

### *FALLACIES AND FANTASIES*

*The Pleiades is the Little Dipper.*

## REVIEW QUESTIONS

_____ 1. Why are open clusters also called galactic clusters?

_____ 2. What type of cluster is found most frequently in the galaxy?

_____ 3. Fill in the following chart with respect to the Pleiades (an open cluster) and the Hercules cluster (M13, a globular cluster).

| | Pleiades | Hercules cluster |
|---|---|---|
| Number of stars | | |
| Presence of gas or dust | | |
| Amount of Symmetry | | |
| Distance | | at least |
| Population type | | |

_____ 4. If a cluster is not close enough for a parallax determination and contains no variable stars, describe a way to find its distance.

_____ 5. A cluster contains a cepheid. How can you use it to find the distance to the cluster?

_____ 6. Describe Shapley's method of finding the position of the galactic center.

_____ 7. What population type does the sun belong to?

_____ 8. Could we obtain a moving cluster parallax for a globular cluster? Explain.

_____ 9. Why does the H-R diagram look different for the two types of clusters?

_____ 10. Describe a way to tell which stars definitely belong to a cluster.

# 6 The Milky Way Galaxy: From the Spiral Arms to the Nucleus



## THE MILKY WAY

### Appearance in the Sky

In the preceding chapter we learned that Harlow Shapley established the **galactocentric** point of view. He gave a good estimate of the distance to the center of our galaxy some eighteen years after the start of the twentieth century. This is a bit of a surprise to many of us because now we feel that the distance to the center must have been known for a long time. In fact, as we shall see, the true extent and nature of the Milky Way galaxy is only now becoming clear some twenty or so years prior to the start of the 21st century.

For centuries the Milky Way (Figure 6.1), seen as a band of light across the sky, was thought to be a nebula or a cloud. The Chinese accurately portrayed its general outline in the eleventh century, but gave it no special meaning. It was not certain that the Milky Way band continued completely around the sky in a great circle. Ferdinand Magellan, the great navigator, was the first to note the continuity of the circle of the Milky Way. Figure 6.2 is a composite photograph of the entire Milky Way. Magellan also observed two cloudlike but permanent celestial objects now referred to as the Large and Small Magellanic Clouds (Chapter 12). He even noted that these clouds were luminous "like the Milky Way which itself has great breadth in Sagittarius." It was difficult for the Chinese and pre-Copernicans to attach any significance at all to the circle of the Milky Way; the important circles for early cultures were the celestial equator and the ecliptic, or band of the zodiac. The sun and the planets seemed to move through the zodiac, and the stars rose and set parallel to the celestial equator. It was only in the 17th century that Galileo, using his first primitive telescope, saw that much of the Milky Way was made up of stars.

At nightfall in the late summer in middle northern latitudes, the Milky Way arches overhead from the northeast to the southwest horizon. It extends through Perseus, Cassiopeia, and Cepheus as a single band of varying width. Beginning in the region of Cygnus, or the Northern Cross, which is overhead in middle northern latitudes, it seems to divide into two parallel bands. The dark area between these bands is called the **Great Rift.** The division is conspicuous as far as Sagittarius and Scorpius. The western band is the broader and brighter one through the Northern Cross. Farther south, in Ophiuchus, this branch fades and nearly vanishes behind dense dust clouds. It emerges again in Scorpius. The eastern branch grows brighter as it goes southward and gathers into great star clouds in Scutum and Sagittarius. Here, in Barnard's words, "the stars pile up in great cumulus masses like summer clouds."

As beautiful as this northern hemisphere spectacle may be, the view in the southern hemisphere's late summer (our late winter) is absolutely awesome. In addition to the Milky Way, Orion is high in the sky and the stars Sirius and Canopus are brilliant. To the east is the region of Crux and Centaurus; to the west are the Magellanic Clouds.

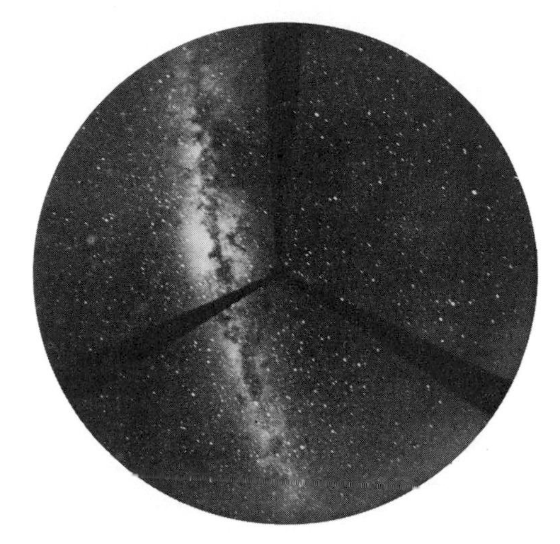

**FIGURE 6.1**
The Milky Way in Sagittarius. This picture was made by placing a camera above a dome-shaped mirror lying on the ground. The camera faced the mirror and was held by three struts which can be seen as dark lines in the photograph. (Washburn Observatory photograph by T. Houck and A. Code)

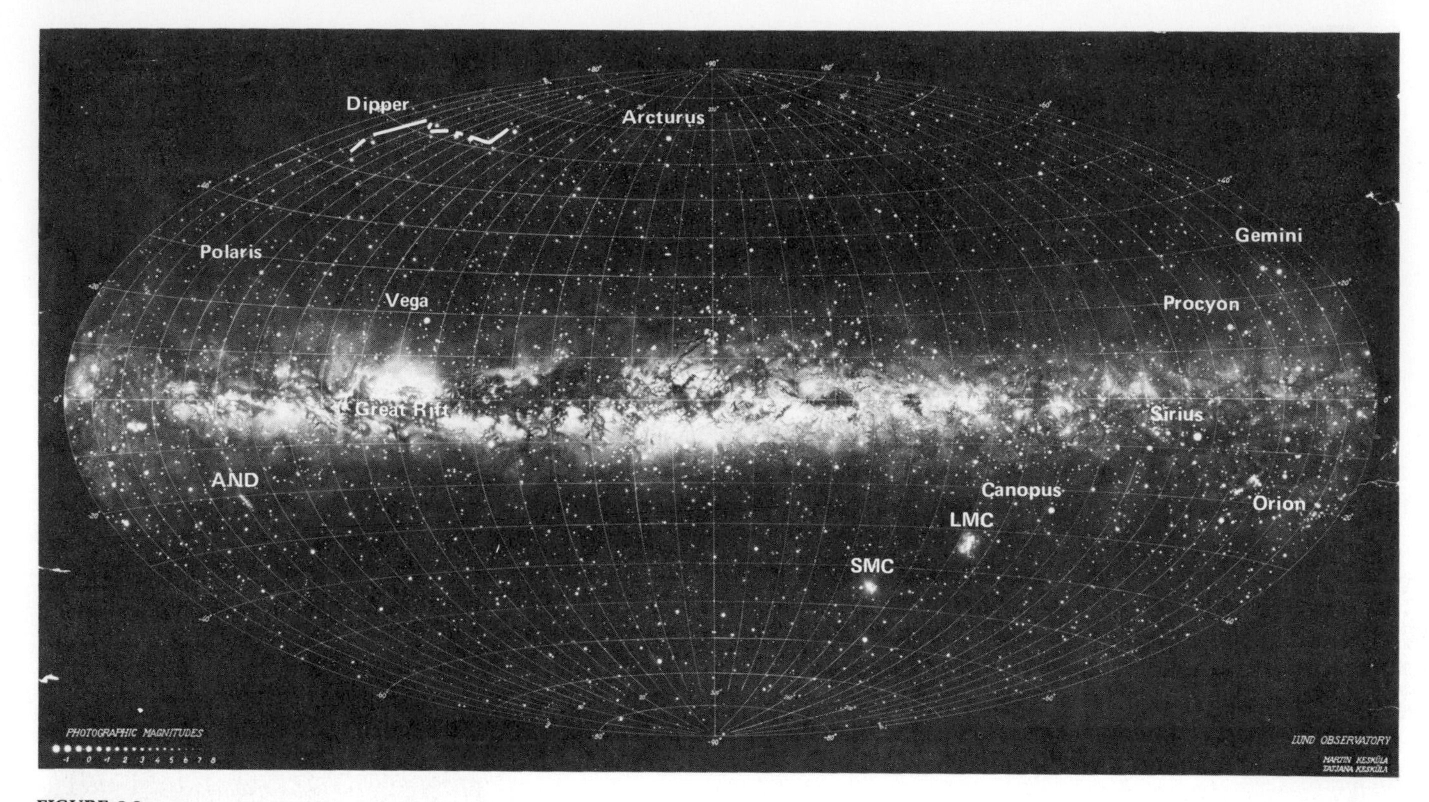


**FIGURE 6.2**
A composite painting of the entire Milky Way with a few well-known celestial objects identified. Three external galaxies are also identified: Andromeda (AND), Small Magellanic Cloud (SMC), Large Magellanic Cloud (LMC). (Lund Observatory)

In the evening skies of the late winter in our northern latitudes, the Milky Way again passes nearly overhead, now from northwest to southeast. The band is thinner than in Cygnus and undivided. From Cassiopeia to Gemini, the Milky Way is narrowed by a series of nearby dust clouds. The Milky Way becomes broader, weaker, and less noticeably obscured as it passes east of Orion and Canis Major down toward Carina.

## Contents

Telescopes show that the Milky Way is made up of great "clouds" of stars. In addition, it had long been known that there were occasional dark areas. These were originally thought to be "holes" in the Universe. With the introduction of photography to astronomy in the early part of the twentieth century, interest in the structure and contents of the Milky Way increased. It did not take observers long to realize that the dark areas previously considered holes were actually

**FIGURE 6.3**
The Southern Coalsack is an excellent example of a dark nebula. To the right is a bright nebula. (Harvard College Observatory photograph)

clouds of opaque material among the stars, which blocked our view of the stars beyond. At the same time it was noted that there were luminous patches or bright clouds of material among the stars. These great clouds of both gas and dust are called the **dark** and **bright nebulae.** For the most part, they are located near the main band of the Milky Way (Figure 6.3).

Astronomers observed one other kind of nebula in the sky. Many small, fuzzy, luminous patches were located in the direction away from the principal band of the Milky Way. To indicate their position and distinguish them from the bright and dark nebulae, they were called **extragalactic nebulae.** These are the same luminous patches studied by Lord Rosse several centuries ago. Many of his sketches show the extragalactic nebulae to be spiral in shape. Following a suggestion by the French astronomer Pierre Laplace (1749–1827), most astronomers originally thought that these spiral nebulae were other solar systems in formation (Figure 6.4). Solving the mysteries of the nebulae occupied the first thirty years of this century.

**FIGURE 6.4**
The Great Andromeda Galaxy was known only as an extra-galactic nebula at the beginning of this century. This photograph was taken at the same resolution and brightness as Lord Rosse was observing and sketching. (Leander McCormick Observatory photograph)

At the beginning of the 20th century, astronomers were trying to determine the nature of the bright and dark nebulae. The task seemed complex and disappointing. Certain bright nebulae were called **emission nebulae** because their spectra contain many bright emission lines. The pattern of hydrogen emission lines was easily identified, as was helium's, but several prominent emission lines could not be identified. The bright Orion nebula (Figure 6.5) was one of the nebulae to have these unidentified lines in its spectrum. It was thought that

these lines must come from a new element, which was called "nebulium." Chemists quickly pointed out that there was no room for such an element in the periodic table of the elements. Nebulium remained a mystery until 1927, when Ira Bowen demonstrated that nebulium was really due to unusual lines of the common elements oxygen and nitrogen when they have had two electrons stripped away. From that observation, it can be concluded that the bright nebulae showing emission lines are in a rather unusual state of temperature, pressure, and density, at least by our earth-based standards.

Other bright nebulae that do not show emission lines do show absorption lines in their spectra. The lines are identical to lines in the spectrum of a nearby bright star. Evidently these nebulae are composed of material reflecting the light of the nearby star. They are called **reflection nebulae.** The classic examples of such reflection nebulae are those around the bright stars of the Pleiades, as shown in Figure 6.6.

In photographing the spectra of distant stars, other thin, dark, interstellar clouds were detected in the Milky Way. These clouds are too thin to block the light of distant stars the way the large, dark nebulae do. But being a cool, low-pressure gas in front of a hot star, the clouds cause very sharp absorption lines to show up in the spectra of the stars.

The bright and dark nebulae, the thin interstellar material, and hundreds of billions of stars make up the Milky Way. In 1924, Edwin Hubble showed beyond any doubt that the extragalactic nebulae are actually great assemblages of stars, gas, and dust that are far from and independent of the Milky Way. To distinguish these "island universes" from nebulae, they are referred to as **galaxies** (Figure 6.7).

The luminous galaxies can be classified into three major types by shape: **spheroidal, spiral,** and **irregular.** The galaxies are too faint to be seen through the Milky Way, so those that were first observed were generally in directions where the Milky Way stars are not very thick. With the confirmation that galaxies do exist came the intriguing question, "Is the Milky Way a galaxy?"

**FIGURE 6.5 (page 116)**
A portion of the Orion Nebula. A perfect example of a bright emission nebula. (Hale Observatories photograph)

## FINDING THE SHAPE OF THE MILKY WAY

### Using Star Counts

Through many different observations, astronomers learned that the Milky Way is a galaxy and discovered something about its extent and shape. An early method of studying the galaxy was the technique of **star counting.** W. Herschel came up with a flattened distribution of

**FIGURE 6.6**
The reflection nebula around the star Merope in the Pleiades. (Hale Observatories photograph)

stars in the galaxy based on counts of stars in all directions. His method was simple, straightforward, and elegant.

Herschel probably became interested in the arrangement of the stars after his discovery that the sun has a preferential space motion toward the Milky Way in the direction of Hercules. Herschel blocked the sky into hundreds of sections and proceeded to count all the stars

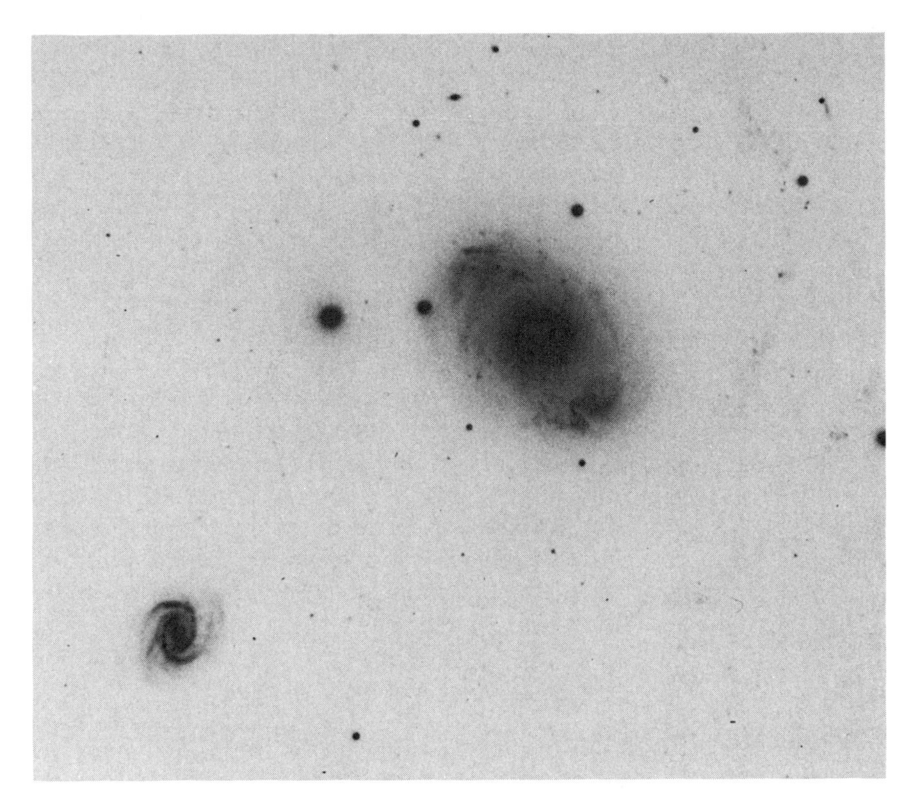

**FIGURE 6.7**
Two galaxies in the same field. The larger galaxy is much nearer than the smaller. Astronomers often prefer negative prints, such as this, as they show details more readily. (Hale Observatories photograph)

he could see in those sections with his 18-in. (46 cm) telescope. Some years later, his son John carried the same telescope to South Africa and made counts in sections of the southern sky not visible from England. Finally, father and son had counts of stars in sample sections all over the sky. What did the numbers mean?

The interpretation of the star counts required some assumptions. Herschel made two basic ones. First, he assumed that all the stars were the same brightness and, therefore, the fainter stars were fainter because they were farther away. Second, he assumed that the stars were spread out with a uniform density in space. These assumptions led him to expect to see four times as many stars at each fainter magnitude. For example, if he counted 10 third-magnitude stars in a section, he should count 40 fourth-magnitude stars and 160 fifth-magnitude stars. The number 4 was not pulled out of a hat; it was the result of mathematical calculations involving Herschel's assumptions and the brightness-magnitude relation discussed in Chapter 3.

**FIGURE 6.8**
Herschel's plot of a cross section of the Universe. Herschel achieved this diagram by counting stars over the sky and then plotting one section of his counts.

In general, the number of stars Herschel counted at each magnitude level increased by the factor of 4 as expected. Occasionally, however, his counts would not increase at all at the next magnitude. This indicated that he had reached the boundary of the Milky Way in that direction. In some directions, the counts increased more slowly than he expected, which indicated a tapering off of stars. When Herschel converted his counts to distances, he arrived at the picture of the galaxy shown in Figure 6.8. He concluded that the sun was at the center of the galaxy, that the galaxy was highly flattened, and that the boundary of the galaxy was irregular. In Herschel's time, the Milky Way was not known to be a galaxy among billions of galaxies, and his map was presumed to be a map of the Universe. The division in the diagram is at the Great Rift between Cygnus and Centaurus.

Herschel's picture of the Milky Way is not bad, considering that both of his assumptions were wrong. The stars are not all the same brightness, nor are they distributed uniformly through space. Also, Herschel could not correct his counts for intervening material dimming the light of the stars. He was aware of the great dark regions but did not know that they were filled with matter.

## Using H-II Regions and Bright O and B stars

Two things happened at the beginning of the 20th century to facilitate the study of the structure of the Milky Way galaxy. First, the extragalactic nebulae were definitely confirmed to be other galaxies. Second, telescope technology advanced to the point that coarse details could be observed in the external galaxies. After discovering something about their structure, we could look for the same things in our Milky Way. For example, if bright emission nebulae are located only in the spiral arms of galaxies, then the bright emission nebulae of the Milky Way must define the arms of the Milky Way galaxy.

Early in the 20th century, the largest telescope in the world was the 100-in. (2.5 m) Hooker telescope located atop Mt. Wilson just north of Pasadena and Los Angeles. By the 1930s, the cities below Mt. Wilson had grown so large that the effectiveness of the great telescope was greatly limited, especially when it was used on very faint objects like other galaxies. But from 1942 to 1945 the area was under wartime blackout conditions, and it was possible to use the telescope on faint stars in the distant galaxies. Walter Baade, using this telescope, studied the Andromeda galaxy (then called the great Andromeda Nebula), Figure 6.9, and made several important discoveries.

One was that there are really two types of cepheid variable stars. Those we described in Chapter 5 are the brighter type. When distance calculations were adjusted to account for the two types of cepheids, the

**FIGURE 6.9**
The Great Andromeda Galaxy and its two easily photographed companion galaxies. Compare this photograph available to Baade to the view available to Lord Rosse in Figure 6.4. (Hale Observatories photograph)

galaxies ended up nearly twice as far away as before. Baade also saw that the spiral arms of the Andromeda galaxy are composed of bright, hot blue stars (O and B type young stars), glowing hydrogen gas clouds (called **H-II regions**), and dust clouds (Figure 6.10). Since the Milky Way has the same kinds of objects, they should trace out any spiral arms that might exist in our galaxy.

Careful studies of the distribution of the bright O and B stars in the Milky Way indicate that there are three streams of these stars. Another study of the glowing H-II regions shows the same streams at the same distances. In Chapter 5 we showed how the galactic clusters also line up in three streams in space. When we superimpose all the information, we see a section of three spiral arms in the sun's vicinity (Figure 6.11). The arms are named for some of the constellations we see in their directions. The sun is located on the inside edge of the **Orion arm.** The **Perseus arm** is farther from the galactic center than we are, and the

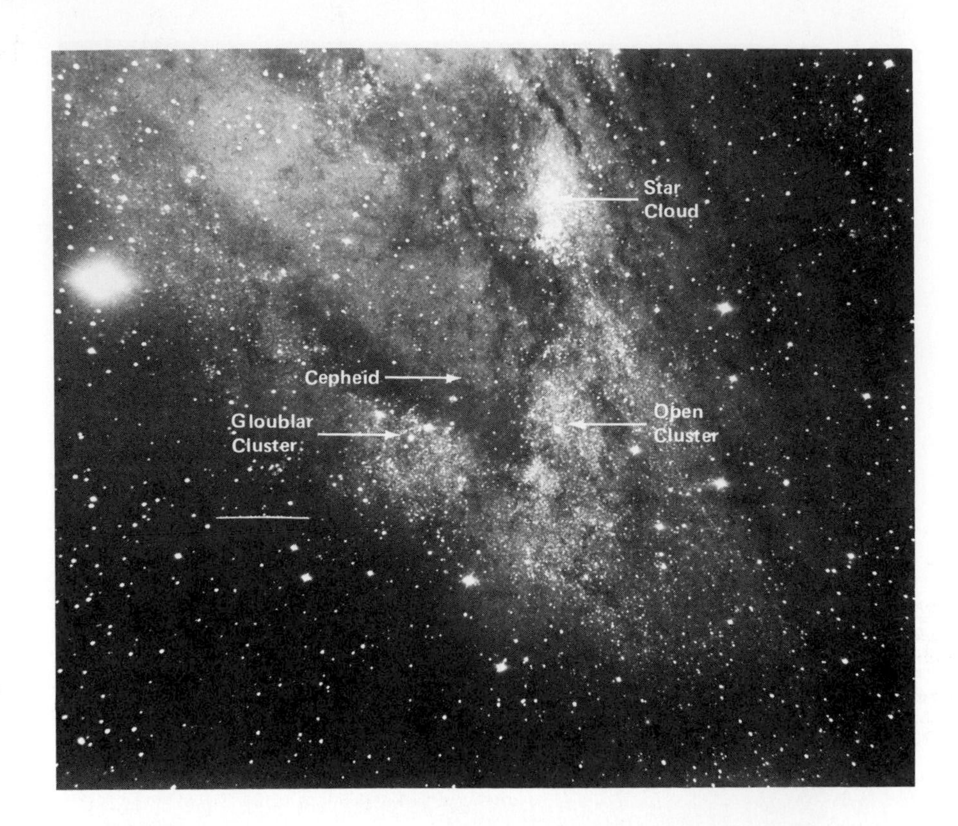


**FIGURE 6.10**
A portion of the Great Andromeda Galaxy viewed by Baade showing great H-II regions, star clusters, and so forth. (Hale Observatories photograph)

**Sagittarius arm** is closer to the center. These three bands of objects confirm that the Milky Way is a **spiral galaxy.** They cannot tell us what type of spiral in detail because they can only be traced to a distance of 6000 lt-yr.

## THE SPIRAL ARMS

### Radio Astronomy

Astronomers knew that the center of the Milky Way was about 30,000 lt-yr away from us. This was fully appreciated after Shapley analyzed the distribution of the globular clusters in space (Chapter 5). Prior to 1951, there was no hope of directly seeing the center of our galaxy or of tracing the spiral arms beyond 6000 lt-yr. There is just too much gas and dust making up the bright and dark nebulae and blocking our view. Then came a major instrumental breakthrough. Astronomers were able to "see" the hydrogen gas in the galaxy to great distances using the new and powerful techniques of **radio astronomy.** Radio antennas receive radiation from the radio region of the electromagnetic spectrum (Chapter 4). The wavelengths of radiation in that region are longer than those of visible light and pass through the gas and dust.

In 1932, Karl Jansky, an engineer with the Bell Telephone Laboratory, was studying atmospheric interference in conventional radio broadcasting. He found two sources of **noise** or interference. One was from local and distant thunderstorm activity, and the other was a steady hiss that appeared every day on his radio apparatus for about an hour. Interestingly enough, the hiss occurred almost 4 min. earlier each day. Since stars seem to rise about 4 min. earlier each day (Chapter 2), it did not take Jansky long to realize that the noise was coming from space. The object that was overhead when the noise was observed was the band of the Milky Way. Quite by accident, a major discovery had been made. Prior to Jansky's discovery of radio noise from the Milky Way, no one had any reason to expect radio waves from the heavens.

In 1938, Grote Reber, with more sensitive radio receivers and a more efficient radio telescope, studied the radiation from the sky at shorter wavelengths than Jansky had used. Reber's wavelengths were still in the radio region of the spectrum and were invisible to our eyes. Compared to radio broadcasts on earth, radio waves from distant space are very weak and require large, sensitive antennas. Radio telescopes can measure the strength of a signal (like measuring the brightness of a star) and the position of a signal in the sky. Reber's radio maps of the sky (Figure 6.12), on which strong signals resemble mountains on a topographical map, clearly show the band of the Milky Way and the bulge in Sagittarius, the suspected center of the galaxy.

## The 21 cm Wavelength of Hydrogen

Jan Oort, long interested in the structure of the Milky Way, also recognized the value of the new science of radio astronomy. He determined to appropriate the German radar antennas located in the Netherlands at the end of World War II and use them as radio telescopes. Before he got the telescopes, Oort began to wonder what he should observe. The radar antennas could be used at any wavelength longer than 10 cm. With that in mind, Oort asked a graduate student to see what might be observed in space. In researching the problem, the student, H. C. van de Hulst, came across an important study on the hydrogen atom by Enrico Fermi.

In the same year that Jansky first identified radio radiation from space, Fermi was studying the hydrogen atom using the newly

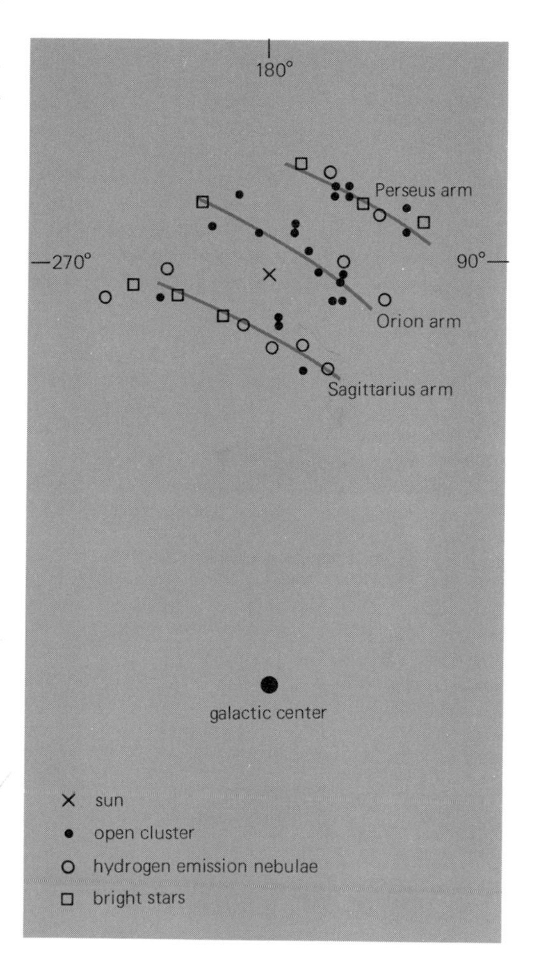


**FIGURE 6.11**
A plot of the open clusters, H-II regions, and the bright O stars on the galactic plane as a function of distance from the sun. Three distinct lines or streams are readily discerned. These streams are associated with spiral structure in the Milky Way.

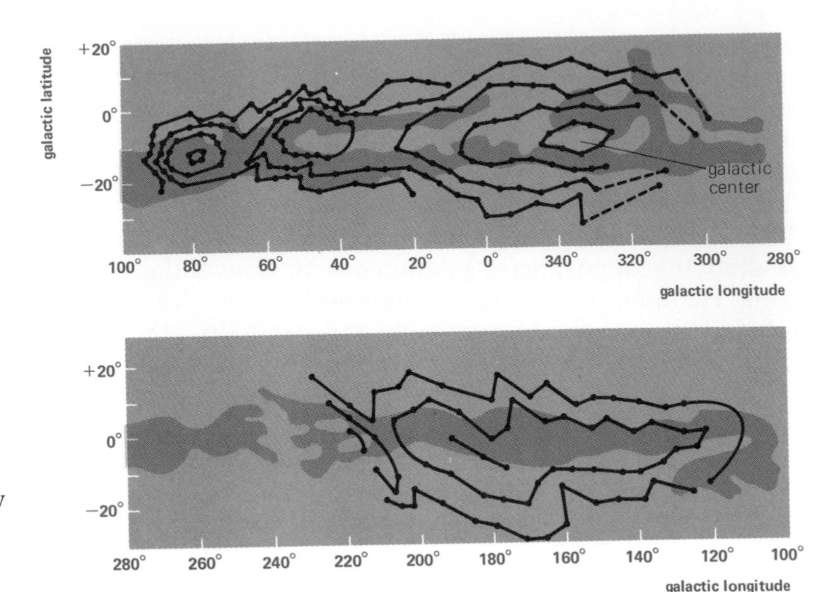


**FIGURE 6.12**
Grote Reber's original radio maps of the sky using the Milky Way as his equator. (Diagram from the David Dunlap Observatory)

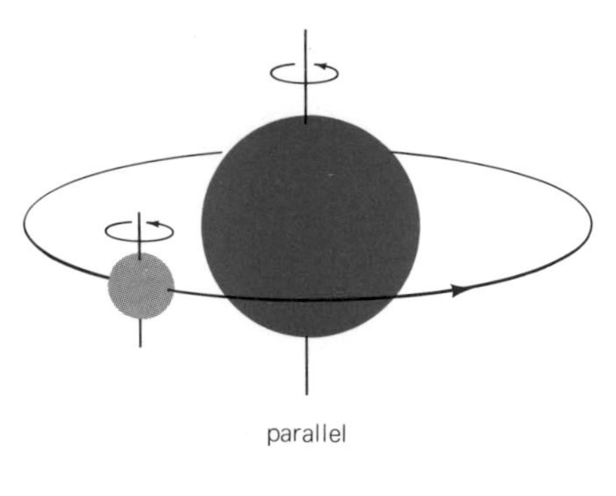


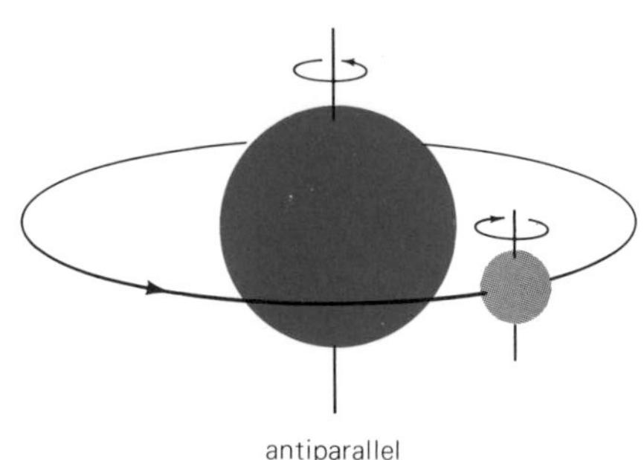


**FIGURE 6.13**
The two possible energy levels of hydrogen in the ground state. The atom has slightly more energy in the parallel state than in the antiparallel state.

developed quantum mechanics. The quantum theory told him that the electron was constantly spinning, but it could have the same spin as the nucleus (parallel) or the opposite spin (antiparallel) (Figure 6.13). He calculated that an atom in the parallel case will, on the average, flip over to the antiparallel condition after 10 million years. When this happens, the hydrogen atom will release some energy. The amount of energy released is so small that physicists showed no interest in this research.

Van de Hulst immediately realized the potential value of Fermi's work. He calculated that the energy released by the hydrogen atom when the electron flipped would have a wavelength of 21 cm. Receivers attached to the radar antennas could be tuned to this wavelength. Indications were that hydrogen was very abundant in the Universe. Van de Hulst proposed that there might be sufficient hydrogen in some of the Milky Way clouds to produce enough flipping from the parallel to the antiparallel spin states to cause a detectable amount of energy at a wavelength of 21 cm. His proposal did not go unnoticed, especially by groups of astronomers having access to the rapidly developing radar technology. In 1951, H. I. Ewen and E. M. Purcell announced the detection of interstellar hydrogen radiation at a wavelength of 21 cm. A few months later, Oort's group confirmed the discovery. From this single important discovery, a new, extensive, and detailed picture of the galaxy would emerge.

As we have noted, the gas clouds in other galaxies follow the spiral arms, and the H-II regions in the Milky Way show the beginnings of spiral arm structures. The new 21-cm radio observations enabled astonomers to trace the hydrogen gas clouds throughout most of the galaxy and by this means map the spiral arms of the Milky Way. By a happy turn of events, even when two gas clouds are lined up in the same direction, we can usually see them both. This is due to the fact that the clouds are moving around the center of the galaxy, as we will discuss below. Since clouds at different distances are moving at different speeds, the Doppler shift of the 21-cm line is different for each cloud.

When astronomers map the nearby hydrogen gas in the galaxy, the results agree reasonably well with the observations of the O and B stars, the H-II regions, and the galactic clusters. At greater distances, the radio map shows a rather chaotic pattern of arms (Figure 6.14). This would seem to say that the Milky Way is a spiral galaxy with very loose branching arms and a rather small center. In Chapter 12 we refer to this type of galaxy as an **Sc galaxy.** Even with radio observations, there are some areas of the galaxy that we cannot map. Notice the missing information directly toward and away from the center of the galaxy in Figure 6.14. In these directions, the gas in the spiral arms is moving across our line of sight and there is no detectable Doppler shift.

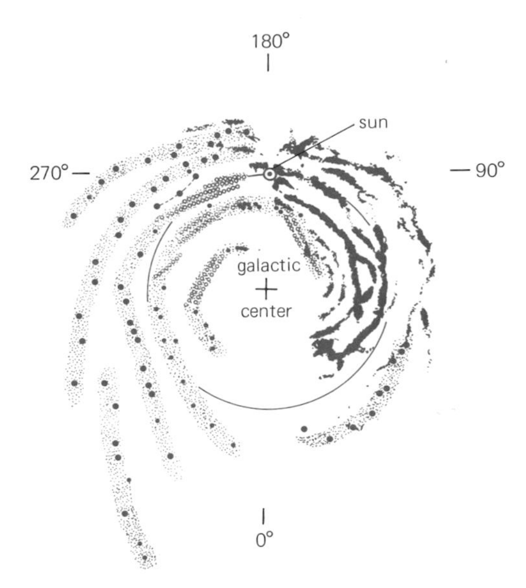


**FIGURE 6.14**
A radio view of the Milky Way Galaxy in the plane of the Milky Way. These observations are made in neutral hydrogen (21 cm) and combined from work by Australian and Dutch astronomers. The somewhat chaotic picture is that of a spiral galaxy.

## Rotation of the Galaxy

When we look at photographs of spiral galaxies, it is only natural to wonder if these galaxies are rotating. Rotation was first observed by V. M. Slipher in 1913. The next step was to search for the rotation of the Milky Way, trying to find the direction and speed of the motion.

The first hint that the stars had an orderly motion was contained in the 18th-century work of W. Herschel (Chapter 3). By the 20th century, Oort reasoned that the stars, gas, and dust of the Milky Way must revolve around the center of the galaxy in keeping with Kepler's laws of motion. This means that objects closer to the galactic center will revolve faster than those farther away. In 1928, Oort predicted the patterns of star motions that we should observe from the sun when we look in all directions. He then proceeded to show that the observations confirmed his predictions. The sun is moving in a nearly circular orbit at a speed of 250 km/sec around the galactic center. Stars closer to the center are moving in about the same direction, but faster than the sun. Stars farther from the center are also moving in about the same direction, but slower than the sun (Figure 6.15).

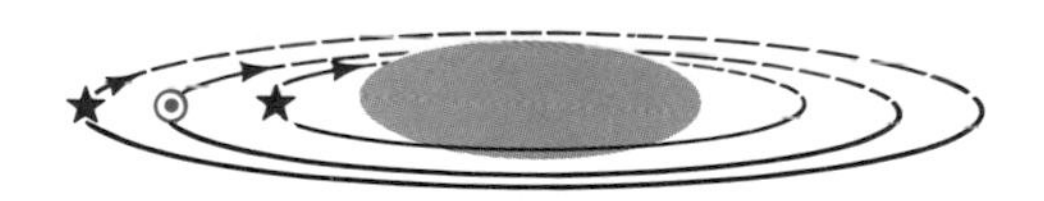

**FIGURE 6.15**
Stars nearer the center of the galaxy have greater orbital velocities than those farther away. This is in accordance with Kepler's Harmonic relation and analogous to the velocities of the planets in the solar system.

Some individual exceptions to these general patterns showed up in the observations. The orbits of all stars around the galactic center are not circular. Some orbits are much more elongated than others. In observing motions of the stars from the sun's point of view, the line-of-sight motion determined by the Doppler shift is really a combination of the sun's motion and the other star's motion. If a star is moving alongside the sun in a similar orbit (nearly circular), we will not detect any line-of-sight motion. All the motion is in a parallel direction. If a star is moving near the sun on a slightly elongated orbit, we will detect a small line-of-sight motion. However, if a star is on an elongated orbit and moving almost directly across the sun's orbit, we will detect a line-of-sight motion nearly equal to the sun's 250 km/sec speed (Figure 6.16). Objects on highly elongated orbits generally reflect the sun's motion around the galaxy.

If we assume the sun is on a nearly circular orbit around the galactic center and we determine the distance to the center, then we can calculate the actual circumference of the sun's orbit around the center of the galaxy. Combining that with the speed of the sun tells us that it takes about 250 million years for the sun to go once around the galaxy.

Interesting situations involving star motions over long periods of time can also be constructed. Suppose a star is located farther from the center of the galaxy than the sun is. The star is traveling on a slightly elongated orbit with the same 250-million-year period as the sun's orbit (Figure 6.17). If we could observe this star from the sun for 250 million years, an unusual effect would result. The star would actually look as though it is *orbiting the sun* once every 250 million years.

Unfortunately, we have only been observing motions of stars around the galaxy for less than a hundred years. Making sense out of what we observe is not easy, but it is being done. Thinking about star motions that take place over long time periods poses new problems and unanswered questions. For example, we know that the sun is located in a spiral arm of the galaxy and we will see in Chapter 10 that the sun is about 5 billion years old. That should mean that the sun has made at least twenty whole trips around the galactic center. The unanswered question is "Why aren't the spiral arms tightly wound up by now?" In the Milky Way, the spiral arms are seen to be fairly loose and disjointed.

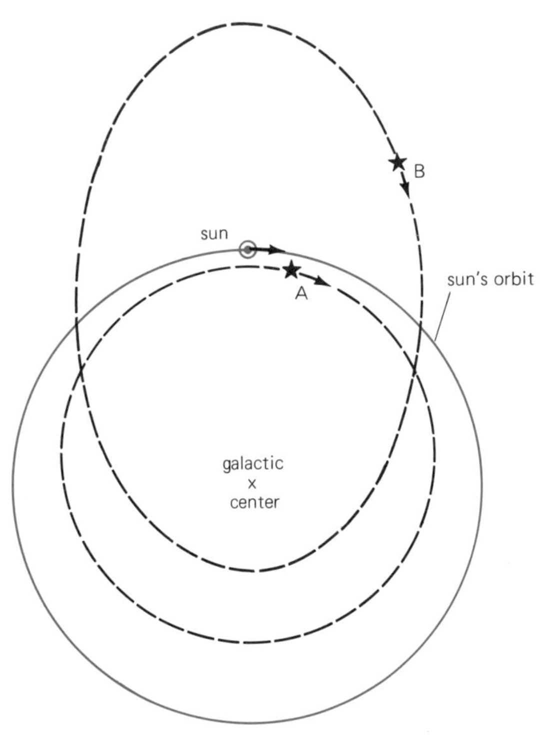


**FIGURE 6.16**
Star A on an orbit very nearly that of the sun shows almost no proper motion nor radial (line-of-sight) velocity. However, star B on a highly elliptical orbit will show a large radial velocity (the reflex of the sun's velocity) and a large proper motion.

## The Spiral Pattern

All spiral galaxies have arms emanating from their central regions. The arms are wound up at most two or three times. It is now believed that the spiral pattern is embedded in the galaxies themselves. The spiral pattern can be thought of as a standing wave or a density wave. The only reason we see the pattern at all is because the gas and dust orbiting around the galaxies tends to pile up in the dense regions. As an example of this kind of pileup, imagine viewing a busy six-lane highway from the air. In one direction a construction crew is repairing guard rails and has one lane of traffic blocked off. As cars pass through the construction area, they will be packed closer together than before and also be slowed down. Once they are past the construction, the cars will spread out and resume a faster speed. We can tell there is something happening just by watching the flow of traffic. The density increase shows up even though there are always different cars in the pattern. Now, if the construction crew finishes one section of guard rail and move farther down the road, that will shift the dense pattern of traffic to a new location.

Something more than just a pileup occurs when gas and dust encounter the density wave in a galaxy. The increased density increases the temperature of the interstellar material and actually triggers star formation. Much of the **density wave theory** is satisfied by our observations. We see just what we expect to see marking the spiral arms in our galaxy: gas and dust clouds; H-II regions; young, hot O and B stars; and galactic clusters. Galaxies where the density wave is strongest show the most pronounced spiral structure. Apparently the density wave in the Milky Way is not terribly strong. Irregular galaxies, which we introduce later, show no evidence of a density wave at all (Figure 6.18).

Density waves as an explanation of the spiral arm structure is the latest of many theories. As we have seen with other astronomical theories, it may eventually be proved correct or another may come along to take its place.

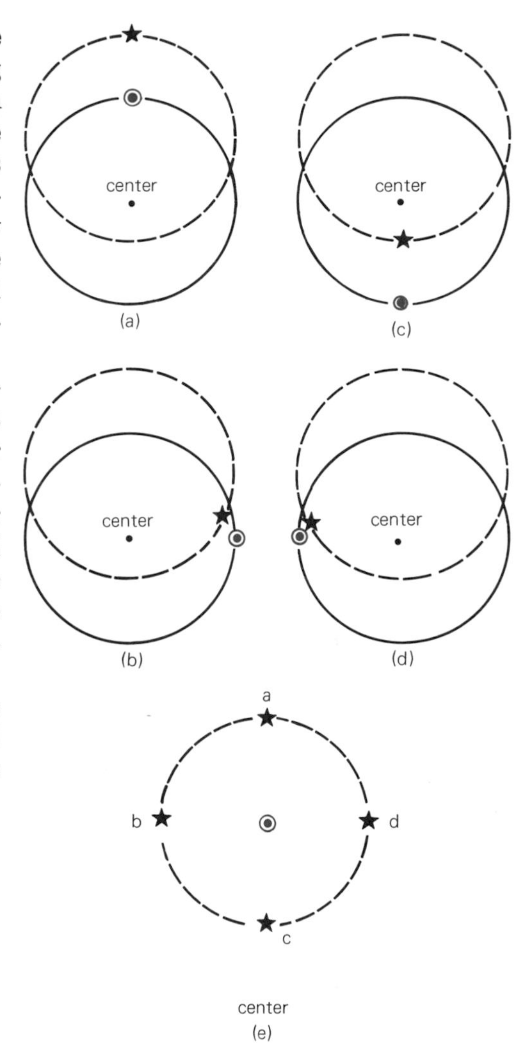


FIGURE 6.17
A star on a slightly elliptical orbit compared to the sun's circular orbit, but having the same period, will appear to perform a circular orbit around the sun. Such an orbit will take 250 million years. In (a) the sun is closer than the star to the center of the galaxy. In (c) the star is closer than the sun to the center of the galaxy. In (b) and (d) the direction to the star from the sun is perpendicular to the direction of the center of the galaxy. From the sun's point of view (e), it appears that the star is orbiting the sun.

## THE NUCLEUS

### Location

For convenience, we often divide the galaxy into three regions: the **halo,** the **disk,** and the **nucleus** (Figure 6.19). We have already described the great spheroidal region or halo containing the outlying older stars and the globular clusters. The diameter of this halo is about 100,000 lt-yr. Within the galactic halo lies the relatively thin disk of stars, gas, and dust where the spiral arms and the sun are located. The diameter of the disk is also about 100,000 lt-yr. It is only about 3000 lt-yr thick. At the center of the disk is the bulging spheroidal concentration of stars, gas, and dust that is the nucleus. The diameter of the nucleus is slightly more than 12,000 lt-yr. The center of the galaxy is located about 30,000 lt-yr away from the sun in the direction of Sagittarius (Figure 6.20).

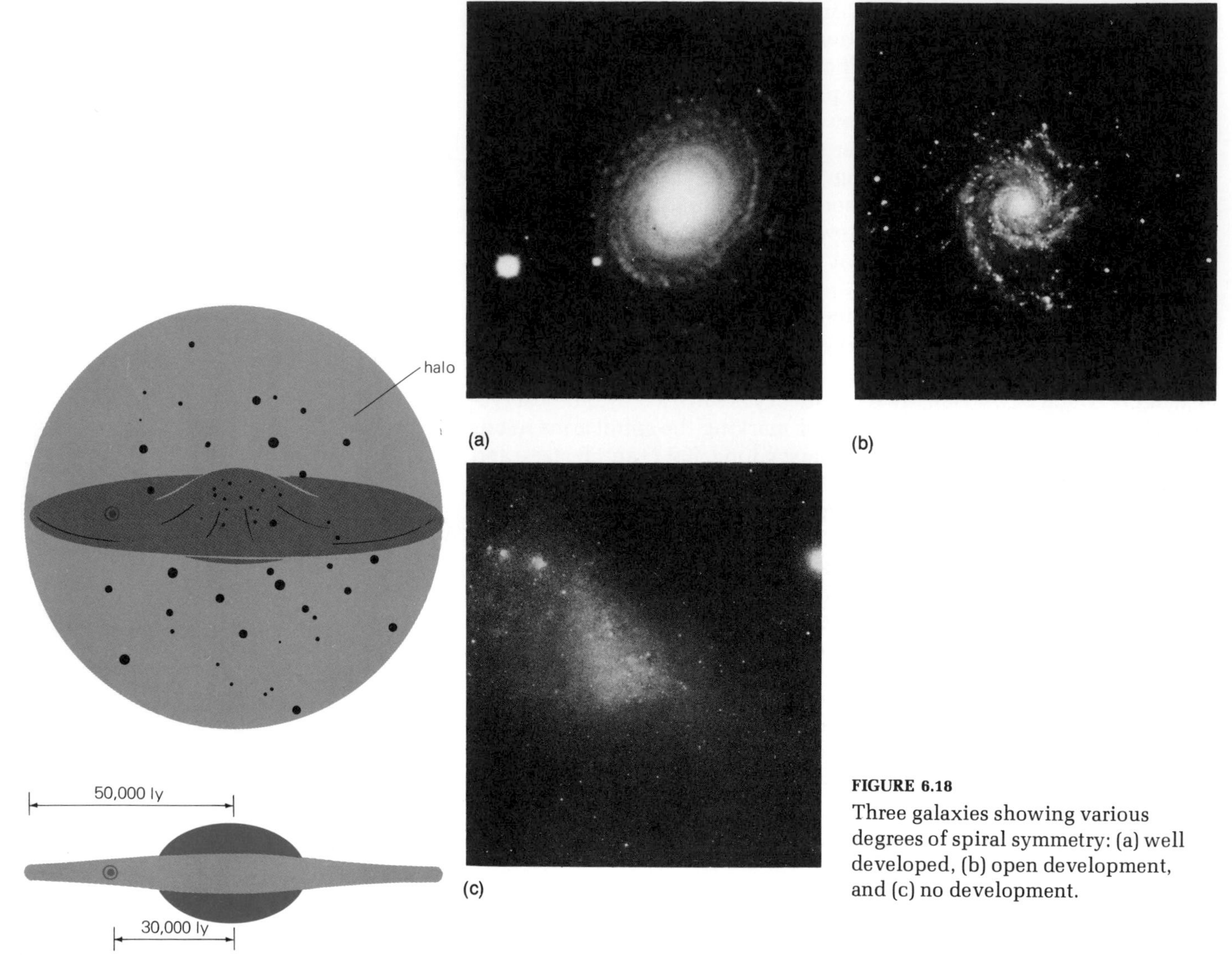


**FIGURE 6.18**
Three galaxies showing various degrees of spiral symmetry: (a) well developed, (b) open development, and (c) no development.

**FIGURE 6.19**
A galaxy is a great spheroidal grouping of stars. The Milky Way galaxy concentrates in a plane with a halo (a). Viewed from the side, the extent of the Galaxy is about 50,000 lt-yr in radius with the sun located in the plane about 30,000 lt-yr from the center (b).

## Makeup

In the dense nucleus of the galaxy, the stars are crowded together two or three times as closely as they are in the sun's vicinity. There are also great clouds of interstellar gas and dust in the nucleus. It is only recently that we have been able to peer into the center of the galaxy. Before 1950 and the application of the techniques of radio astronomy, the nucleus was hidden from our view by vast clouds of obscuring dust. With radio telescopes, we can "see" through the dust and examine the center. In general, we do not see stars (they have very little energy at radio wavelengths), but rather we see the hydrogen gas and several strong radio sources located in the nucleus.

At the very center of the galaxy is the radio source called **Sagittarius A.** It is no more than 3 lt-yr across and radiates tremendous amounts of energy. In addition to several other individual radio sources, there is evidence for two rotating and expanding rings of gas (or partial rings of gas) in the outer regions of the nucleus. Closer to the center is a rotating disk of gas. Within that disk, a ring of dense clouds is rotating and expanding outward (Figure 6.21).

Why gas appears to be moving away from the center of our galaxy at around 100 km/sec is one of the many mysteries of the nucleus. Some astronomers have suggested that the rings of gas are the aftermath of a violent explosion which occurred in the center at least 10 million years ago. If so, what triggered the explosion? Great speculation surrounds the activity in the hub of our galaxy. It will take many more years of careful observations coupled with advanced technology to learn its secrets.

## SUMMARY

All the stars visible in the night sky are part of the Milky Way galaxy. The band of light in the sky called the Milky Way is the densest part of the galaxy. We first learned that the Milky Way was a galaxy in the early part of this century. In addition to billions of stars, the galaxy contains great clouds of gas and dust called nebulae, as well as some thin interstellar material. Most of the stars and nebulae are located in the disk and nucleus of the galaxy. The thin galactic disk containing the spiral arms is about 100,000 lt-yr in diameter and 3000 lt-yr thick. At the center of the disk is the bulging nucleus, 12,000 lt-yr in diameter.

Maps of the positions of bright O and B stars, H-II regions, and galactic star clusters show portions of the three spiral arms near the sun. The sun is located on the inner edge of the Orion arm. Using the 21 cm spectral line of hydrogen (in the radio region), maps of the hydrogen gas in the galaxy were made. These also showed the spiral structure and confirmed that the center of the galaxy is in the direction of Sagittarius, about 30,000 lt-yr from the sun.

The spiral arms of the Milky Way galaxy appear loose and disjointed. The density wave theory for spiral patterns in galaxies says that the pattern is fixed to the galaxy. As stars orbit the center of the galaxy, they tend to pile up in the density wave defining the spiral pat-

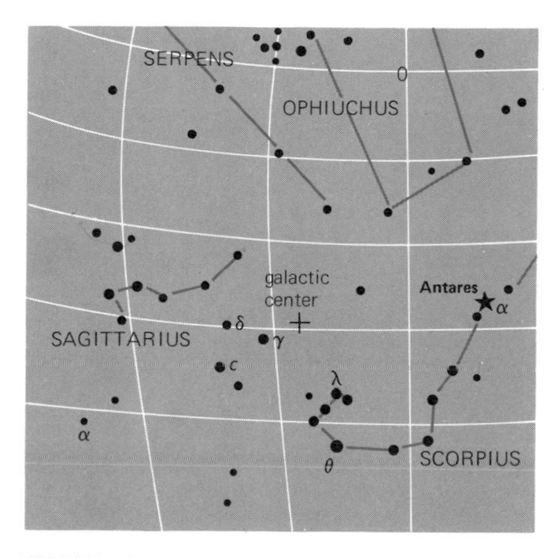


**FIGURE 6.20**
The center of the galaxy marked on a star map. View this region on any clear dark summer night and you will be convinced that the center of the Milky Way galaxy is in that direction.

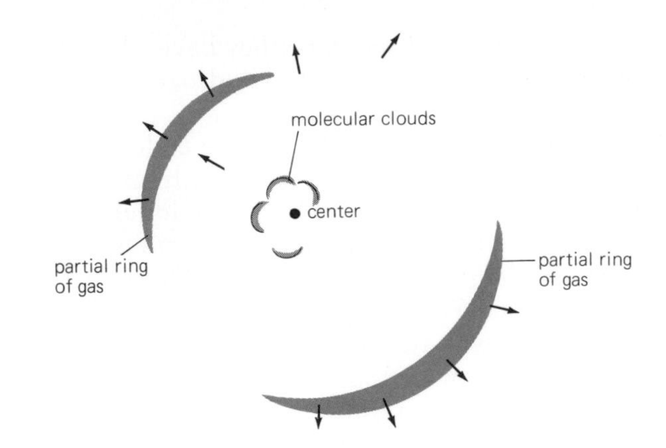


**FIGURE 6.21**
Molecular clouds and rings of gas are expanding from the center of the Milky Way.

tern. Stars obey Kepler's laws in their orbits around the galactic center. At the sun's distance, one orbit takes 250 million years.

Studying the nucleus of our galaxy with radio telescopes shows that it contains some expanding clouds and rings of gas that might have been ejected from the center 10 million years ago.

### *FALLACIES AND FANTASIES*

*The sun is at the center of the galaxy.*
*Visible stars in the sky that are not in the main band of the Milky Way are not part of the galaxy.*

## REVIEW QUESTIONS

_____ 1. Compare the spectra of emission and reflection nebulae. Why are they different?

_____ 2. What simple observation is a reasonable indication that the solar system is located in a flat system of stars?

_____ 3. What objects are good tracers of spiral arms?

_____ 4. List all the different kinds of objects you can think of that constitute the Milky Way galaxy.

_____ 5. Why are radio telescope observations so valuable for studying the galaxy?

______ 6. Sketch and label the disk, nucleus, and halo of the Milky Way galaxy. Indicate the sun's position.

______ 7. What is an example of a halo object?

______ 8. Describe the motion of the sun in the galaxy.

______ 9. Now that you know what makes up the galaxy, why does the Milky Way look like a faint band of light in the sky?

______ 10. Why are we able to see reflection nebulae?

# 7 Interstellar Material: From Galactic Clouds to the Birth of Stars



## INTERSTELLAR CLOUDS

### Between the Stars

We have already mentioned that large amounts of gas and dust pile up in the spiral arms of galaxies. Most of the material between the stars in a galaxy is in a gaseous state and, like the rest of the Universe, consists primarily of atoms of hydrogen and helium. Dust grains represent about one or two percent of the mass of the interstellar material.

Astronomers know very little about the nature of the dust grains. From the way they reflect light, it can be concluded that they are no bigger than .00001 cm (house dust particles are bigger), that they are about as reflective as snow, and that they are most likely elongated in shape. Some astronomers have speculated that the particles of dust are ice-covered graphite. Others suggest that silicates make up the dust. They are most likely a mixture of graphite, silicate, and iron grains. Their origin is also a mystery. Do the dust particles form from the interstellar gas or are they formed in stars and then blown out into the surrounding space? No matter what their origin or composition, examining the dust clouds gives us clues to physical conditions in space.

The real excitement in studying the interstellar material has been the detection of small amounts of atoms other than hydrogen and helium, the detection of **molecules,*** even some complex molecules, and the detection of traces of various minerals. The interstellar molecules are densest where the interstellar dust is densest. The famous Orion Nebula (Figure 7.1) is a favorite spot for detecting molecules.

*Combinations of two or more atoms, molecules are the smallest particles of an element or compound that can retain chemical identity with the substance.

**FIGURE 7.1**
The visible portion of the Orion Nebula. In this region many different species of molecules have been discovered. (Hale Observatories photograph)

The pictures of the great bright and dark clouds of the Milky Way convey various impressions. Some clouds (Figure 7.2) are like cumulus clouds, while others are chaotic (Figure 7.3). Some bright clouds show long, dark intrusions called **elephant trunks** (Figure 7.4). Some are reasonably symmetrical globs (Figure 7.5). Others are highly localized clouds, as if they resulted from a very local event, for example the explosion of a star.

For the past two centuries we have speculated that the beautiful, timeless gas and dust clouds of the Milky Way are the site of star formation. With our better understanding of the energy and life cycles of stars, we can see that stars of all different ages are coexisting in the galaxy. Rigel (a bright star in Orion) can be expected to shine for only

**FIGURE 7.2**
The beautiful Horsehead Nebula gives the impression of terrestrial cumulus clouds. (Hale Observatories photograph)

**FIGURE 7.3**
Some nebulae appear quite chaotic as we see in 30 Doradus. (European Southern Observatories photograph)

**FIGURE 7.4**
Some nebulae seem to have high velocity objects penetrating them giving rise to what astronomers call elephant trunks. (Hale Observatories photograph)

**FIGURE 7.5**
The Trifid Nebula is a large symmetrical nebula showing deep absorption lanes. (Hale Observatories photograph)

about 10 million years, whereas the sun has already been shining at least 4.5 billion years. Rigel must have formed after the sun. Because it is located in one of the spiral arms of our galaxy where we also find much gas and dust, the evidence points to the fact that Rigel and many other stars have been and are forming out of the interstellar medium.

What is it like in regions of star formation? What are the physical conditions inside the great gas and dust clouds of the galaxy? Astronomers are just beginning to know the answers to these questions. Exploration of the space between the stars is almost exclusively a story of 20th-century astronomy.

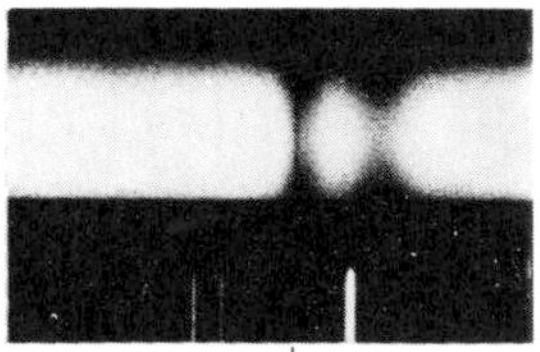

Ca II

**FIGURE 7.6**
Two very sharp interstellar lines of calcium well displaced from the same absorption line in the star. The duplicity of the calcium line tells us there are two interstellar clouds between us and this star. (Hale Observatories photograph)

## Discoveries of Interstellar Matter and Molecules

The first recognition that the interstellar material is everywhere in the galaxy (not just in the great, easily visible bright and dark clouds) came in 1904. Johannes Hartmann was observing the spectra of bright stars. He noted some very sharp absorption lines in the stellar spectra (Figure 7.6). The sharp lines were identified as being produced by calcium, sodium, and a few other elements. When these lines appeared in the spectra of double stars, they remained in the same position while the lines of the stars showed a Doppler shift back and forth. Hartmann suggested that the lines were produced in interstellar space, between us and the bright star. Since the material absorbed the light of the stars, it must be cooler than the stars. Indeed, from the sharpness of the lines, the material must be very cold, ranging from 40°K to 100°K. The broadening of a spectral line depends on the speed of the atoms, which is a measure of the temperature of the gas. When the temperature is low, the velocity is low and hence the line is sharp.

The first interstellar molecule was found in 1937 by T. Dunham. It was a combination of carbon (C) and hydrogen (H) called methyladyne (CH). Methyladyne produced an extremely sharp absorption line on the spectrum of a more distant bright star. Two other interstellar molecules were identified during the next few years. In general, the molecules were located in the same places as the interstellar calcium and sodium. These observations gave direct evidence that the material between the stars occurs in clumps and coexists.

At this time, in the early 1940s, it was doubtful that more complex molecules could be detected even if they existed. Optical telescopes can "see" only two or three thousand light years into space before the obscuring dust blocks the view. Molecules are not easily formed and are very fragile once they are made. High-energy ultraviolet and x-ray radiation from nearby stars might destroy a molecule or keep it from forming a more complex one. Although molecules can form by collisions of atoms, it was not known if the interstellar material was dense enough for collisions to be common.

Then came radio astronomy, which changed the whole picture. One of the first major breakthroughs of the new radio technology was the detection of interstellar hydrogen—lots of it! The neutral hydrogen atoms produced an emission line at a wavelength of 21 cm. The maps of the locations of the interstellar hydrogen (Figure 6.14) confirmed that it was also clumped in the same locations as the other material. But with radio astronomy, the hydrogen, and presumably the other material, is seen to be spread throughout the galaxy from edge to edge. Also it tends to collect along the spiral arms.

In 1963 came the discovery of a fourth molecule in space. It was a combination of oxygen (O) and hydrogen (H) called hydroxyl (OH). Hydroxyl produces emissions at a wavelength of 18 cm. This was followed in 1968 by the discovery of ammonia ($NH_3$) and water vapor ($H_2O$). These discoveries served to show that more complex molecules do exist in the interstellar medium. Then, in early 1969, came the remarkable discovery of formaldehyde ($H_2CO$), an organic molecule. An **organic molecule** is one in which hydrogen is attached to carbon.

**TABLE 7.1 Interstellar Molecules compiled by L. E. Snyder**

Diatomic

| Inorganic | Organic |
|---|---|
| $H_2$—hydrogen | CH—methyladyne radical |
| HD—heavy hydrogen | $CH^+$—methyladyne ion |
| OH—hydroxyl radical | CN—cyanogen radical |
| NS—nitrogen sulfide | CO—carbon monoxide |
| SiO—silicon monoxide | CS—carbon monosulfide |
| SO—sulfur monoxide | |
| SiS—silicon sulfide | |

Triatomic

| Inorganic | Organic |
|---|---|
| $H_2O$—water | CCH—ethyl radical |
| $N_2H+$ | HCN—hydrogen cyanide |
| $H_2S$—hydrogen sulfide | HNC—hydrogen isocyanide |
| $SO_2$—sulfur dioxide | DNC—deuterium isocyanide |
| | HCO—formyl radical |
| | $HCO^+$—formyl ion (X-ogen) |
| | OCS—carbonyl sulfide |

4—Atomic

| Inorganic | Organic |
|---|---|
| $NH_3$—ammonia | $H_2CO$—formaldehyde |
| | HNCO—isocyanic acid |
| | $H_2CS$—thioformaldehyde |

5—Atomic

| Inorganic | Organic |
|---|---|
| None | $H_2CNH$—methanimine |
| | $H_2NCN$—cyanamide |
| | HCOOH—formic acid |
| | $HC_3N$—cyanoacetylene |

6-Atomic

| Inorganic | Organic |
|---|---|
| None | $CH_3OH$—methyl alcohol |
| | $CH_3CN$—methyl cyanide |
| | $HCONH_2$—formamide |

7—Atomic

| Inorganic | Organic |
|---|---|
| None | $CH_3NH_2$—methylamine |
| | $CH_3C_2H$—methylacetylene |
| | $HCOCH_3$—acetaldehyde |
| | $H_2CCHCN$—vinyl cyanide |
| | $HC_5N$—cyanodiacetylene |

| 8—Atomic | |
|---|---|
| Inorganic | Organic |
| None | $HCOOCH_3$—methyl formate |
| **9-Atomic** | |
| Inorganic | Organic |
| None | $(CH_3)_2O$—dimethyl ether |
| | $CH_3CH_2OH$—ethyl alcohol |

The term organic originates from the fact that such molecules are found in living organisms. The discovery of formaldehyde marked the beginning of a flood of molecule discoveries that still goes on. To date, over forty molecules have been detected in space. All of them involve hydrogen, carbon, nitrogen, oxygen, silicon, sulfur, and a few other atoms. Most of the 40 plus molecules are organic. It is tempting to think that a carbon chemistry such as is found on earth is possible, if not common, throughout the Milky Way. This possibility has great consequences for our ideas of the origin and evolution of life. Table 7.1 is a complete listing of interstellar molecules known to date. The table is arranged in order of increasing number of atoms forming the molecules beginning with 2 atom molecules (diatomic) and ending with 9 atom molecules. It also lists the molecules as organic or inorganic. Organic molecules are those found in living organisms and involve hydrogen and/or carbon combined with other elements. No inorganic molecules have been found made up of 5, 6, 7, 8, or 9 atoms.

It is not yet clear exactly how the molecules form, but it is evident that the most complex ones are found in the densest regions of interstellar space. It is in the interiors of the great dust clouds that astronomers are searching for even more complex molecules. It is also in these regions that the mechanism of new star formation must operate.

## Physical Conditions and Motions in Prestellar Clouds

Interstellar clouds must pull together and collapse in order for stars to form, but this is not easy. Consider the standard conditions of the material between the stars. On the average, we will find only one atom of hydrogen per cubic centimeter ($cm^3$) of space. Even in some of the denser areas of the spiral arms, the clouds of gas will have only 100 to 10,000 atoms per cubic centimeter. That may sound like a lot, but it represents a better vacuum than we can produce on earth. The interstellar clouds are composed of the basic elements in roughly these proportions: for every 10,000 hydrogen atoms, there are about 1,200 helium atoms, 2 carbon atoms, 1 or 2 nitrogen atoms, 3 or 4 oxygen atoms,

1 neon atom, 1 sulfur atom, and traces of some heavier atoms. Astronomers refer to these proportions as the **cosmic abundance** since this is the proportion we find everywhere. Some of these elements are bound up in molecules. The dust grains are intricately mixed up with the interstellar gas.

Two forces are constantly at work in the clouds. Mutual gravitation works toward collapse and heating effects work against collapse. As mutual gravitation draws the material together, it also causes parts of the material to bump into other parts and produce heat. The additional heat causes the material to move faster and resist the gravitational pull. An example of this collisional heating effect is the constant beating of a hammer on a cold anvil. After only a few minutes of hammering, the anvil is quite warm.

The relative internal motions of the material in the interstellar clouds have been measured. Typically they are 4 to 10 km/sec—not very large by astronomical standards. However, those speeds are high enough to keep the temperature at a level that prevents the clouds from collapsing to form stars. Since normal interstellar conditions prevent the collapse of clouds, we must look for an external way to trigger the collapse.

Like all other matter in the Milky Way, the interstellar material is orbiting the center of the galaxy. The interstellar clouds are on essentially circular orbits with periods appropriate to their distance from the center. Clouds as far from the center as the sun have orbital periods of 250 million years. When an interstellar cloud passes through the galactic density wave (Chapter 6) it is compressed. We could assume that this compression causes the cloud to collapse and form stars. If compression is all that is happening, the cloud would get denser but it would also get hotter and not collapse. The secret to star formation is to compress the cloud and cool it at the same time. Then the gravitation can take over. To understand Nature's way of cooling the cloud, we have to turn to chemistry. The key is in the formation of a few special molecules.

## STAR FORMATION

### Molecules and Collapse

Photographs of clouds of interstellar material give the impression of timelessness and lack of activity. If we were to follow a single atom this might be true, but when we examine all the billions of atoms in giant clouds, we find teeming activity. Under the standard conditions of the interstellar material, magnificent chemical changes occur. As the cloud orbits the galaxy, atoms combine with other atoms to form molecules, dissociate, combine with other atoms, stick to dust grains, get bumped off, and so on. The molecules that form depend upon the mixture of materials forming the cloud.

If we assume the standard cosmic abundance mentioned earlier, we can ask what molecules form. The answer depends upon density and temperature. At the density of normal interstellar clouds, only a few hardy molecules can form, for example molecular hydrogen, silicon monoxide, methyladyne, cyanogen, and carbon monoxide. More complex molecules are easily dissociated by cosmic rays, x-rays, ultraviolet radiation, or collisions with another molecule or atom.

Now imagine the interstellar cloud orbiting the center of the galaxy and repeatedly passing through the density wave. After one passage into the density wave (spiral arm), the density of the cloud increases and it heats up slightly. Although the density goes up only a little, it may be enough to prevent x-rays or ultraviolet rays or even light from penetrating the cloud. The longer infrared and radio wavelengths, however, can still pass through the cloud (Figure 7.7). The increase in density provides some natural shielding from the destructive x-rays and ultraviolet light so that additional molecules can form and survive inside the cloud. If enough carbon monoxide is formed, the cloud will actually cool and contract and become even denser. We will use carbon monoxide as our example of how certain molecules cool the interstellar clouds.

When a carbon monoxide molecule is bumped, it absorbs some of the collision energy and is then in a higher energy state. Normal molecules in this condition move around until they bump into other molecules, sharing their excess energy with these molecules. As the gas is compressed, the molecules will bump together more often, and the temperature of the gas increases. However, carbon monoxide radiates its excess energy as an infrared photon in a very short time, often before it bumps into another molecule. The infrared photon passes out of the cloud. Since the cloud has lost energy, it is now cooler. When this process is repeated by the billions of carbon monoxide molecules present, the cloud cools. Then gravity may cause the cloud to become denser.

At the higher densities and cooler temperatures, even more complex molecules may form and survive since they are shielded from disruptive outside radiation. Deep inside the cloud we can find ammonia, hydrogen cyanide, and many other molecules. In the central regions we can find formaldehyde, methyl alcohol, and ethyl alcohol. Indeed, we find organic molecules in abundance. At the temperatures observed,

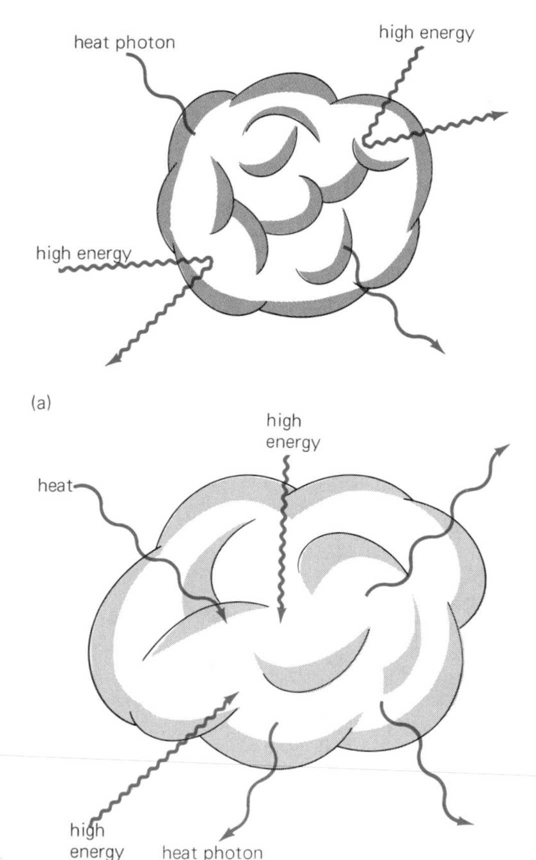


**FIGURE 7.7**
A very dense interstellar cloud (a) allows long wavelength radiation to penetrate and pass through while high-energy radiation is immediately reflected. A thin cloud (b) will generally absorb high-energy radiation while heat photons or long wavelength radiation passes through.

minerals and ices form as well. Now, when the cloud moves into the density wave again, the little extra compression triggers its collapse, and star formation results.

We have just described one of several theories of beginning star formation. Some astronomers feel that the trigger for the collapse of the interstellar clouds may be the compression wave or blast wave generated by exploding stars called **supernovae.** These stars are discussed in Chapter 11. Other astronomers feel that star formation occurs before galaxy formation. At this time the observational evidence seems to favor the theory of a collapsing interstellar cloud triggered by the density wave. We observe the youngest stars in the Milky Way galaxy just where they would be expected to form, along the inner edges of the spiral arms.

## Observations of Dark Clouds, Globules, and Protostars

In general, interstellar dust grains scatter starlight and make it appear redder than normal. When enough dust is present, total blocking out of starlight results. This produces the dark clouds of dust that look like "holes" in photographs of background stars (Figure 7.8). About a dozen **dark clouds** are observed within 1000 lt-yr from the sun. These

**FIGURE 7.8**
Two dark globules relatively nearby on the sky. (Photograph by B. Bok)

dark clouds are each about 12 lt-yr in diameter and have a quantity of dust equal to 20 times the mass of our sun. From the molecules that are observed in the clouds (hydroxyl, ammonia, formaldehyde, carbon monoxide, and so on) it can be estimated that there is about 100 times more gas than dust in the clouds. The total mass of each cloud (gas plus dust) is about 2000 solar masses. The density of the gas is between a few hundred and ten thousand particles per cubic centimeter. The temperatures are between 4°K and 25°K. Calculations show that these dark clouds should be gravitationally collapsing.

Within 1000 lt-yr of the sun we also see about 100 **globules.** These are regions of dust and gas smaller and denser than the dark clouds. A large globule may be 3 lt-yr in diameter and contain material equal to 60 solar masses. Globules can sometimes be seen against the background of bright emission nebulae (Figure 7.9). They look like tiny holes in the sky. Globules are probably one of the final stages in the collapse of an interstellar cloud into protostars.

A **protostar** is the term given to an even denser concentration of gas and dust about the size of the solar system. It is very close to becoming an actual star. A protostar can be expected to have a density and temperature sufficient for it to radiate energy in the infrared region of the spectrum. Astronomers have detected what they believe to be protostars embedded in a dark molecular cloud behind the Orion Nebula.

Dark clouds that are ripe for star formation will probably fragment into many globules. As the globules collapse, they can in turn break up to form many protostars (Figure 7.10 and color plates). Some protostars will collapse much faster near their centers than in their outer regions. When that happens, these newly born stars will be surrounded by a large region of gas and dust. At this stage in the formation of stars, conditions may lead to the formation of planets from the surrounding gas and dust.

## BIRTH OF THE SUN AND THE SOLAR SYSTEM

### The Solar Nebula

One fragment of our parent interstellar cloud was almost like any other. In general the motions in this fragment, which we call the **solar nebula** for convenience, were random. Gradually, as the solar nebula collapsed, one direction of motion became more prevalent than any other. After a time, the entire nebula was rotating more or less in one direction around a central axis (Figure 7.11). The molecules, ices, dust, and heavier particles quickly settled out into a very thin disk. The hy-

**FIGURE 7.9**
The Lagoon Nebula shows dark globules superposed on the bright emission nebula. (Lick Observatory photograph)

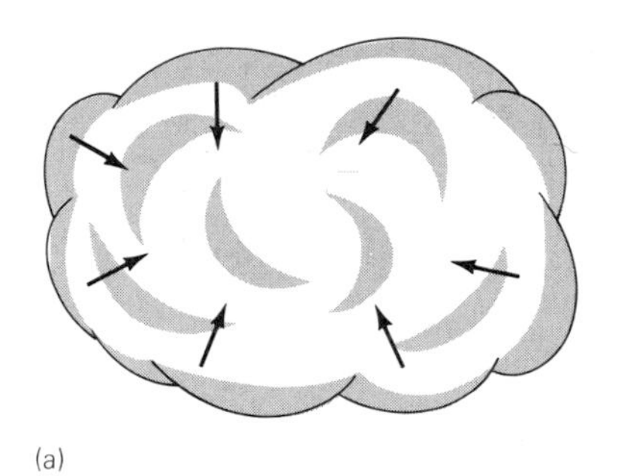


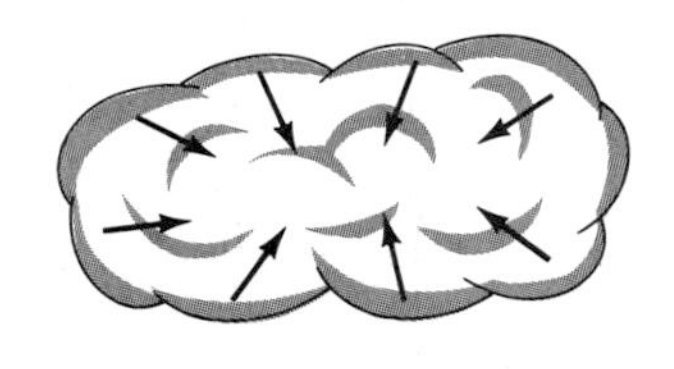

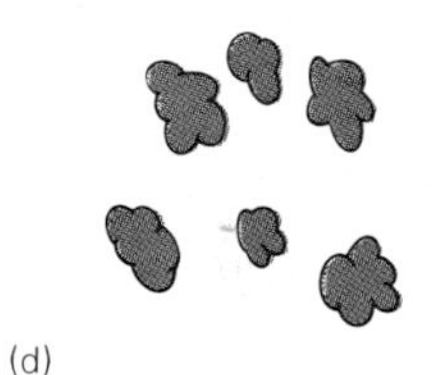


**FIGURE 7.10**
A typical interstellar cloud cools and collapses (a) and (b) and fragments (c). The fragments in turn collapse and fragment still farther (d).

drogen and helium, and other gaseous components, remained in a more or less spheroidal distribution with only a slight tendency to concentrate toward the disk (Figure 7.12). In the disk, ices bumped into other ices and gradually accreted or collected material into lumps. The largest lump formed at or near the central axis and would become the sun.

As the solar nebula collapsed, it heated up. But by this time the increased density of the nebula was such that the heating from collisions could not override the mutual gravitation of the material. The sun and solar system were destined to form.

## The Protosun and Planets

When the center of the nebula began to glow due to the heating described above, we really had a **protosun.** It was 400 times larger than the sun is now. The heat from the protosun flowed out through the solar nebula. According to one theory, the planets formed out of the remaining solar nebula. The central regions of the nebula near the protosun were relatively hot, while the edges were relatively cold. The temperature ranged from several thousand degrees K near the center of the nebula to about 40°K at the outermost edge. Because the temperature varied so greatly across the nebula, the composition of the nebula

**FIGURE 7.11**
A cloud fragment or protostar nebula has a preferred axis of rotation.

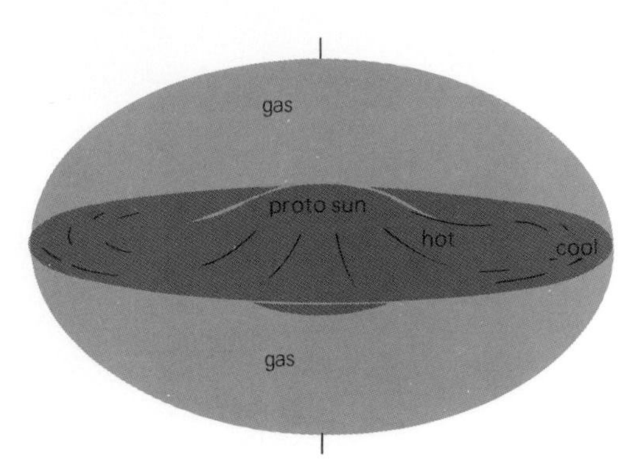


**FIGURE 7.12**
Nongaseous material in the solar nebula forms a disk which is hot near the central protosun and cool near the edge.

also varied (Figure 7.13). Next to the protosun, all the materials were totally evaporated. At greater distances, some elements condensed into solid particles, minerals, and dust. Farther from the protosun, some of the volatile elements froze and formed ices. Just by looking at the temperature of the nebula at various distances, we can predict the composition and densities of the planets.

At a distance of 60 million km from the center, the temperature was about 1200°K and only heat-resistant materials continued to exist. Thus the clump accreting materials at this point collected only minerals containing iron-nickel alloys, aluminum oxides, and compounds of magnesium. Any planet of a reasonable size at this distance would be composed of this material and have a mean density of about 5.4 g/cm$^3$. This is what we observe with Mercury.

At a distance of 108 million km, the temperature of the nebula was slightly hotter than 800°K. At this temperature, the major condensates were the heat-resistant materials mentioned above plus compounds of sodium, sulfur, potassium, and so on. We would expect a planet at this distance to be rich in these compounds and have a mean density near 5.0 g/cm$^3$. We observe exactly this with Venus.

At the earth's distance, the nebula was just under 700°K. The condensates present on the earth and the density of the earth (5.5 g/cm$^3$) agree with the theory. All the small bodies of the solar system seem to fit into this picture of planet formation. The snag comes with the giant planets—Jupiter, Saturn, Uranus, and Neptune. They do not fit into the theory, although most of their moons do. The giant planets

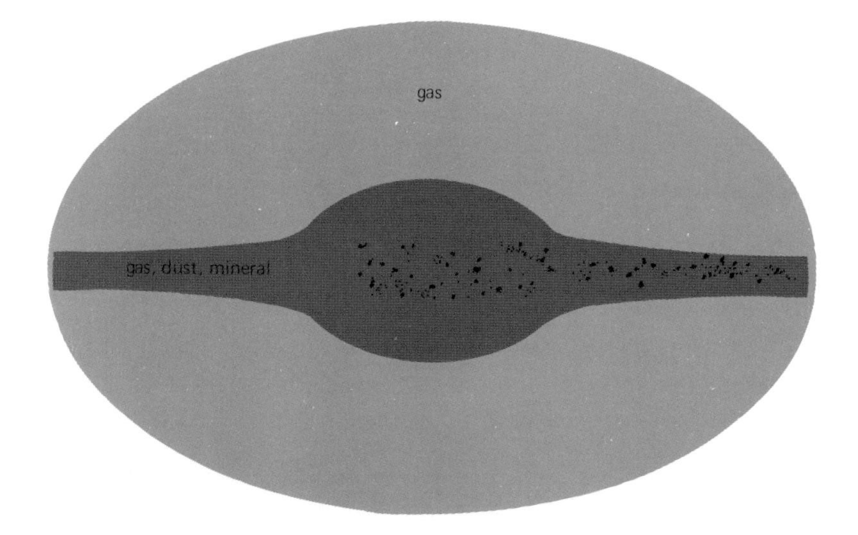


**FIGURE 7.13**
A cross section of the solar nebula shows that the inner portion of the disk contains minerals while the outer portion contains gas, dust, and ices of volatile materials.

have compositions very similar to the sun's. It has been suggested that these planets might have begun with extremely large rocky, icy cores. These large masses could then have captured tremendous amounts of the original gaseous material. This would have given them a composition similar to what we observe.

All the while the planets were forming, the protosun was accreting additional material. It became so massive that it began to collect outlying gaseous material as well as dust and solids from the disk. This great mass of material continued to contract under its own gravitation until its central pressure caused hydrogen fusion to begin. Deep in the interior of the protosun, hydrogen atoms fused into helium atoms and created energy. Details of the sun's energy production will be discussed in Chapter 10. The protosun had to adjust to this new energy source. During a relatively brief period of time the sun went through various flashes and flickerings as atomic energy generation began. We call this phase of star formation the **T Tauri phase,** after a well-observed star in Taurus that is flashing and flickering right now.

The sun's flash of intense radiation pushed the remnants of the solar nebula out of the solar system forever (Figure 7.14). The young sun and planets began evolving to their ultimate fate. The sun settled down very quickly to a long period of steady shining as a main sequence star. In Chapter 11 we will continue the story of the life of stars. In a star's declining years, some of the most spectacular events of nature occur.

(a)

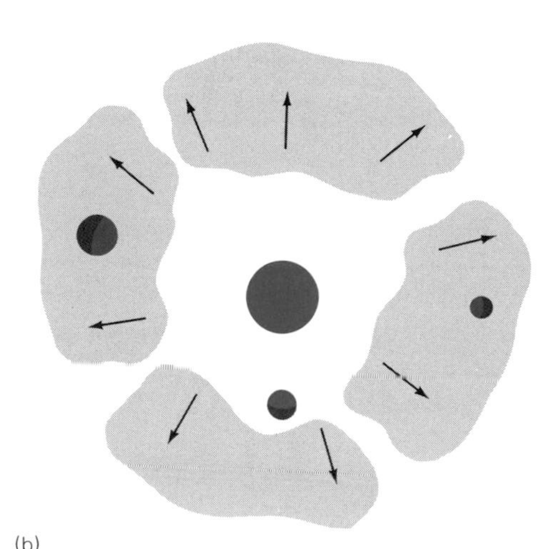

(b)

**FIGURE 7.14**
The sun and planets form deep in the solar nebula (a). A bright hot flash or flashes from the young sun blast away the small material of the solar nebula (b).

## SUMMARY

In our galaxy, the space between the stars contains hydrogen and helium gas. It is spread thinly (only one hydrogen atom per cubic centimeter) and is quite cold (near absolute zero). Along the spiral arms of the galaxy, some of this gas has piled up into great clouds. A small amount of dust is mixed up in the gas clouds. Occasionally the dust will obscure our view of what lies beyond. Inside these great gas and dust clouds, where they are protected from high-energy photons, molecules, even organic molecules, are forming and surviving. More than 40 interstellar molecules have been discovered.

The great interstellar clouds are thought to be the birthplace of new stars. A current theory says that as the interstellar clouds orbit the center of the galaxy, they move in and out of the density wave (spiral arms), which triggers their collapse. Some clouds continue to collapse because certain molecules are radiating energy and cooling the interiors of the clouds. The clouds fragment, and gravity eventually takes over as the dominant force in each fragment. Collapsing clouds called globules have been observed in space. A globule is probably the last stage before the formation of a protostar.

The interstellar cloud that collapsed to form the sun and planets is called the solar nebula. As it collapsed, it rotated. The protosun formed at the center, attracting most of the material. In the outer part of the nebula some lumps of material added more material and became the planets. The solar nebula was hottest in the center and coldest at the edges. This temperature difference affected the composition of material throughout the nebula and is responsible for the different makeup of each planet. When the center of the protosun got hot enough, it began producing atomic energy. The initial burst of energy cleared out the remains of the solar nebula between the planets. Our star was born.

### *FALLACIES AND FANTASIES*

*The space between the stars is empty.*
*Stars make up most of the mass of the Universe.*

## REVIEW QUESTIONS

______ 1. Even though most of the interstellar material is located in clouds, there is some small amount between the clouds. How do we know this?

______ 2. Where are the best places to search for complex organic molecules in space? Why?

______ 3. What is the cosmic abundance?

______ 4. We cannot see dark nebulae beyond about 5000 lt-yr. Why not?

______ 5. What is a protostar?

______ 6. All the planets orbit the sun in the same direction. Why do you think this is so?

______ 7. What must happen to a normal interstellar cloud in order for it to form stars?

_____ 8. Why is radio astronomy so valuable for studying the interstellar medium?

_____ 9. In what parts of the galaxy would you search for new stars?

_____ 10. You have probably heard that interstellar space is a vacuum. Is it? Explain.

# 8 The Solar System: From Giant Gaseous Worlds to Small Rocky Worlds



## INTRODUCTION

### Inventory

Perhaps the only interesting thing about the medium-sized star of average brightness which we call the sun is that in its development a system of small bodies formed around it, and on one of these life developed. Our solar system is made up of the sun and all the celestial objects that are gravitationally bound to and revolve around the sun. These include the nine planets, their thirty-five moons, thousands of minor planets called asteroids, and billions of small bits of debris called meteoroids, comets, and dust. In Chapters 8 and 9 we take a close look at 1% of our solar system, everything but the sun. In Chapter 10, we will learn about the other 99%, the sun.

## Perspective

In this age of space travel, when it takes space probes about 9 months to reach the nearby planet Mars, the solar system may seem vast and remote. Pluto, the outermost planet, is almost 6 billion km from the sun. If the entire solar system out to the orbit of Pluto could be placed on the Pentagon, the next nearest star or star system would be located in Dallas, about 2000 km away (Figure 8.1). Thus when distances within the solar system are compared to the enormous distances between the stars, the members of our solar system turn out to be quite close to each other.

If we were on the star closest to the sun (about 4.5 lt-yr or 40 trillion km away), looking through our largest telescope, we would not be able to see any of the planets orbiting the sun. Their feeble reflected light would be undetectable in the glare of the sun. If we were on a star 40 lt-yr away from the sun, we would not be able to see the sun with the unaided eye. This distance is not even one-tenth of one percent of the distance across the Milky Way galaxy. In this chapter, we are focusing on a very minute region of the Universe.

## Regularities and Similarities

Some of the most precise information we have about the planets is on their orbits and their motions. When data are compiled, certain regularities of the solar system stand out. The most notable are these:

1. The orbits of the planets are coplanar (except perhaps Pluto's). That is, all the orbits are in nearly the same plane.

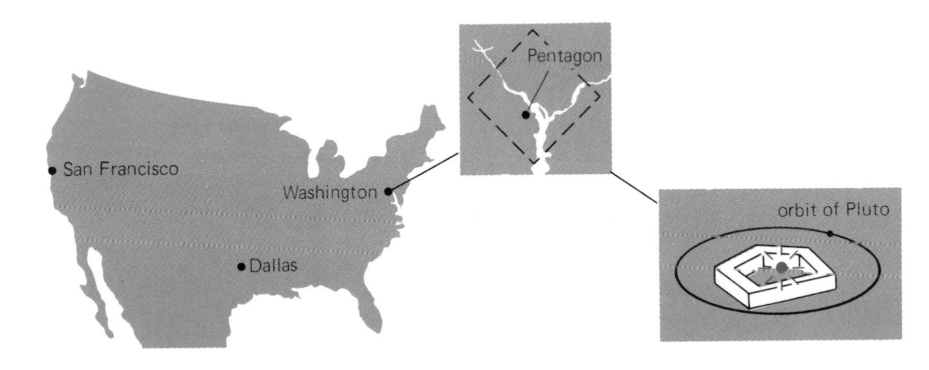


**FIGURE 8.1**
A schematic drawing showing that if Pluto's orbit were about the size of the Pentagon, the distance of the nearest star to the sun would be the distance between Washington and Dallas.

**TABLE 8.1 Characteristics of Members of the Solar System**

| *Characteristics* | *Sun* | *Mercury* | *Venus* | *Earth* | *Moon* | *Mars* | *Ceres* | *Jupiter* | *Saturn* | *Uranus* | *Neptune* | *Pluto* |
|---|---|---|---|---|---|---|---|---|---|---|---|---|
| Symbol | ☉ | ☿ | ♀ | ⊕ | ☾ | ♂ | ① | ♃ | ♄ | ♅ | ♆ | ♇ |
| Distance from Sun (Astronomical Units) | — | 0.387 | 0.723 | 1.000 | 1.000 | 1.524 | 2.767 | 5.203 | 9.539 | 19.182 | 30.058 | 39.439 |
| Distance from Sun (millions of kilometers) | — | 57.91 | 108.20 | 149.60 | 149.60 | 227.94 | 413.98 | 778.33 | 1429.99 | 2869.57 | 4496.60 | 5900.00 |
| Period of revolution (years) | — | 0.241 | 0.615 | 1.000 | 0.075 | 1.881 | 4.604 | 11.862 | 29.458 | 84.013 | 164.794 | 247.686 |
| Orbital inclination to ecliptic | — | 7° | 3°24′ | 0.0 | 5°9′ | 1°51′ | 10°37′ | 1°18′ | 2°29′ | 0°46′ | 1°46′ | 17°10′ |
| Period of rotation (days) | 25.38 | 58.65 | −243 | 0.997 | 27.322 | 1.026 | — | 0.413 | 0.426 | 1.042 | 0.917 | 6.39 |
| Inclination of equator to orbit | 7°15′ | 0° | 3°18′ | 23°27′ | 6°41′ | 23°59′ | — | 3°4′ | 26°44′ | 97°53′ | 28°48′ | — |
| Diameter (kilometers) | 1,392,000 | 4868 | 12,112 | 12,756 | 3476 | 6787 | 780 | 143,200 | 120,000 | 50,800 | 49,500 | 5800: |
| Mass (earth = 1) | 332,960 | 0.05 | 0.82 | 1.0 | 0.012 | 0.11 | — | 317.9 | 95.12 | 14.6 | 17.2 | 0.11: |
| Density (water = 1) | 1.41 | 5.44 | 5.26 | 5.52 | 3.34 | 3.94 | — | 1.314 | 0.704 | 1.31 | 1.66 | 4.9: |
| Surface temperature (daytime °K) | 6000 | 700 | 740 | 295 | 400 | 250 | — | 123 | 93 | 63 | 53 | 40: |
| Main atmospheric constituent | Hydrogen Helium | — | Carbon dioxide | Nitrogen Oxygen | | Carbon dioxide | — | Hydrogen Helium | Hydrogen Helium | Hydrogen Helium | Hydrogen Helium | — |
| Number of moons | — | 0 | 0 | 1 | — | 2 | — | 14 | 11 | 5 | 2 | — |

: designates uncertain values

2. The orbits of the planets are only slightly flattened ellipses and may be regarded as essentially circular.
3. The planets orbit the sun in the same direction that the sun rotates.
4. The planets rotate on their axes in the same direction as the sun (except Venus and perhaps Uranus).
5. Planetary densities decrease with distance from the sun (with a few exceptions).
6. The spacings of the orbits follows a geometric relation called the Titius-Bode relation (except Neptune and Pluto).

Almost all these regularities can be explained by the theory of the origin of the solar system discussed in Chapter 7. Though there are similarities among the planets, each has its own unique and exclusive features. The origin of the planets was similar, but the course of evolution of each planet was quite different because of their different distances from the sun.

Collecting data about the planets was slow and laborious until twenty years ago and the beginning of the space age. Ground-based optical telescopes are of limited value, because the atmosphere of the earth constantly smears and blurs the images of the planets. Radar and radio telescopes are not greatly affected by the earth's atmosphere. They can even penetrate some of the planetary atmospheres giving us valuable information. But, the latest and best information on the planets comes to us from above the earth's atmosphere—from satellites orbiting above the earth, from six manned landings on the moon, from the space probes that have flown past the majority of the planets, and from the two Viking orbiters and landers on Mars.

New and significant discoveries about the planets have become commonplace. Before we look at the planets individually, we summarize various characteristics of the planets and their orbits in Table 8.1. A reasonable fraction of the entries have been improved recently and a few have changed dramatically. If the recent history of planetary exploration is a guide, the information we present in this chapter will probably need to be revised, corrected and added to more quickly than any other in the book.

## Two Types of Planets

For convenience, we divide the planets at the asteroid region (Figure 8.2). Planets closer to the sun than that region are called inner or **terrestrial planets.** Planets outside that region (except Pluto) are called Jovian or gaseous or **giant planets.** This seemingly arbitrary division is striking when we consider the physical compositions, sizes, and positions of the planets. With the exception of Pluto, the giant planets occupy the region of the solar system farthest from the sun. The temperatures in their upper atmospheres are never warmer than 146°K (−127°C). In terms of size, any one of the giant planets could easily swallow up all the tiny terrestrial planets. The composition of Jupiter and the other giant planets is much like that of the sun, over 90% hydrogen and helium.

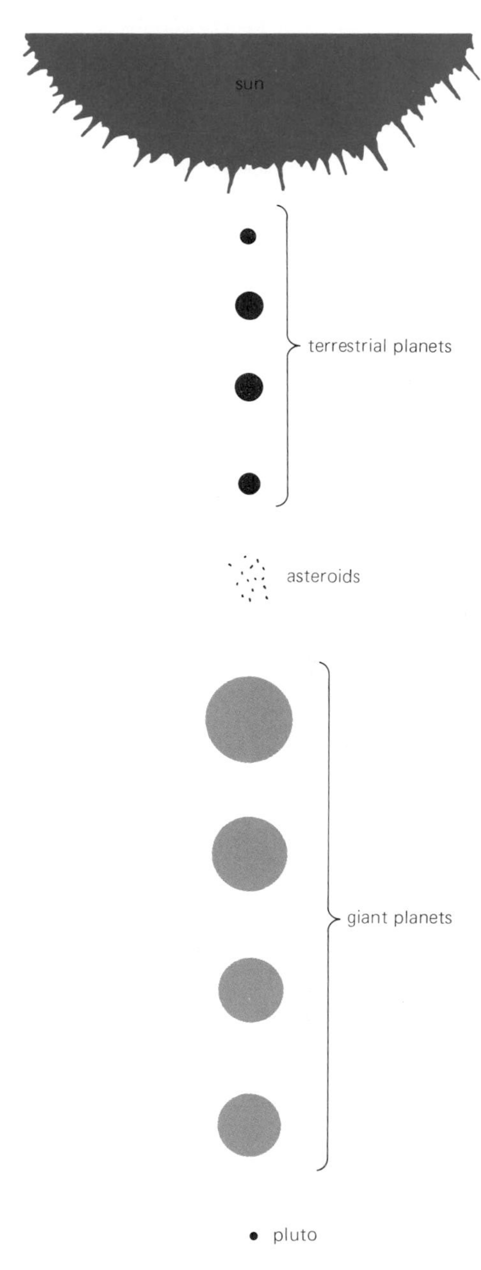


**FIGURE 8.2**
The division of planets into inner, terrestrial planets and outer, giant planets. Pluto is an outer planet but may be terrestrial. The asteroid belt is the division point. Sizes and distances are not to scale.

## THE GIANT PLANETS

### General Properties and Jupiter

After a journey through space of nearly two years, the Pioneer 10 and Pioneer 11 space probes flew by Jupiter in 1973 and 1974 (Figure 8.3). Information from those probes greatly enhanced our knowledge of the largest planet. Spacecraft photographs of Jupiter show its familiar atmospheric cloud bands (Figure 8.4). Jupiter's grey-white zones are weather cells of rising gases. The highest gas clouds in these zones are thought to be crystals of ammonia. The dark red-brown belts are areas of descending gases. Observations of the upper atmosphere of Jupiter have shown traces of methane and ammonia, gases retained from the solar nebula.

The **Great Red Spot** on Jupiter is now thought to be the top of a massive, hurricane like storm center that has been raging for over three hundred years. It was first observed in the 17th century. The spot, which has varied in size, shape, and hue, is several times larger than the earth. Spacecraft photographs (Figure 8.5) show other smaller "spots" and swirls in Jupiter's turbulent atmosphere.

Astonomers cannot really define a surface for any of the giant planets. As we go down through the clouds, we can deduce that their atmospheres get denser and slushier and that their interiors consist of liquid molecular hydrogen over liquid metallic hydrogen directly above a small, solid metallic core (Figure 8.6). Deep inside Jupiter, incredibly high pressures exist along with temperatures hotter than the surface of the sun (6000°K). Solids at high pressure are not understood well enough for this to be any more than a theory of what the interiors of the four giants planets are like. Jupiter has some internal heat in excess of that expected from theory. The source of the heat is a topic of current research and is believed to result from the continuing gravitational contraction of Jupiter.

Jupiter is sometimes described as a star that failed. Apparently, the accreted core of Jupiter did not form as fast as the sun did. When the sun began to flash and flicker and sweep the solar system clear of excess debris, Jupiter could not compete with it. But this planet was

**FIGURE 8.3**
An artists conception of Pioneer 10 flying past Jupiter. (Courtesy of the National Aeronautics and Space Administration)

massive enough to retain gases from the solar nebula. Had its orbit not been so nearly circular, Jupiter may have dominated the region around the sun and swept up enough material to become a star. In that case, the solar system would have become a binary star system. Studies show that the circular orbits of the major planetary bodies seem essential to the formation of a solar system like ours.

**FIGURE 8.4**
A Pioneer 10 photograph of the cloud bands on Jupiter. (National Aeronautics and Space Administration photograph)

All the planets rotate around an axis. Jupiter spins, or rotates, faster than any other solar system planet. It completes one rotation in less than 10 hours. The result of such rapid rotation is to cause the planet to bulge at its equator. This is very evident in photographs of the planet (Figure 8.7). Jupiter also has an extensive **magnetic field** that rotates with the planet. The magnetic field of Jupiter is stronger than the earth's and it extends to a much greater distance from the planet. Like the earth's magnetic axis, the magnetic north-south axis of Jupiter is tipped compared to the rotational axis. The effect is much more extreme on Jupiter than on the earth. The intersection of these two axes is actually outside the surface of the planet. As Jupiter rotates, the magnetic pole alternately sweeps into view and out of view (Figure 8.8). This is an example of what is called an **oblique rotator.**

Jupiter has fourteen moons or **satellites.** The inner five (four of which were discovered by Galileo) are regular in that they have nearly circular orbits, are regularly spaced, and are of reasonable size compared to Jupiter. Pioneer 10 returned a dramatic picture of the largest of these, Ganymede. In Figure 8.9 we see that Ganymede greatly resembles our moon with its **maria** (dark plains) and craters. Ganymede is larger than the planet Mercury. The next four moons beyond the first five form a second group. Their orbits are not quite coplanar and are distinctly elliptical. The outer five moons are clearly irregular. Their orbits have high inclinations, and their orbital motion is retrograde,

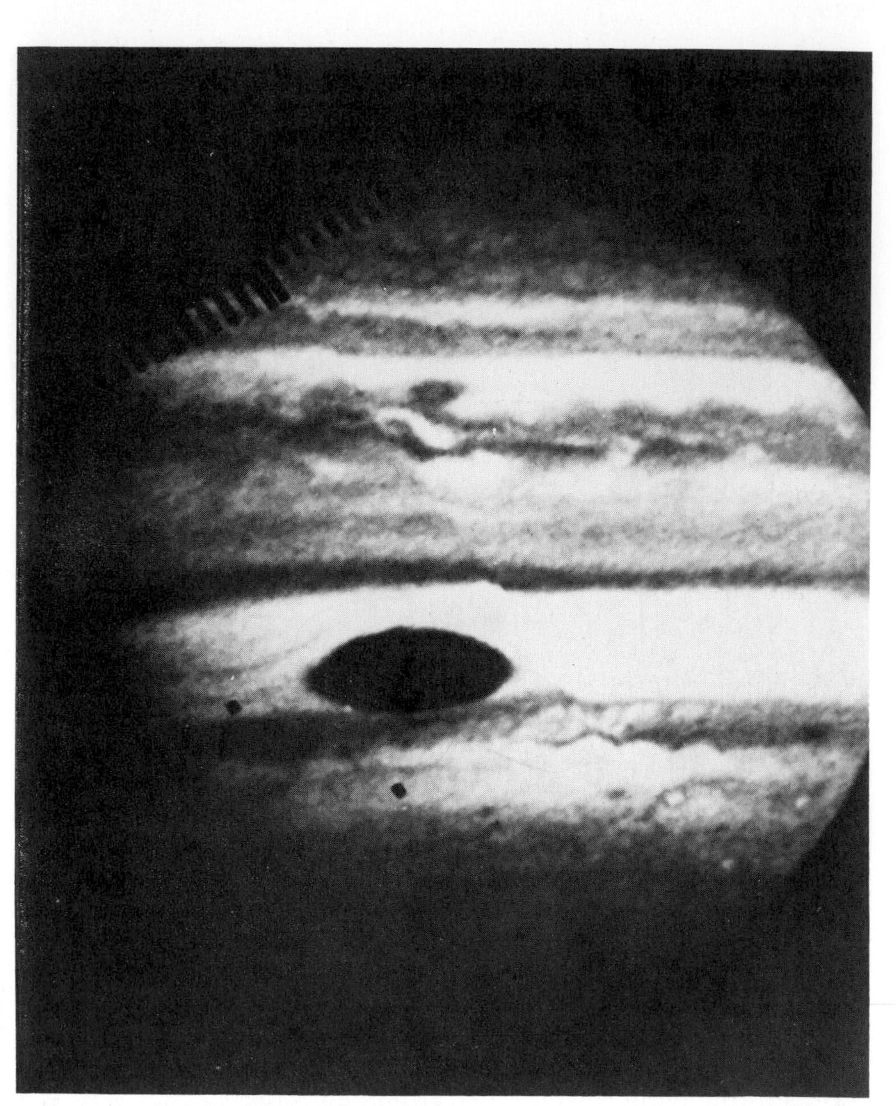

**FIGURE 8.5**
Jupiter from Pioneer 10. The great Red spot stands out. (National Aeronautics and Space Administration photograph)

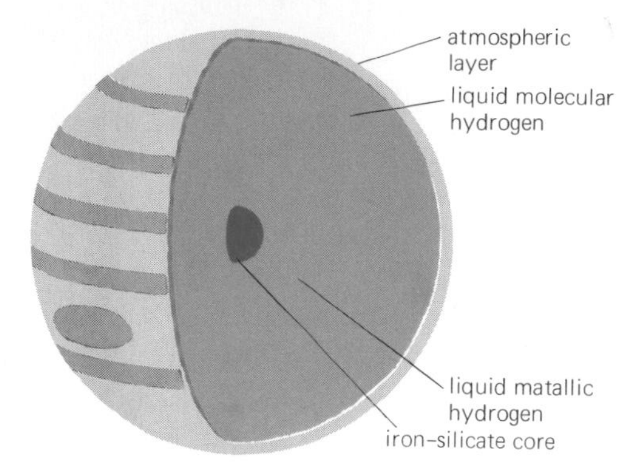


**FIGURE 8.6**
A cut-away view of what the interior of Jupiter may be like.

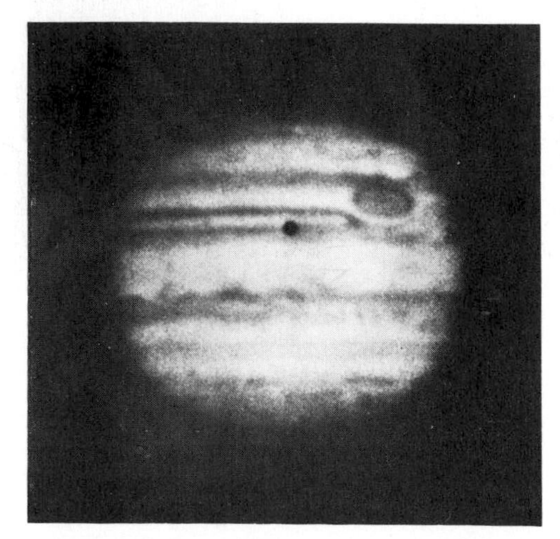

**FIGURE 8.7**
Jupiter shows slight oblateness. Its polar diameter is 6% shorter than its equatorial diameter. (Lowell Observatory photograph)

that is, in the opposite direction from the other moons. They are small and resemble asteroids. Once in a while a sharp, black spot will appear on a photograph of Jupiter (Figure 8.10). This is merely a shadow cast on the clouds of Jupiter by one of its nearby moons. (See color plates.)

## Saturn

Saturn is the second largest planet in the solar system and is distinguished by its very low density and spectacular ring system (Figure 8.11 and color plate). We would expect a density of about 2 gm/cm³ for

Saturn based on the theory of the origin of the solar system described in Chapter 7. Instead, the density of Saturn is only 0.9 gm/cm³—less dense than water. It appears that, like Jupiter, the condensation that formed Saturn grew large enough to retain a significant amount of the original hydrogen and helium of the solar nebula. Titan, one of Saturn's 11 moons, has a density of 2.1 gm/cm³, which agrees well with the theory. Titan is one of the largest moons in the solar system. It is almost twice as massive as our moon and is known to have an atmosphere.

Saturn's atmosphere and interior must be very much like Jupiter's. Hydrogen and helium are the main components of the atmosphere, followed by methane and ammonia.

Saturn's brilliant rings are made up of millions of small particles. The particles are most likely ice crystals and bits of ice-covered dust resembling hailstones. At times, distant stars can be seen right through the rings, which demonstrates that they are neither solid nor very thick. Each particle in the rings may be thought of as a tiny moon orbiting Saturn. The origin of the rings is still uncertain. One theory says that the material in the rings was once a moon of Saturn. The moon got too

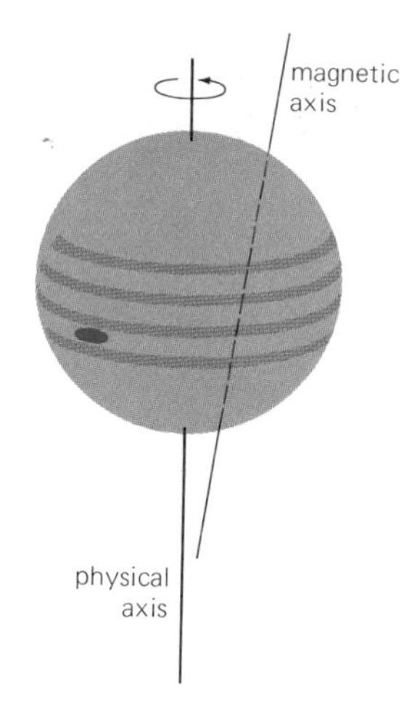


**FIGURE 8.8**
The magnetic pole of Jupiter is tilted with respect to the pole of rotation. The magnetic axis intersects the rotation axis well below the planet.

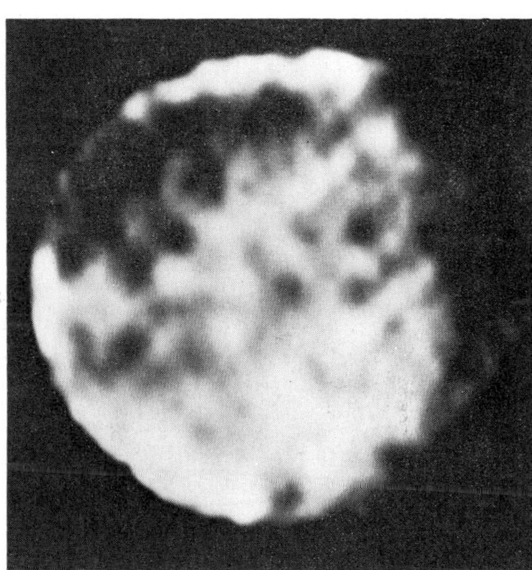

**FIGURE 8.9**
A photograph of Ganymede from Pioneer 10. Note the dark mare-like areas. (NASA photograph)

**FIGURE 8.10**
A Pioneer 10 photograph of Jupiter. The shadow of the satellite Io falls on the upper clouds. (National Aeronautics and Space Administration photograph)

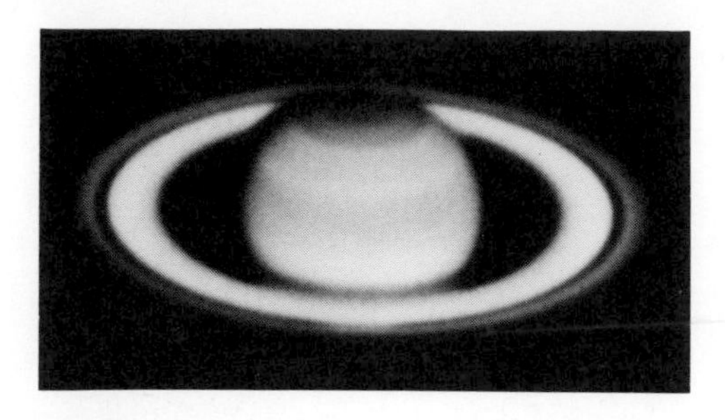

**FIGURE 8.11**
Saturn the ringed planet. Saturn shows a smooth banded structure. The poles of the planet are darker than the equatorial regions. Note that the ball of the planet can be seen through the outer ring. (Lowell Observatory photograph)

close to Saturn and was shattered into millions of small particles by strong tidal forces, caused by the gravitational attraction of the planet. Another theory says that the particles of the rings have been there since the formation of Saturn, but are located too close to the planet to pull together to form a moon.

Pioneer 11, which passed Jupiter in 1974, is now on its way to Saturn (Figure 8.12). It will approach and photograph the planet in 1979. Some clues to the origin of the rings may come from that brief encounter; we may also learn whether or not Saturn has a magnetic field. There is some indirect evidence for a magnetic field, but it is far from conclusive. Pioneer 11 will then leave the solar system to travel endlessly through outer space, as is the case with its predecessor, Pioneer 10.

## Uranus

In a telescope, Uranus looks like a greenish disk with no distinct markings. When telescopes are sent above the earth's atmosphere in balloons or satellites, faint cloud bands can be photographed on Uranus (Figure 8.13). Its atmosphere consists of hydrogen, helium, and methane; any ammonia that may have been there has frozen out.

The most unusual characteristic of Uranus is that its north-south axis is tilted almost 90° compared to its orbit plane. That is a greater tilt than any other planet. We can imagine Uranus "rolling" around the sun rather than spinning upright, like the other planets (Figure 8.14). This 90° tilt of Uranus gives it extreme seasonal variations. Actually, Uranus is so far from the sun that it is always cold. In addition, some locations on the planet would experience the effect of constant sunlight at some times and constant darkness at other times. Since a year on Uranus is about 84 of our years, the north pole of the planet would face toward the sun for 42 years and face away from the sun for another 42 years.

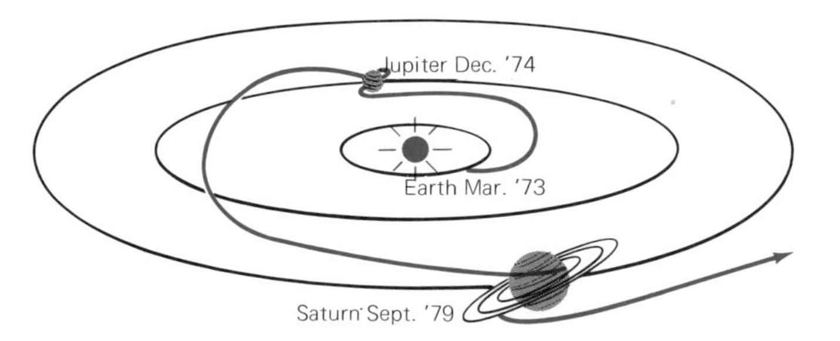


**FIGURE 8.12**
The trajectory of Pioneer 11. Between Jupiter and Saturn the spacecraft is high out of the plane of the solar system. After encountering Saturn and making its observations Pioneer 11 will leave the solar system.

Because of the extreme tilt of Uranus, there is a lack of agreement as to which pole is the north pole. Some astronomers consider the pole presently facing the earth to be the north pole. Since it points slightly below the orbit, they say that Uranus is rotating backward (east to west). In forty years, the other pole will be facing the earth. If we consider that pole as the north pole of Uranus, then the rotation is direct (west to east). For a long time, the rotation period of Uranus was thought to be 11 hours. Just recently that was corrected to 25 hours (such large changes in our knowledge of basic planetary data today are not uncommon).

Another of the most recent and major discoveries in the solar system is that Uranus has a system of rings similar to Saturn's. The rings were discovered while astronomers watched Uranus pass directly in front of a star. When a planet or the moon eclipses or passes in front of a star, it is called an **occultation.** Measuring the time from when the star disappears behind the planet until it appears again can provide accurate information about the size of the planet. In the case of Uranus, the starlight faded off and on five times before it disappeared behind the planet. After it reappeared from behind the planet, it again faded off and on five times. The observations can be explained by at least five rings surrounding Uranus. Figure 8.15 shows the path of the planet and its rings in front of the star. The rings are too faint to be seen with telescopes. We will only see these rings when a space probe flies by Uranus and sends back photographs. Outside the rings are Uranus' five satellites (Figure 8.16).

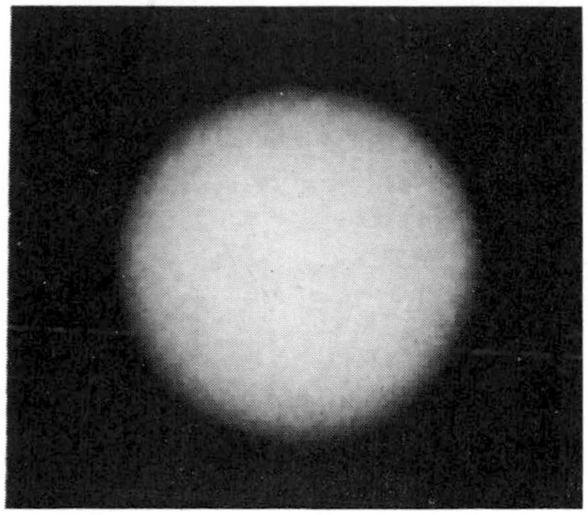

**FIGURE 8.13**
Uranus photographed by Stratoscope. This picture relayed from the balloon-borne telescope shows Uranus to be essentially featureless. (Princeton University photograph)

## Neptune

Neptune (Figure 8.17) like Uranus, presents a featureless, blue-green image in a telescope. We believe it is a thick, cloud-covered planet similar to the other giant planets. Its atmosphere is made up of hydrogen, helium, and methane.

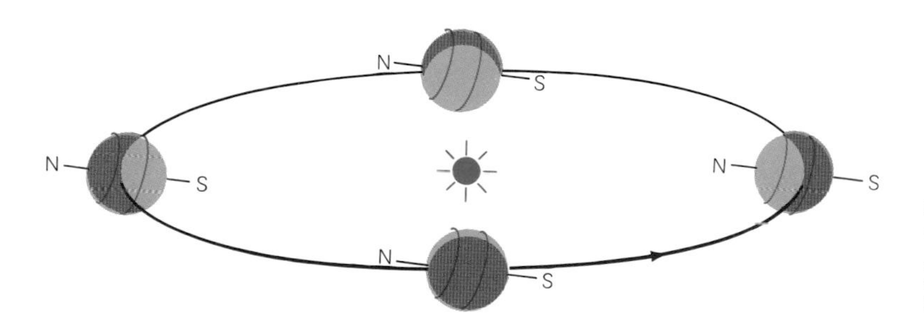


**FIGURE 8.14**
The axis of Uranus lies almost in the orbital plane. The north pole alternately points to and away from the sun.

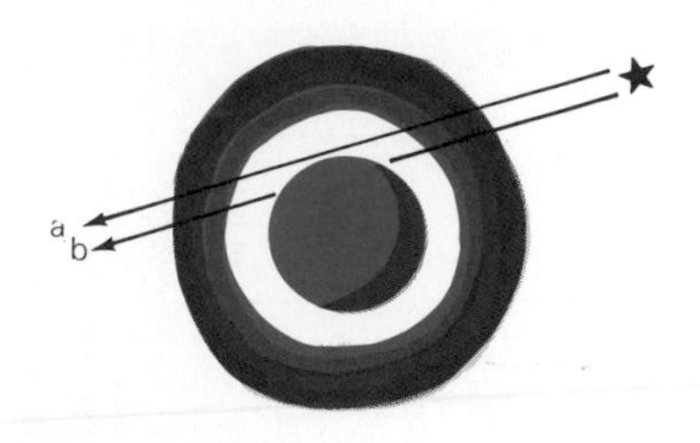


**FIGURE 8.15**
When the axis of Uranus points toward the sun its rings are open for us. The rings were detected by observing a star about to be occulted by Uranus. The star was not covered by the planet's disk for observers at Perth (a). Observers on an airplane much farther south observed the occultation (b).

Since its discovery in 1846, Neptune has not even completed one orbit around the sun. It did occult a star once, and from that event, its size has been determined (Figure 8.18). Neptune is slightly smaller and denser than Uranus. Its rotation period has been revised upward from 15 to 22 hours. Neptune has two moons, one of which, Triton, is probably the largest moon in the solar system.

Two Voyager spacecraft were launched in 1977. Their mission is to study Jupiter and its satellites and Saturn and its satellites. If all goes well and the mission is accomplished, one of the spacecraft will be redirected to visit Uranus and the other to visit Neptune. This latter mission, if successful, will greatly enhance our knowledge of Neptune.

## THE TERRESTRIAL PLANETS

### General Properties

In addition to being much smaller than the giant planets, the terrestrial planets share other characteristics. With the exception of Pluto, they are fairly close to the sun, within 225 million km. Only two of the terrestrial planets have moons: the earth has one, and Mars has two. Unlike the giant planets, these small planets are made up of very little hydrogen and helium. They have shallow atmospheres, if any at all. The interior structure of the terrestrial planets must be similar to the earth's (Figure 8.19). They are characterized by hot, molten, metallic-rocky cores surrounded by a thick rocky mantle and a surface composed of a relatively thin crust of lighter rocks.

**FIGURE 8.16**
Uranus and four of its satellites. The satellites are left to right: Titania, Umbriel, Ariel, and Oberon. Uranus is greatly overexposed. (Lunar and Planetary Laboratory photograph by E. Roemer)

Several common events have left their imprint on the terrestrial planets. If these planets have any atmosphere at all, it is not their original one. The original atmospheres of hydrogen, helium, methane, ammonia, and other gases were stripped away in their early history by violent energy bursts from the sun, probably by a single exceptionally hot blast. Then volcanos appeared and spewed out masses of rock and gas. From these gases, a second atmosphere built up. On the earth, this was composed of carbon dioxide, water vapor, methane, nitrogen, and gases containing sulfur. Temperatures on the earth were such that the water vapor condensed out as a fluid. The heavier gases dissolved in the water, fell to earth as rain, and left an atmosphere primarily of nitrogen and oxygen.

Whether or not the terrestrial planets kept their second atmospheres depended on two factors. First, if the planet was small, with a weak gravitational pull, it would have trouble holding on to the gases in the atmosphere. Second, if the planet had a high surface tempera-

**FIGURE 8.17**
Neptune (overexposed) and its closest satellite Triton. (Lick Observatory photograph)

ture, that would cause the atoms of the gases to be more energetic and more likely to escape from the planet. Mercury is an example of a planet that could not hold on to its atmosphere. Both its small size and high surface temperature worked against retaining its atmosphere.

After the gases and small particles had been cleaned out from between the planets, there were still some miscellaneous clumps of material in the solar system. All the planets were bombarded by these remaining clumps. The giant planets collected them without much effect. Some of the terrestrial planets and the moons of the solar system show permanent scars from their early bombardment. The large dark areas on the moon (now called maria) were produced by impact of very sizable clumps of material. The impacts formed huge basins which were then filled in by lava flows (Figure 8.20). Almost all the miscellaneous material was collected or accreted during the first billion years of the solar system. When the accretion phase ended, the small planets were covered with craters of all sizes. For the past 3.5 billion years, the number of impacts has been small and sporadic. On the earth the record of bombardment has been almost erased by erosion, volcanic activity, and movement of the earth's crust.

## Earth

The earth, the largest of the terrestrial planets, is not even one-tenth the size of Jupiter. Since it is composed primarily of the solid material of the solar nebula, it has a high density. Its density is 5.5 g/cm$^3$, the largest of any of the planets. The earth is plastic enough that its 24-hour rotation causes it to be slightly flattened at the poles. Clouds often cover the entire earth. At such times its reflectivity is extremely high,

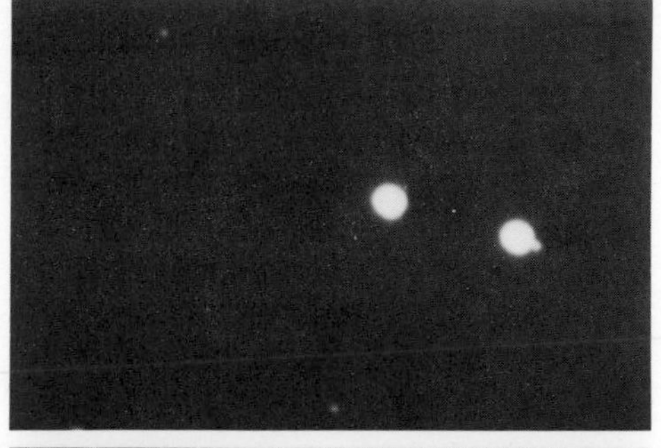

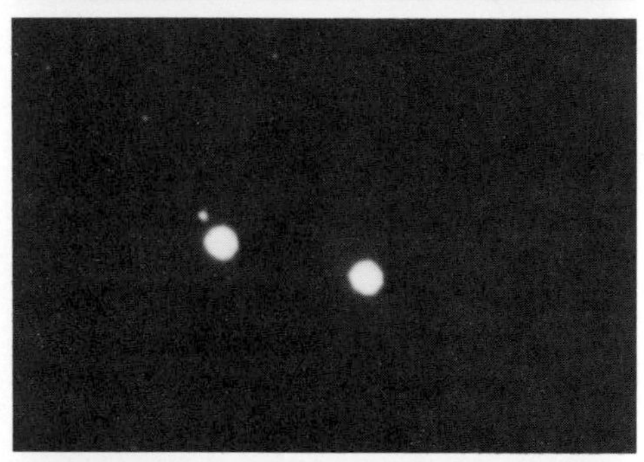

**FIGURE 8.18**
Neptune occults a star. The sequence was taken on four successive days and runs from (a) through (d). Neptune moves only about 21.5 seconds of arc per day. Note the motion of Triton around Neptune. (Photograph courtesy of Tersch Enterprises)

almost 80% (see color plate). As viewed at opposition from the location of Venus, it would be the brightest object in the night sky.

The earth has a varied surface about 60% covered with water and ice. Much of the land mass is covered with plant life. It is the only planet where water exists as a solid, liquid, and vapor. Water, along with the wind, has been responsible for rapid and continued erosion. Equatorial temperatures often reach 330°K (57°C) and seldom drop below 270°K ( −3°C). In other regions, temperature variations are more extreme. General temperature patterns are affected by the tilt of the earth's pole of rotation, which is 23½° from the perpendicular to the plane of its orbit.

The earth stands alone among the terrestrial planets in the fact that it has an appreciable magnetic field. It is also unique in that it has a very large natural satellite. It appears to be the only planet sustaining active plant and animal life. You might skim through an earth science book for a more complete picture of the geology of the earth. The geological processes on earth are dynamic and continuous and are fascinating.

## Mercury

Mercury, the small planet nearest the sun, is a little more than 1000 km larger in diameter than our moon. It revolves around the sun once in only 88 of our earth days. Not long ago it was thought that Mercury always kept one side toward the sun. Radar astronomy observations have since proved that notion false. The true period of Mercury's rotation is about 58.65 earth days, two-thirds of its period of revolution. The rotation and revolution of Mercury combine to give a very long Mercurian day. If we were on Mercury, the time between one sunrise and the next would be about 176 earth days. As a result, daytime temperatures reach 700°K (427°C) and nighttime temperatures drop to 100°K ( −173°C).

Long before the Mariner 10 spacecraft mission photographed Mercury, it was thought that Mercury looked like the moon. The diameter of Mercury could be measured on its occasional **transits** across the face of the sun. The diameter of Mercury is 4868 km. By knowing the size of Mercury and its distance and then measuring its brightness, we can calculate its reflectivity. Mercury reflects only 7% of the sunlight it receives. This is very nearly the same as the moon's reflectivity. But, few, if any, astronomers expected Mercury to look as much like the moon as it does (Figure 8.21).

Mercury is covered with impact craters and marelike basins. In its early development, as it cooled internally, it contracted and cracked. At some point after the heavy bombardment stage it underwent a period of volcanism. Mercury had and probably still has a molten core. On the surface, there is no evidence of erosion. Mercury has no atmosphere, and none was expected; any atmosphere that it may have had must have escaped quite early.

## Venus

In size and mass, Venus is nearly a twin of the earth. After the sun and moon, Venus can be the most brilliant object in our sky. Venus is perpetually covered by a thick layer of clouds that obscures all surface features from our view (Figure 8.22). With telescopes we can observe only the tops of the clouds and watch as the planet goes through phases as it orbits the sun. There was much speculation about the surface of Venus in the past. It was imagined to have everything from tropical rain forests with swamps inhabited by reptiles to one immense ocean of carbonated water. Today, radar and radio astronomy techniques allow us to study and learn about this mysterious surface. Information has also come to us from a few Mariner spacecraft that have flown by the planet (see color plate) and several Soviet Venera spacecraft that have landed on Venus.

The atmosphere of Venus, far from being the life-bearing Garden of Eden depicted by poets and science fiction writers, is a hot, oppressive, acid-containing atmosphere dominated by carbon dioxide. There is only a trace of water on Venus, and it is in the form of vapor. The surface pressure of the atmosphere is ninety times that on the earth. To experience such immense pressure, you would have to descend an impossible 2.5 km into the earth's oceans. The surface temperature of Venus is 740°K (476°C), higher than the melting point of lead. High above the surface are clouds. These are not water clouds but clouds that rain sulfuric acid which never reaches the ground because of the intense heat. The rain just vaporizes and goes back up into the clouds.

Two of the Venera landers managed to survive this atmosphere and the severe surface conditions for 50 minutes. During that time they each sent back a picture showing that Venus has rocks and "soil" on its surface (Figure 8.23). Erosion appears to be present but not dominant. Radar observations from the earth have characterized some of the surface areas of Venus as rough or smooth and high or low. Venus has several large craters with very shallow rims, presumably eroded.

Recent studies have indicated what might be an enormous volcano. On earth, a feature as large as the suspected Venus volcano would cover all of New Mexico and spill over into Colorado, Arizona, and Texas. Elsewhere on Venus we find what could be a second major volcano and a long, curving mountain range. For the most part, however,

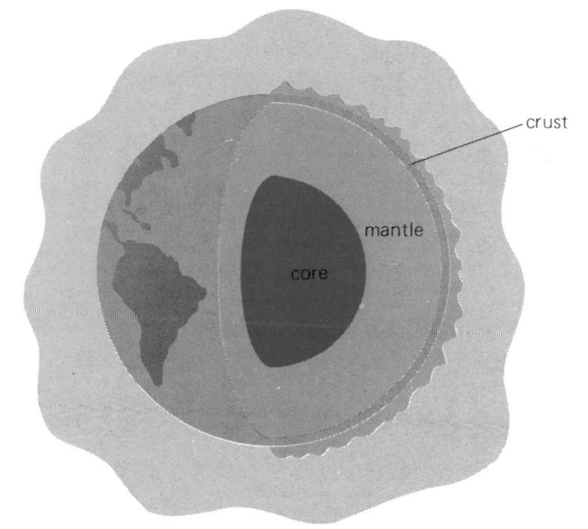


**FIGURE 8.19**
A cut away drawing of the earth showing the three principal divisions of the interior. Typical of the terrestrial planets is the molten core.

**FIGURE 8.20 (page 164)**
Full moon. Some of the major features visible to the unaided eye or through small binoculars can be seen. (Yerkes Observatory photograph)

**FIGURE 8.21**
Mercury photographed by Mariner 10. This is a mosaic pieced together from 18 pictures. (National aeronautics and Space Administration photograph)

**FIGURE 8.22**
The clouds of Venus photographed by Mariner 10. The photograph was made in blue light to enhance the visibility of the cloud structure. (National Aeronautics and Space Administration photograph)

the surface of Venus is much smoother than the moon or Mars or Mercury. Few features appear to be more than several kilometers high. Some of this is undoubtedly due to the weathering produced by the hot, dense, and corrosive atmosphere.

Why is Venus so much hotter than the earth? In part the answer comes directly from Venus' position closer to the sun. In addition, there is an indirect effect due to its atmosphere. When Venus built up its second atmosphere, being close to the sun, it was warmer than the earth. On Venus, carbon dioxide and other molecules remained in the vapor phase. The vapors in this condition actually trapped the sun's energy and caused the surface temperature to rise some more. This process is called the **greenhouse effect,** since it is analogous to the trapping of heat inside a greenhouse (Figure 8.24). The clouds on Venus continue to act as a greenhouse, keeping the planet's surface incredibly hot.

Venus exhibits an exception to one of the regularities of the solar system mentioned earlier. It rotates in the opposite direction (east to west) from the other planets. Its rotation period is very slow, 243 days.

The rotation and revolution of Venus produce a curious effect. Every time the planet is between the earth and sun, the same side is facing the earth. This tidal lock between the earth and Venus is unexplained.

## Mars

Mars has been called the "newspaper" planet because through the years so much has been written about it. From a lively discussion at the turn of the century about possible canals on Mars to many science fiction stories about "little green men," Mars has spurred the imaginations of scientists and nonscientists alike.

Mars does bear some striking similarities to earth. Polar caps made up of frozen water shrink in the Martian summer and grow large in the Martian winter. Dark markings on the planet change color and shape in the different seasons. A day on Mars, like that of the earth, is almost 24 hours long. The axis of Mars is tipped about 23°, giving places on the surface a change of seasons that is almost earthlike. Each season lasts almost 6 months instead of 3 because of Mars' longer period of revolution. The equatorial temperatures on Mars can get to 300°K (27°C) in the daytime and down to 173°K (−100°C) at night.

Several Mariner spacecraft have flown by as well as orbited Mars. They returned thousands of pictures of the Martian surface. In the summer of 1976, two Viking spacecraft arrived at Mars and went into orbit around the planet. Each Viking sent a lander safely to the surface. Viking orbiter photographs have shown that Mars' surface is cratered and bleak, very similar to the moon's (Figure 8.25). Close examination of the pictures shows some outstanding differences between the features of Mars and the moon and indicate a lively past for Mars.

The rims of many Martian craters are smoother than those of lunar craters. Over millions of years, the thin Martian atmosphere, with its occasional global dust storms, has worn away the crater rims. Once Mars cooled it was too cold to sustain water in the fluid phase, so water erosion no longer plays a role. Now water is generally trapped in the polar icecaps, although some water ice (water in the ice phase) may be present under the surface at many places. Because of Mars' small size and mass, much of its atmosphere has been lost. The primary constituent of the thin Martian atmosphere is carbon dioxide. As with the other terrestrial planets, Mars lost its orginal atmosphere, which was replaced with its present secondary one by volcanic activity.

Olympus Mons is a giant dormant volcano on Mars (Figure 8.26). Its summit reaches nearly 25 km above the surface of Mars—five times

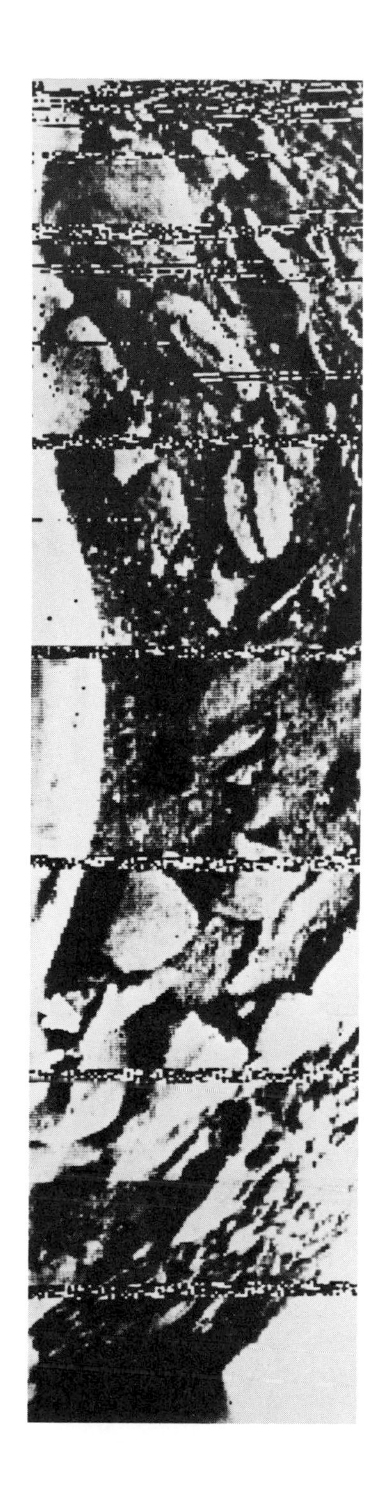

**FIGURE 8.23** The surface of Venus from Venera 9. Broken and eroded rocks are visible. (Novosti from Sovfoto)

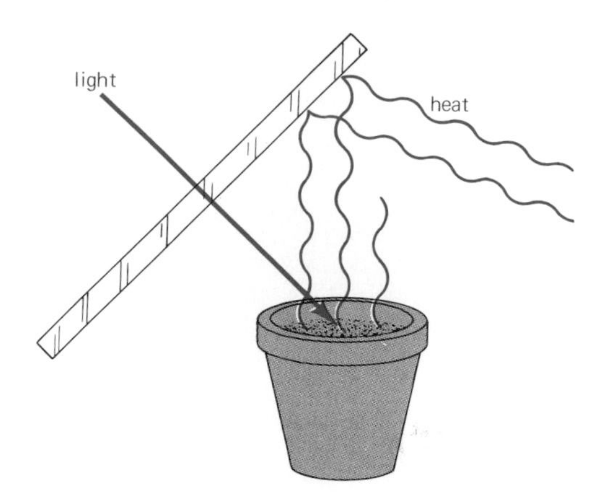


**FIGURE 8.24**
Greenhouse effect develops when light passes through a window and warms the interior. The interior radiates long wavelength energy which is reflected by the glass and hence trapped in the greenhouse.

higher than Mt. Everest. The base of the volcano spans over 500 km. In contrast, the entire volcanic island of Hawaii only spans about 120 km at the sea floor. Near the center of Olympus Mons are circular structures that were once vents for the molten lava. The volcano has been inactive for a long time.

There is a region of enormous canyons on Mars (Figure 8.27). The Valle Marineris canyon system would stretch from one side of the United States to the other. The immensity of this completely dwarfs our own Grand Canyon. Probably the most remarkable discovery on Mars is the existence of long, sinuous channels that appear to have been formed by running water (Figure 8.28). The channels must have been cut fairly early in Mars' history, perhaps after the major bombardment ceased but before the final cooling of the planet.

The first closeup color pictures of Mars were returned by the Viking landers. (See color plate.) The red color is very vivid. It is thought to be due to limonite (hydrated ferric oxide, which is simply a form of rust). The terrain is very similar to the desert regions of the western United States (Figure 8.29). Sand dunes swept by the global

**FIGURE 8.25**
The surface of Mars' southern hemisphere. The southern hemisphere of Mars is more cratered than the northern hemisphere and resembles the moon. (National Aeronautics and Space Administration photograph)

dust storms where winds may reach speeds of 350 km/hr are evident in the distance. The rocks in the picture range in size from several centimeters to about one meter. The Viking missions photographed Mars and analyzed the atmosphere and soil for more than a year. Three experiments designed to test the soil for possible signs of life have given negative but intriguing results.

Mars has two moons. When Mariner 9 arrived at Mars in the midst of a global dust storm, it turned its cameras to the moons. Small Phobos (Figure 8.30) has an irregular shape full of craters. The other moon, Deimos, looks the same. They resemble certain asteroids and meteors leading to speculation that both of Mars' moons are asteroids captured by the planet.

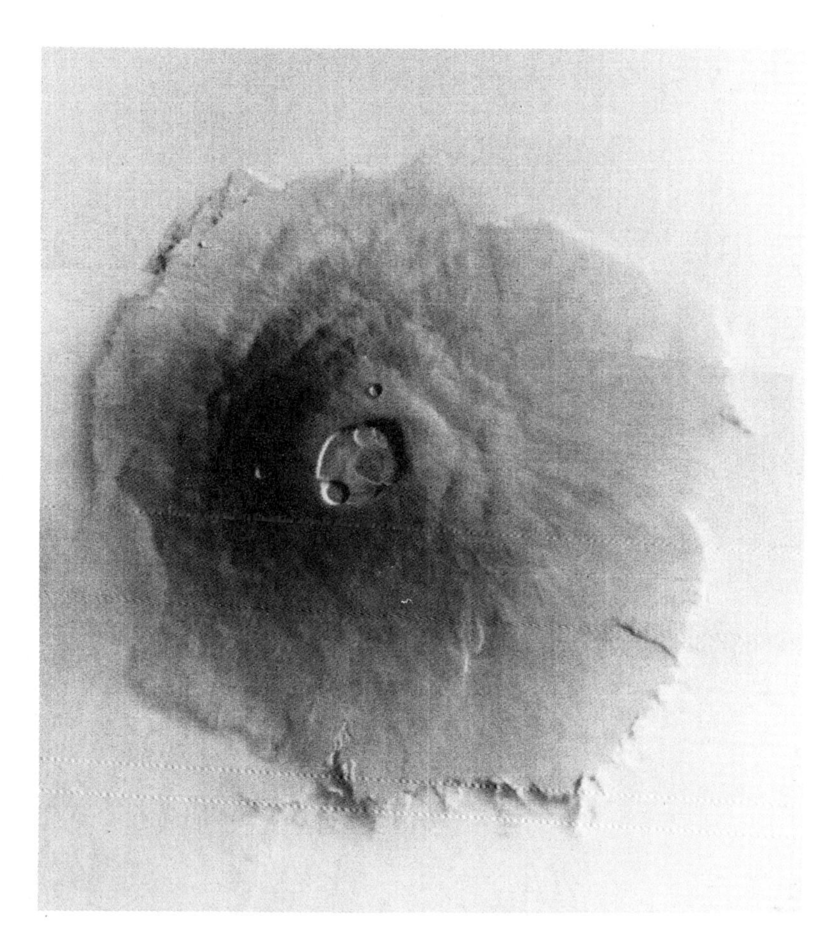

**FIGURE 8.26**
The great Martian volcano Olympus Mons. The volcano is 25 km high and the diameter of its shield is greater than 500 km. The shield shows only two major impact craters indicating that the volcano is younger than the surrounding planes. (National Aeronautics and Space Administration photograph)

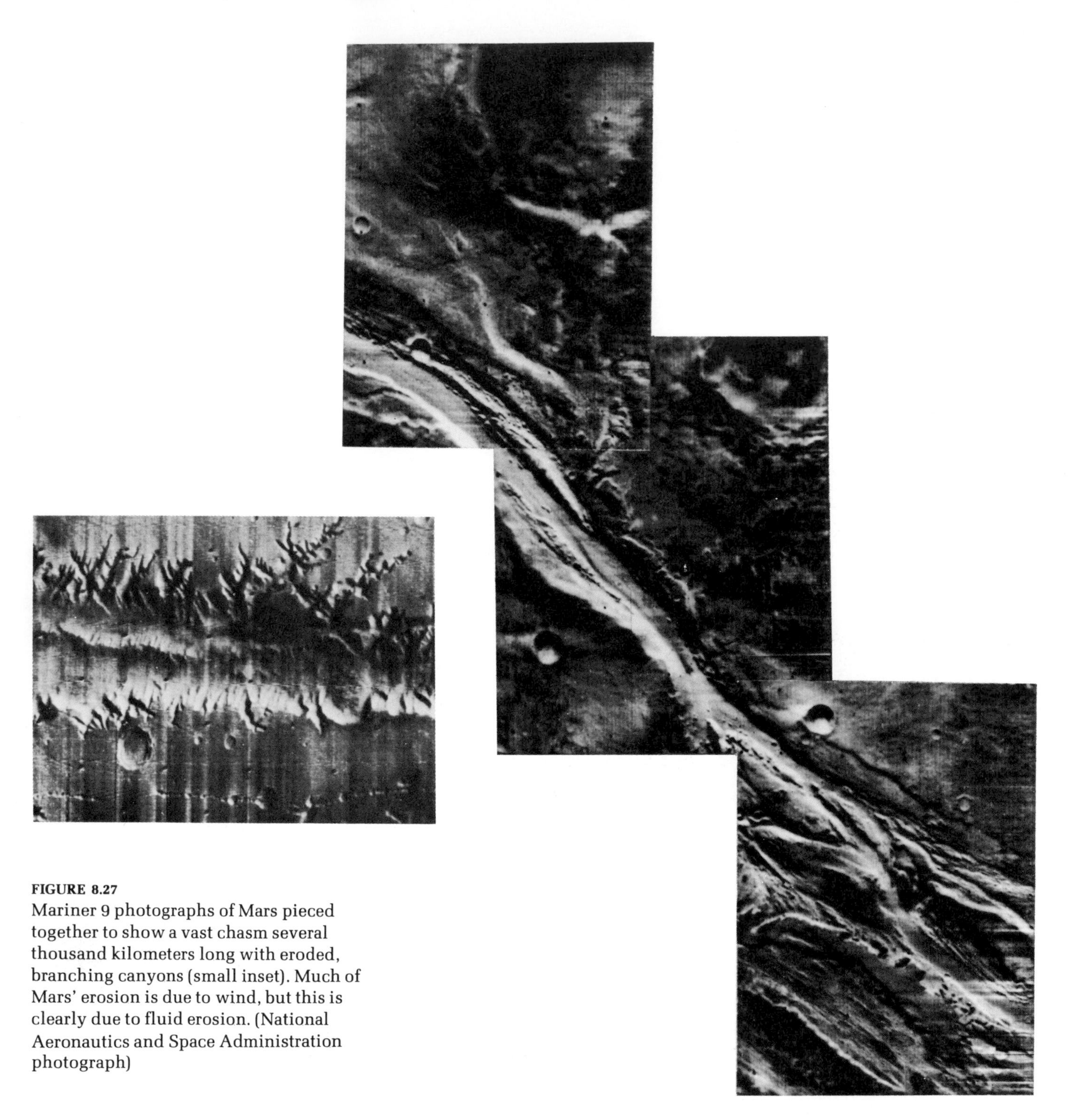

**FIGURE 8.27**
Mariner 9 photographs of Mars pieced together to show a vast chasm several thousand kilometers long with eroded, branching canyons (small inset). Much of Mars' erosion is due to wind, but this is clearly due to fluid erosion. (National Aeronautics and Space Administration photograph)

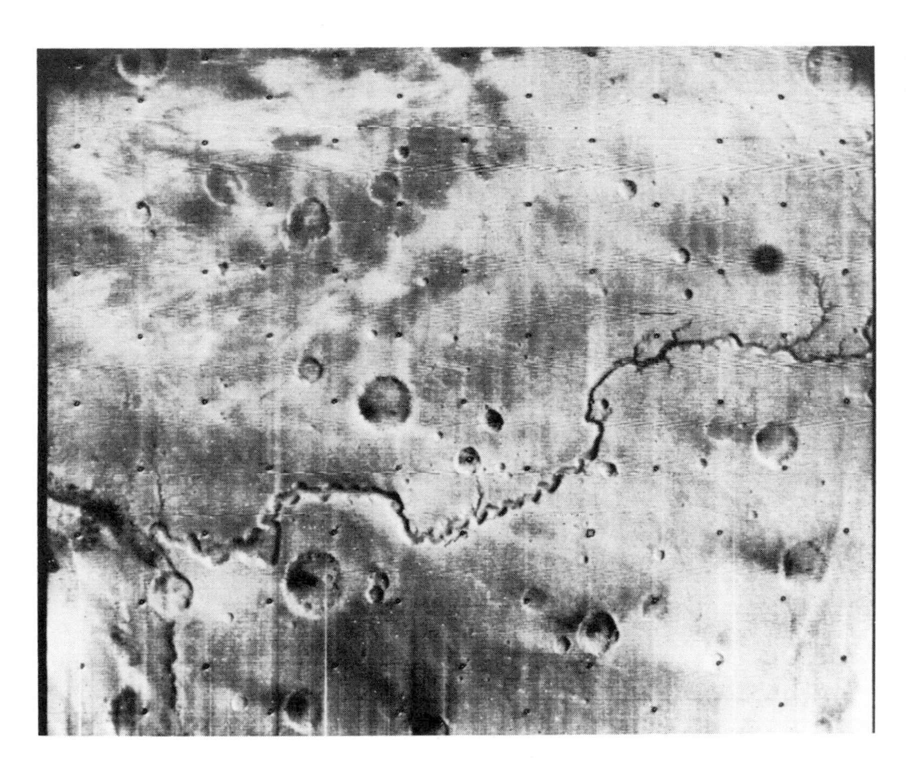

**FIGURE 8.28**
A long sinuous rille on Mars resembling an arroyo or water cut gully commonly found in desert regions on the earth. (National Aeronautics and Space Administration photograph)

**FIGURE 8.29**
The surface of Mars as seen by Viking 1 lander. The wind swept sand dune area is remarkably similar to desert areas on earth. (National Aeronautics and Space Administration photograph)

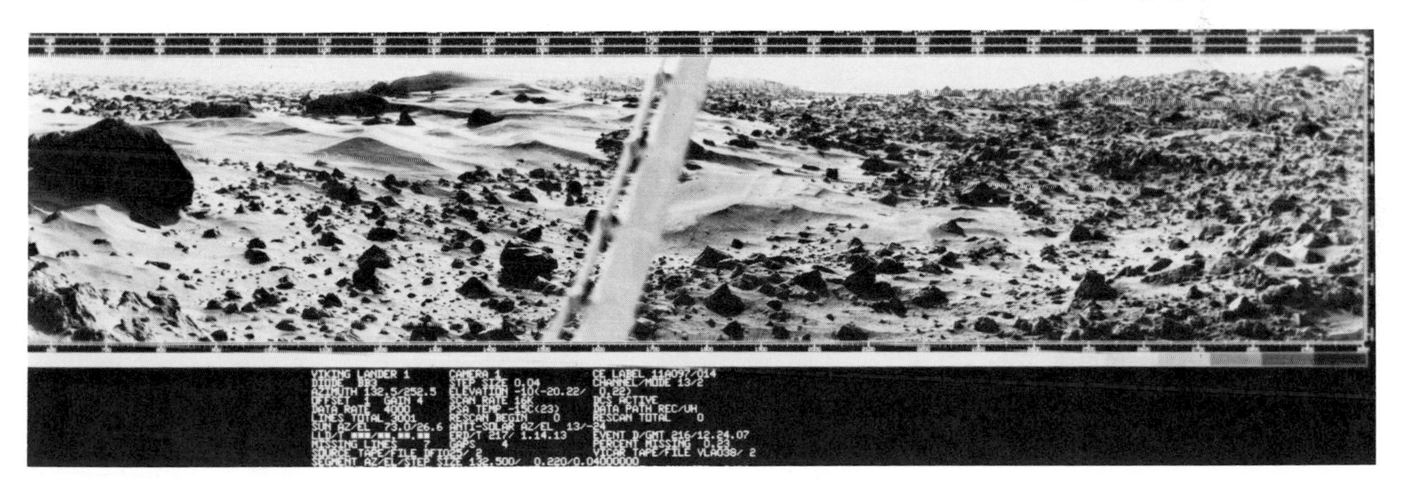

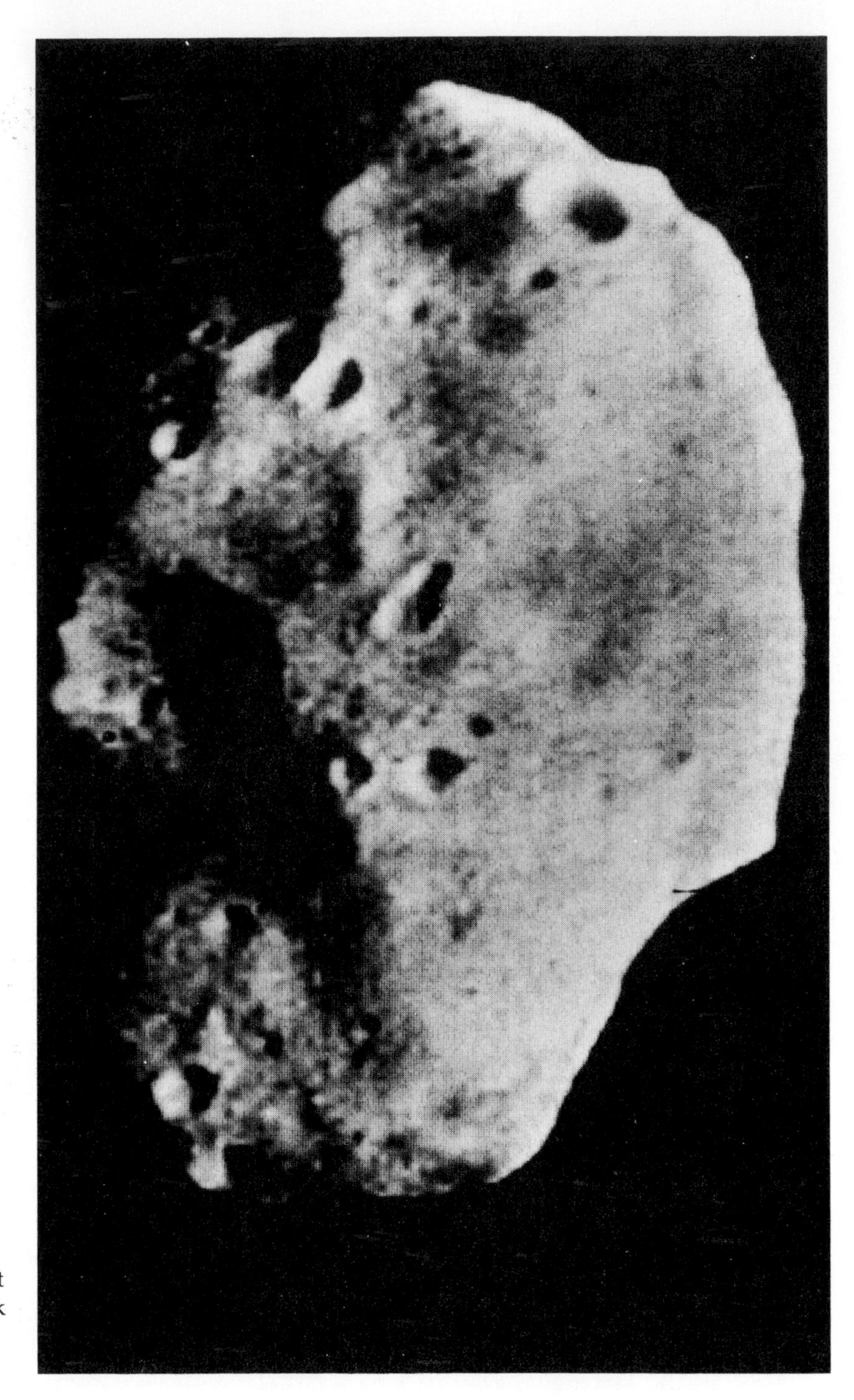

**FIGURE 8.30**
Phobos photographed by Mariner 9. Impact craters are evident. Asteroids probably look like this. (National Aeronautics and Space Administration photograph)

## Pluto

Very little is known about Pluto. It is a small planet. We do not have a good measure of its size because it has not occulted a star recently. A near occultation in 1966, however, makes it possible to say that its largest possible diameter is 5800 km. Many astronomers think that Pluto is in fact only about half that size, making it smaller than six of the moons in the solar system. On a photograph made with the 200-in. (5 m) Hale telescope, Pluto looks like a star (Figure 8.31).

The light reflected from the surface of Pluto varies regularly every 6.39 days. This variation is probably due to the rotation of the planet and the way that different features on its surface reflect light. There is speculation that its surface is covered with ices of ammonia and pools of liquid methane.

Pluto's orbit is interesting. It is the most eccentric of all the planetary orbits. It is also the most inclined to the ecliptic. Although the orbit of Pluto never actually intersects the orbit of Neptune, the inclination and shape of Pluto's orbit is such that Pluto can actually be closer to the sun than Neptune. Pluto will be closer to the sun than

**FIGURE 8.31**
Two photographs of Pluto taken one day apart. Pluto appears to move less than 2 min of arc per day, which is readily visible on photographs. (Hale Observatories photograph)

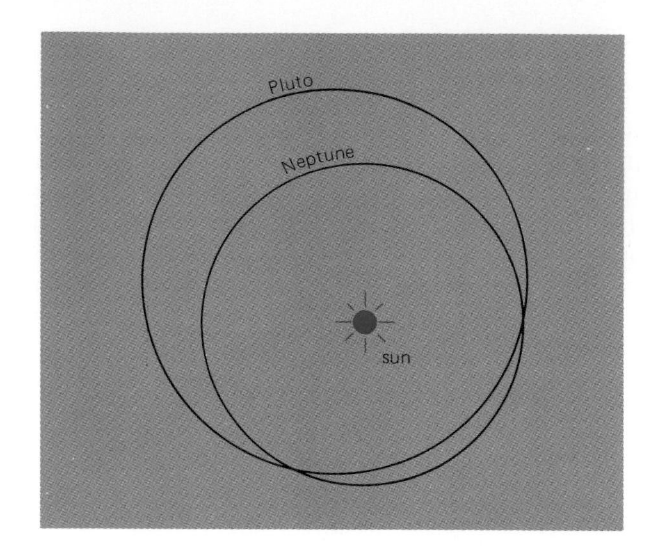


**FIGURE 8.32**
The orbits of Neptune and Pluto.

Neptune from about 1979 until 2000 (Figure 8.32). The sun as viewed from Pluto would look like a very bright star against the background of other stars.

## SUMMARY

Ninety-nine percent of the material in the solar system is in the sun. The rest is contained in the small bodies that orbit the sun: the major planets, their moons, the asteroids, comets, and meteors. There are many regularities in the planet motions. For example, they all move in the same direction around the sun.

The planets can be divided easily into two groups: the small terrestrial planets and the giant gaseous planets. The gaseous planets trapped more of the solar nebula material and are like the sun in composition. The terrestrial planets are similar to the earth in composition. With the exception of Pluto, they are relatively close to the sun. Extensive bombardment by interplanetary debris early in the development of the solar system has left the small planets and moons with cratered surfaces. Their original atmospheres were stripped away by bursts of energy from the young sun. Only two of the terrestrial planets have moons. All the giant planets have moons, and two of them also have systems of rings. Pluto is regarded as a terrestrial planet because of its size, but very little is known about this distant world.

The most up-to-date information on the planets has been obtained during spacecraft missions. Spacecraft have flown past Mercury, Venus, the moon, Mars, and Jupiter. There have been manned landings on the moon and unmanned landings on Venus and Mars.

### *FALLACIES AND FANTASIES*

*The earth is a large planet.*
*Mercury keeps the same side toward the sun always.*

## REVIEW QUESTIONS

______ 1. The surface temperature of Mercury and Venus are nearly equal even though Venus is almost twice as far from the sun as Mercury. What differences in the planets would account for this?

______ 2. What do the low densities of Jupiter, Saturn, Uranus, and Neptune imply about the composition of these planets?

______ 3. What is a transit? Which planets might transit the sun?

______ 4. List one unusual characteristic about the following planets: (a) Venus, (b) Saturn, (c) Uranus, (d) Pluto.

______ 5. Why do some terrestrial planets have atmospheres while others do not? Explain.

______ 6. Explain the striped appearance of Jupiter.

______ 7. What is an occultation? Why is it useful for planetary studies?

______ 8. Why do we not find as many impact craters on the earth as on Mercury and Mars?

______ 9. How are the Martian craters different from lunar craters?

______ 10. Compare the interior structure of the giant planets and the terrestrial planets.

# 9 Between the Planets: From Moons to Meteors



## MOONS AND RINGS

We refer to the moons of planets, the asteroids, the comets, and the meteors as lesser bodies of the solar system. Such a statement is, however, relative. Four moons are larger than Mercury and, if our guess about the size of Pluto is correct, at least six moons including our moon are larger than Pluto. In Table 9.1 we summarize the current data on the moons of the planets.

### Our Moon

Our constant companion as we move through space is the moon. It is the fifth largest moon in the solar system. The moon is about one-fourth the size of the earth. This size relationship makes it the largest

**TABLE 9.1 The Moon and Other Satellites**

| Name | Discovery | Mean Distance from Primary (kilometers) | Mean Period of Revolution (days) | Diameter (kilometers) | Stellar Magnitude at Mean Opposition |
|---|---|---|---|---|---|
| Moon | | 384,397 | 27.322 | 3476 | −12.6 |
| | | Satellites of Mars | | | |
| Phobos | Hall, 1877 | 9,400 | 0.319 | 21.8 | +11.6 |
| Deimos | Hall, 1877 | 23,500 | 1.262 | 11.4 | 12.7 |
| | | Satellites of Jupiter | | | |
| Fifth | Barnard, 1892 | 181,300 | 0.418 | 193 | +13.0 |
| I Io | Galileo, 1610 | 421,600 | 1.769 | 3658 | 5.0 |
| II Europa | Galileo, 1610 | 670,800 | 3.551 | 2840 | 5.2 |
| III Ganymede | Galileo, 1610 | 1,070,000 | 7.155 | 5280 | 4.6 |
| IV Callisto | Galileo, 1610 | 1,880,000 | 16.689 | 4720 | 5.6 |
| Thirteenth | Kowal, 1974 | 10,170,000 | 210.6 | 15: | 20: |
| Sixth | Perrine, 1904 | 11,470,000 | 250.58 | 96: | 14.7 |
| Tenth | Nicholson, 1938 | 11,710,000 | 259.21 | 19: | 18.6 |
| Seventh | Perrine, 1905 | 11,740,000 | 259.67 | 30: | 16.0 |
| Twelfth | Nicholson, 1951 | 20,700,000 | 631 | 19: | 18.8 |
| Eleventh | Nicholson, 1938 | 22,350,000 | 692 | 19: | 18.1 |
| Eighth | Melotte, 1908 | 23,300,000 | 735 | 19: | 18.8 |
| Ninth | Nicholson, 1919 | 23,700,000 | 758 | 19: | 18.3 |
| Fourteenth | Kowal, 1975 | ? | ? | ? | ? |
| | | Satellites of Saturn | | | |
| — | Larsen & Fountain, 1977 | 151,000 | 0.702 | 200: | 16.0 |
| Janus | Dollfus, 1967 | 157,700 | 0.749 | 300 | 14.0 |
| Mimas | Herschel, 1789 | 185,600 | 0.942 | 480 | 12.1 |
| Enceladus | Herschel, 1789 | 238,200 | 1.370 | 560 | 11.8 |
| Tethys | Cassini, 1684 | 294,800 | 1.888 | 960 | 10.3 |
| Dione | Cassini, 1684 | 377,500 | 2.737 | 960 | 10.4 |
| Rhea | Cassini, 1672 | 527,200 | 4.417 | 1287 | 9.8 |
| Titan | Huygens, 1655 | 1,221,000 | 15.945 | 4830 | 8.4 |
| Hyperion | Bond, 1848 | 1,483,000 | 21.276 | 480: | 14.2 |
| Iapetus | Cassini, 1671 | 3,560,000 | 79.331 | 1090: | 11.0 |
| Phoebe | Pickering, 1898 | 12,952,000 | 550.333 | 190: | 16.5 |
| | | Satellites of Uranus | | | |
| Miranda | Kuiper, 1948 | 130,500 | 1.413 | 300: | 16.5 |
| Ariel | Lassell, 1851 | 191,800 | 2.520 | 800: | 14.4 |
| Umbriel | Lassell, 1851 | 267,200 | 4.144 | 650: | 15.3 |
| Titania | Herschel, 1787 | 438,400 | 8.706 | 1130: | 14.0 |
| Oberon | Herschel, 1787 | 586,300 | 13.463 | 960: | 14.2 |
| | | Satellites of Neptune | | | |
| Triton | Lassell, 1846 | 355,200 | 5.876 | 3700 | 13.5 |
| Nereid | Kuiper, 1949 | 5,562,000 | 359.875 | 320: | 18.7 |

: denotes uncertain values.

**FIGURE 9.1**
The barren, rocky surface of the moon. (National Aeronautics and Space Administration photograph)

moon in comparison to its primary planet for any moon in the solar system. The earth-moon system has sometimes been regarded as a double planet. With the exception of the earth and an occasional meteorite, the moon is the only object in the entire Universe for which we have samples on earth. Over 400 kg of lunar rocks and soil were brought back by the astronauts during six successful Apollo missions.

The moon is a barren world without atmosphere or water. The rugged surface of the moon, covered with craters and mountainous areas and strewn with rocks, has become a familiar sight (Figure 9.1). The moon has undoubtedly looked exactly as it does now for billions of years. Without wind or water, there can be no erosion of the type we are used to on the earth.

Craters on the moon have been produced in two ways. Most are due to direct impacts of chunks of space debris. A few are the result of volcanic action on the moon. Moon rocks are mostly basalt of the sort that forms from molten lava as it cools. Information from lunar seismometers measuring moonquakes indicates that the moon is somewhat layered, similar to a terrestrial planet, with a semi-molten core, a mantle, and a crust.

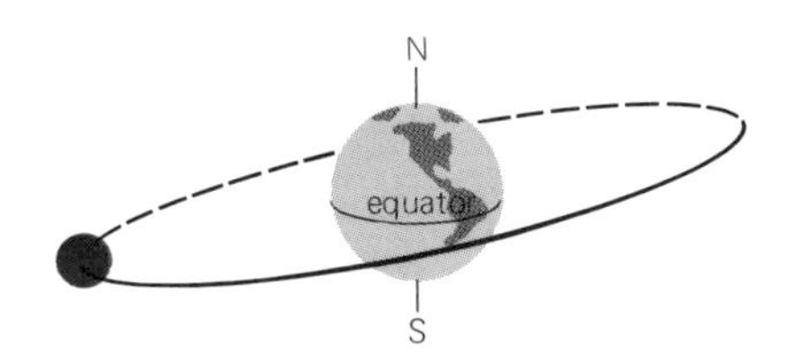


**FIGURE 9.2**
The moon's orbit is more nearly in the plane of the earth's orbit than in the plane of the earth's equator.

Rocks brought back from the moon by the Apollo astronauts show that the moon is about 4.5 billion years old and, therefore, formed along with the earth and the rest of the solar system. The moon's general composition and density, however, suggest that it may have formed elsewhere in the solar nebula, perhaps near the orbit of Mars. The orientation of the moon's orbit supports this idea. If the moon had formed near the orbit of the earth, it probably would be in orbit around the equator of the earth. Instead, the moon's orbit is tipped (Figure 9.2) so that it is more nearly in the plane of the earth's orbit rather than that of the earth's equator. The exact origin of the moon is still in doubt. Many theories have been proposed, from the moon breaking away from the earth from what is now the Pacific Ocean to the Moon being captured from another planet. None of the theories is without problems.

The moon is slightly egg-shaped, and such a shape is stable in rotation and revolution about a more massive body if its long axis points toward the massive body. The earth's strong, steady gravitational pull on the moon has resulted in equalizing the moon's rotation and revolution. Both these moon motions require about one month. The result of this equalization is that, from the earth, only one side of the moon is visible. The changing illumination of that one side from full to just a thin crescent is due to the moon's orbiting the earth and hence continually changing its angle to the sun as viewed from the earth. Figure 9.3 shows part of the orbits of the earth and moon around the sun. Since the moon has no atmosphere, there is no twilight, and the line separating daytime from night is very sharp. One lunar day is about four earth weeks long. Any one place on the moon will have about two weeks of daytime when the temperatures reach 100°C followed by two weeks of night when temperatures plunge to −150°C.

The moon has several effects on the earth. The ocean tides were known from earliest times to be caused by the moon. Newton correctly ascribed the cause of the tides to the attractions of both the moon and the sun. Since gravitational pull depends strongly on the distance to an object, our nearby moon, even though it is smaller than the sun, has the greatest effect on our tides. To simplify the explanation of the tides, we may imagine that the whole earth is covered by very deep water, as in Figure 9.4. The moon's attraction is greatest for the part of the ocean directly beneath it, intermediate for the solid earth, and is least for the part of the ocean farthest away, on the opposite side of the earth. This force draws the ocean into an **ellipsoid.** The tides will be high along a line connecting the earth and moon, and low at the in-between points.

The earth is constantly rotating eastward under the tides, since the earth's period is 24 hours, whereas that of the moon and hence the tides is 1 month. Any place on the earth passes through two high-tide regions and two low-tide regions each day. Since the moon is also moving, the interval between one high tide and the next is a little more than half a day, 12 hr 25 min. In reality, the friction of the water against the

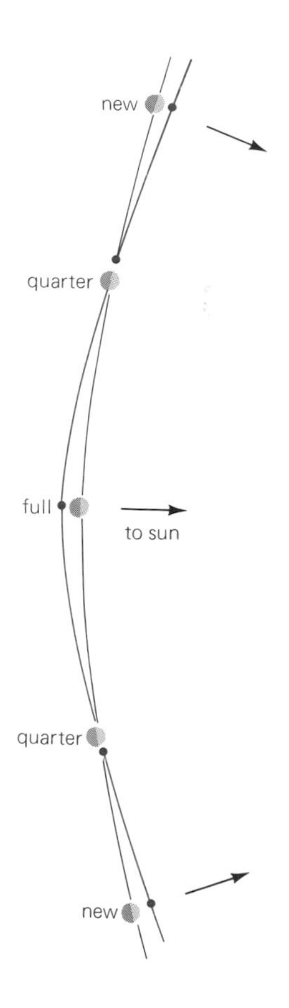


**FIGURE 9.3**
The path of the moon around the sun is everywhere concave despite the fact that the moon orbits the earth.

ocean floor and the contour of the land near the shore determine exactly when a high or low tide will occur and how high or low it will be. There will always be some delay between the time the moon is overhead and the high tide. Local observations at any particular location are the best way to determine the times and extent of the tides.

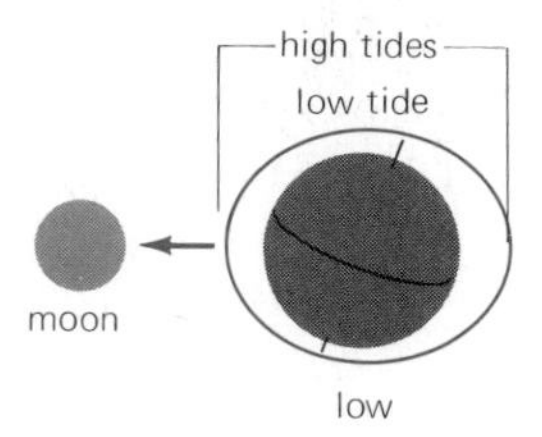


**FIGURE 9.4**
High tides occur in line with the moon on both sides of the earth. Low tides occur in a band around the earth 90° between the high tides.

Another effect of the moon on the earth is to stabilize the tilt of the earth's axis with repect to its orbital plane. This in turn results in the smooth, constant cyclic changes of the seasons on the earth. Without the moon, the earth's tilt would oscillate and result in wild fluctuations in the seasons, which would have adverse effects on the development of life on the earth.

It has also been suggested that the moon played a role in the establishment of the earth's magnetic field. Our magnetic field is important to our survival on the earth. Without the magnetic field, some particles, which currently get trapped in the field, could penetrate to the surface of the earth and damage many life forms. The moon is thus far more than just a "counter light" to the sun or an object for poetic dreaming.

## Other Moons and Rings

Moons are very useful in helping us study the planets. The existence of a moon around a planet makes it possible for us to determine the mass of the planet. Since any object in orbit around another obeys Kepler's laws, by applying Kepler's Harmonic Law as modified by Newton (Chapter 3) we can obtain the mass of the planet plus its moon. To calculate the mass, all we need to measure is the period of the moon's orbit and the average separation of the moon and the planet. In the case of a large planet and a small moon, the mass of the planet plus its moon is practically the same as the mass of the planet alone. In addition to moons, artificial satellites orbiting the planets can be used to find their masses.

*Mars* Kepler was the first to propose that Mars had two moons. In the 17th century it was known that the earth had one moon and Jupiter had four. Struck by numerology, Kepler reasoned that Mars should have two moons. This gave an orderly progression (1, 2, 4, . . .) for the number of moons of the successively more distant planets. Jonathan Swift must have been aware of Kepler's predictions when he wrote his famous satire, *Gulliver's Travels,* in 1726. When Gulliver visited the Island of Laputa, he met some of the King's astronomers who had

Jupiter Pioneer 10 took this photograph of Jupiter. The great Red Spot is prominent. Note the equally red cloud bands and the shadow of Io on the disk of the planet. (National Aeronautics and Space Administration photograph)

Saturn The smooth banded cloud pattern on the ball of the planet is striking. (Lick Observatory photograph)

Crab Nebula The Crab Nebula is the remains of the recent supernova in 1054. The twisted filamentary structure attests to the violence of the event. (Hale Observatories photograph)

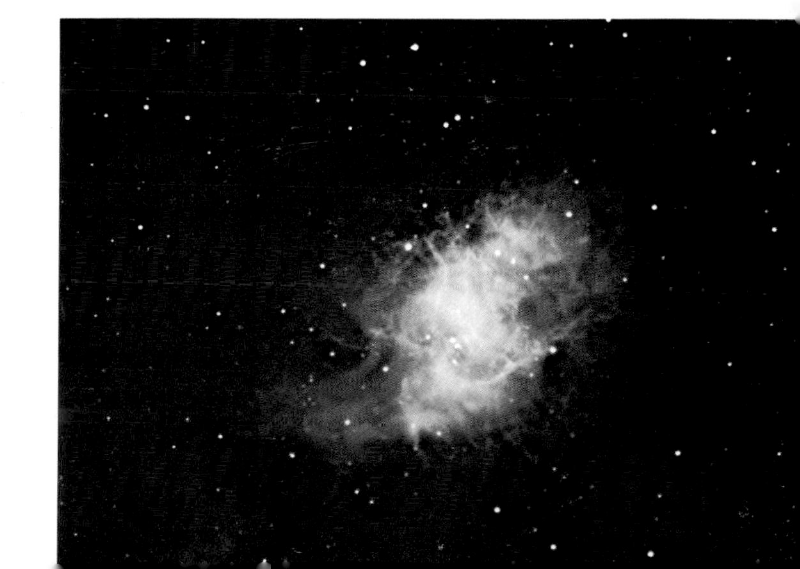

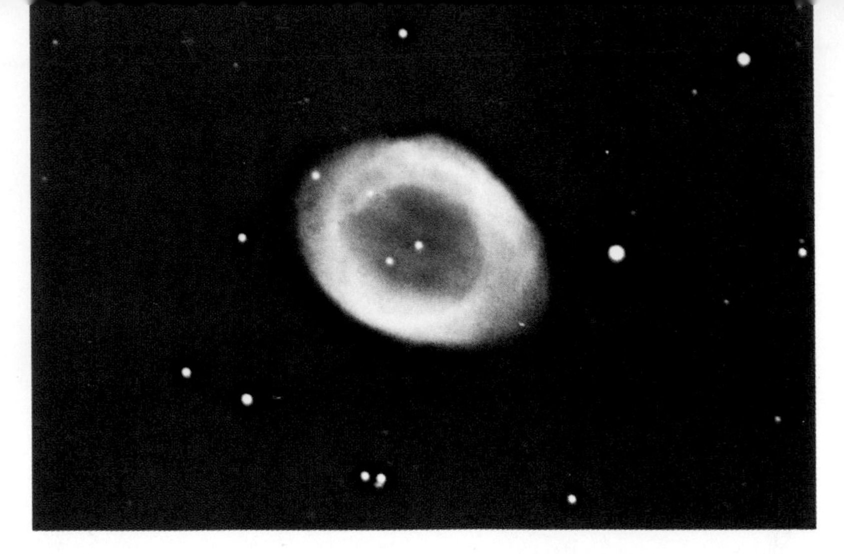

Ring Nebula This famous nebula, the Ring Nebula in Lyra, is an example of a star losing mass by ejecting an outer shell. A small, hot, blue star remains at the center. (Hale Observatories photograph)

Helix in Aquarius The Helix Nebula is an example of a common type of planetary nebula. Release of material as the central star rotated is suggested. (Hale Observatories photograph)

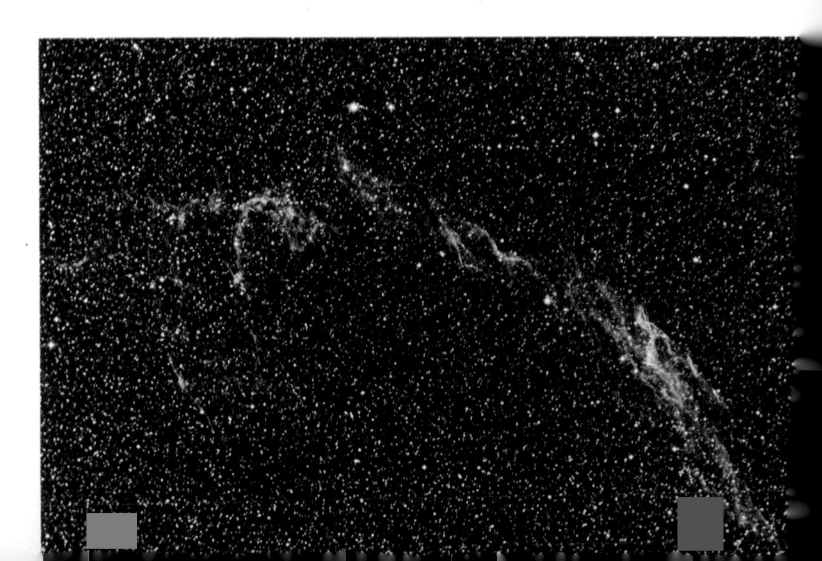

Veil The Veil Nebula is a small portion of a wreath-like structure in Cygnus. It is the remains of a supernova that occurred 50,000 years ago. (Hale Observatories photograph)

M31 The Great Andromeda Galaxy. Faintly visible to the unaided eye, this galaxy along with its elliptical companions is quite impressive in a photograph. Note that the spiral structure is blue while the central region is distinctly yellow. (Hale Observatories photograph)

NGC 7331 An Sb-type spiral galaxy in Pegasus. The blue knots in the spiral arms are open clusters and great glowing gas clouds. (Hale Observatories photograph)

NGC 598 This Sc-type galaxy is seen almost face on. Note the loose spiral structure and also the very bright star-like central source in the nucleus. (Hale Observatories photograph)

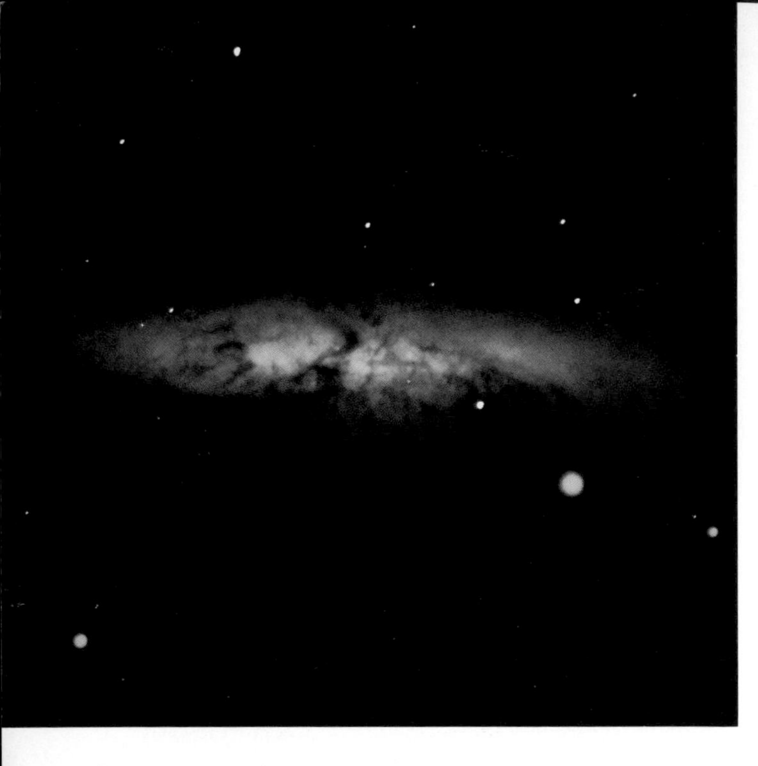

NGC 3034 The galaxy M82 is undergoing a violent disruption in its central region. (Hale Observatories photograph)

NGC 5128 The galaxy NGC 5128. This galaxy shows evidence of a violent event in the spiral structure between us and its nucleus. It also has invisible radio emitting regions extending 15 times its diameter above and below it. (Cerro Tololo InterAmerican Observatory photograph. © 1975 by AURA)

Rosette Nebula The large Rosette Nebula is the apparent remains of star formation as the beautiful, centrally located open cluster readily attests. (Hale Observatories photograph)

Trifid Nebula The Trifid Nebula is a good example of the colors found in emission nebulae. Note that one portion is pink and the other blue. Stars are forming in this nebula. (Hale Observatories photograph)

Lagoon Nebula The Lagoon Nebula is a nebula where star formation has and is taking place. Note the fine open cluster associated with the nebula. (Kitt Peak National Observatory photograph. © 1974 by AURA)

Pleiades The relatively young open cluster of the Pleiades. Some of the nebulosity remains visible by reflected light from the bright, hot, young stars. (Hale Observatories photograph)

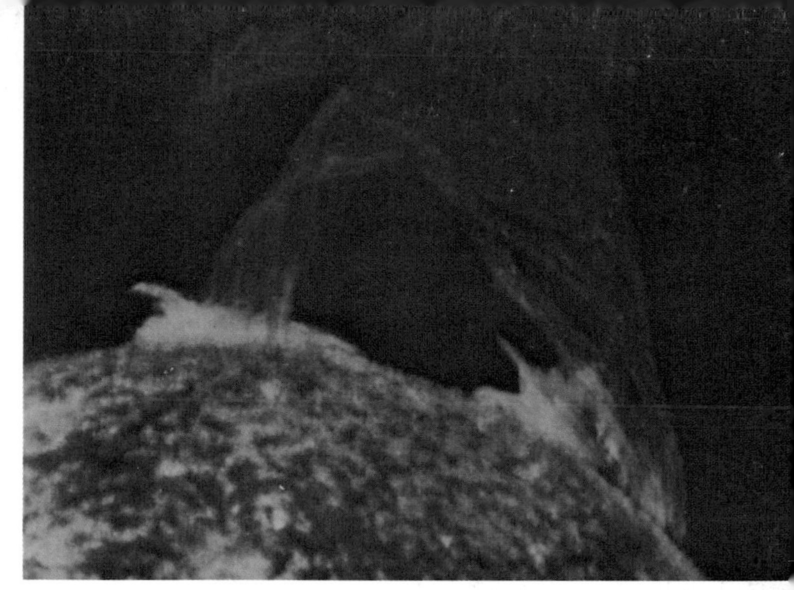

Sun A great twisting prominence on the limb of the sun as photographed from Skylab. (National Aeronautics and Space Administration photograph)

Earth The earth as a celestial object. The earth is a colorful ball and goes through all phases as viewed from the moon. (National Aeronautics and Space Administration photograph)

**Venus** Streaked cloud banks dominate the upper atmosphere of Venus in this ultraviolet photograph from the spacecraft Mariner 10. (National Aeronautics and Space Administration photograph)

**Mars** From the Viking 1 lander the surface of Mars is rust colored and the sky is distinctly pink. This region resembles a desert on earth. Note the wind blown dust on the spacecraft and on the foreground rock. (National Aeronautics and Space Administration photograph)

> . . . discovered two lesser stars, or satellites, which revolve about Mars, whereof the innermost is distant from the centre of the primary planet exactly three of the diameters, and the outermost five; the former revolves in the space of ten hours, and the latter in twenty-one and an half; so that the squares of their periodical times are very near in the same proportion with the cubes of their distance from the centre of Mars, which evidently shows them to be governed by the same law of gravitation, that influences the other heavenly bodies.

In the 19th century two small moons were finally discovered circling around Mars. It is almost uncanny how close they are to Swift's description. The moons, named **Deimos** and **Phobos** (meaning "panic" and "fear"), greatly resemble asteroids. Both moons show cratering (Figure 9.5), indicating that they have suffered bombardment like the rest of the planets and moons in the solar system. A Viking Orbiter 2 photograph of Phobos shows a series of nearly parallel lines. It is thought that these lines may be stress fractures. The evidence seems to indicate that both Deimos and Phobos are captured moons. As seen from the surface of Mars, Phobos rises in the west and sets in the east. Among natural satellites this is true only for Phobos, Amalthea, the innermost moon of Jupiter, and parts of the rings of Saturn.

**FIGURE 9.5**
Deimos (top) and Phobos (bottom), the well cratered moons of Mars. The streaks on Phobos are believed to originate from a large impact crater on the satellite. (National Aeronautics and Space Administration photographs)

*Jupiter* Jupiter has the largest family of moons. The fourteen Jovian moons may be divided into three groups, as we mentioned in Chapter 8. The first group consists of the five satellites closest to the planet—**Almathea, Io, Europa, Callisto,** and **Ganymede.** The last four were discovered by Galileo and are Jupiter's largest moons. Ganymede is larger than Mercury, and all the Galilean satellites may be larger than Pluto. Together with Jupiter they resemble a miniature solar system. These five satellites have nearly circular orbits around the equator of Jupiter. The three inner satellites (Almathea, Io, and Europa) are smaller and have higher densities than do the outer two (Callisto and Ganymede). Their orbits are coplanar and lay in Jupiter's equatorial plane. Their revolutions are in the direction of Jupiter's rotation. The formation of Jupiter and its inner satellites appears to be a scaled down version of the formation of the solar system.

Io is a most interesting moon because it is the reddest object in the solar system. It gets its color from a surface that appears to be covered by sodium salts, except at its poles, where there may be sulfur deposits. As Io orbits, it disturbs Jupiter's magnetic field, causing bursts of radio energy from the planet.

The next two groups of moons of Jupiter are very different. These moons are quite small, and the orbits are tilted compared to Jupiter's

**FIGURE 9.6**
Saturn's rings. The outer and bright rings separated by Cassini's division are most readily evident in photographs. Encke's division of the outer ring can be seen in the original photograph. (Lowell Observatory photograph)

equator. It has been suggested that since their orbits are very large and elliptical these moons were captured by Jupiter. It has also been suggested that they represent fragments of two larger moons that were shattered during the early history of the solar system, when heavy bombardment was prevalent.

*Saturn* Saturn has 11 identified moons, one of which, **Titan,** is quite large and interesting. Titan's diameter of 5800 km makes it the second largest satellite in the solar system. Titan has a thick atmosphere, which is unusual among moons. Its atmosphere is composed of methane and hydrogen and a dustlike material. Such an atmosphere is analogous to smog and tends to trap the feeble solar radiation, which warms the surface slightly above the expected temperature. The surface of Titan is probably methane and ice. If there is volcanism, however, there may also be some warm pools of water present. This is one of the objects in the solar system where scientists have speculated we might someday find evidence of life.

Saturn is best known for its beautiful ring system (Figure 9.6). There are four rings: the outer ring, the main or bright ring, the Crepe-ring, and the inner ring. Photographs only show three of them because the contrast between the planet and the inner ring is too great. The outer edge of the outer ring is 137,000 km from the center of Saturn. It has a trace of a gap in its center referred to as Encke's Division. The outer ring is separated from the main ring by a very noticeable dark gap called Cassini's Division. On photographs, the darker Crepe ring seems to blend right into the inner edge of the main ring. Inside the Crepe ring there is a fairly large gap, and then we see the main body of the planet. The tenuous inner ring, which is too faint to be seen photographically, is closest to Saturn. This is summarized in Figure 9.7. The gaps in the rings are due to tidal interactions with Saturn's moons, especially Titan.

The rings are not solid. As we mentioned earlier, they are most likely composed of billions of ice crystals and bits of ice-covered dust. The light from stars occulted by the rings does not completely fade out. Spectroscopic observations show that the particles at the outer edge of the rings are moving more slowly than the particles at the inner edge, as we would expect for objects obeying Kepler's laws. Recent radar observations show that the particles range in size from 3 to 30 cm. Saturn's rings are only about 2 km thick. The ratio of thickness to width is far less than that of a gummed reinforcement for a ring binder.

We see the rings from different angles because they lie in the equatorial plane of Saturn. Since Saturn's poles are tilted by almost 27° from the vertical to the plane of its orbit, we alternately see the rings opened-out and edge-on (Figure 9.8). Over a period of about 14 years, we see both the northern and southern sides of the rings. The total brightness of Saturn in our skies varies markedly, depending on whether we are seeing the planet when its rings are most exposed or when they are edge-on. Sometimes the rings of Saturn appear dull and indistinct and not as bright as normal. There is no explanation as yet for this curious behavior.

*Uranus* Saturn is no longer the only planet with rings. In 1977, astronomers discovered at least five rings around Uranus. We described their detection in Chapter 8. The rings of Uranus are not as wide or nearly as bright as Saturn's. They may be composed of somewhat larger particles than those making up Saturn's rings. Since the particles in the rings are orbiting in the equatorial plane of Uranus and since Uranus has such an extreme tilt, we get a slightly different perspective on the rings. When we view the planet pole-on, there will be very little foreshortening, and we will see the entire ring system. ("See" is a misleading word, because Uranus' rings are too faint to be seen visually even with a telescope.)

Uranus has five moons located outside the rings (Figure 8.16). They are also orbiting the planet in the equatorial plane. Because of this, it is likely that their formation is tied to the rotation of Uranus rather than to the solar system in general. The usual tilt of Uranus is variously attributed to the loss of Pluto as a moon, or to a near collision with a large chunk of space debris since ejected from the solar system.

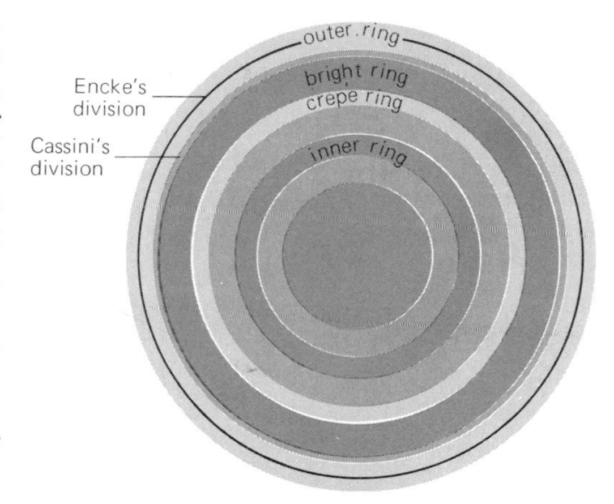


**FIGURE 9.7**
A polar view of the rings of Saturn for identification purposes.

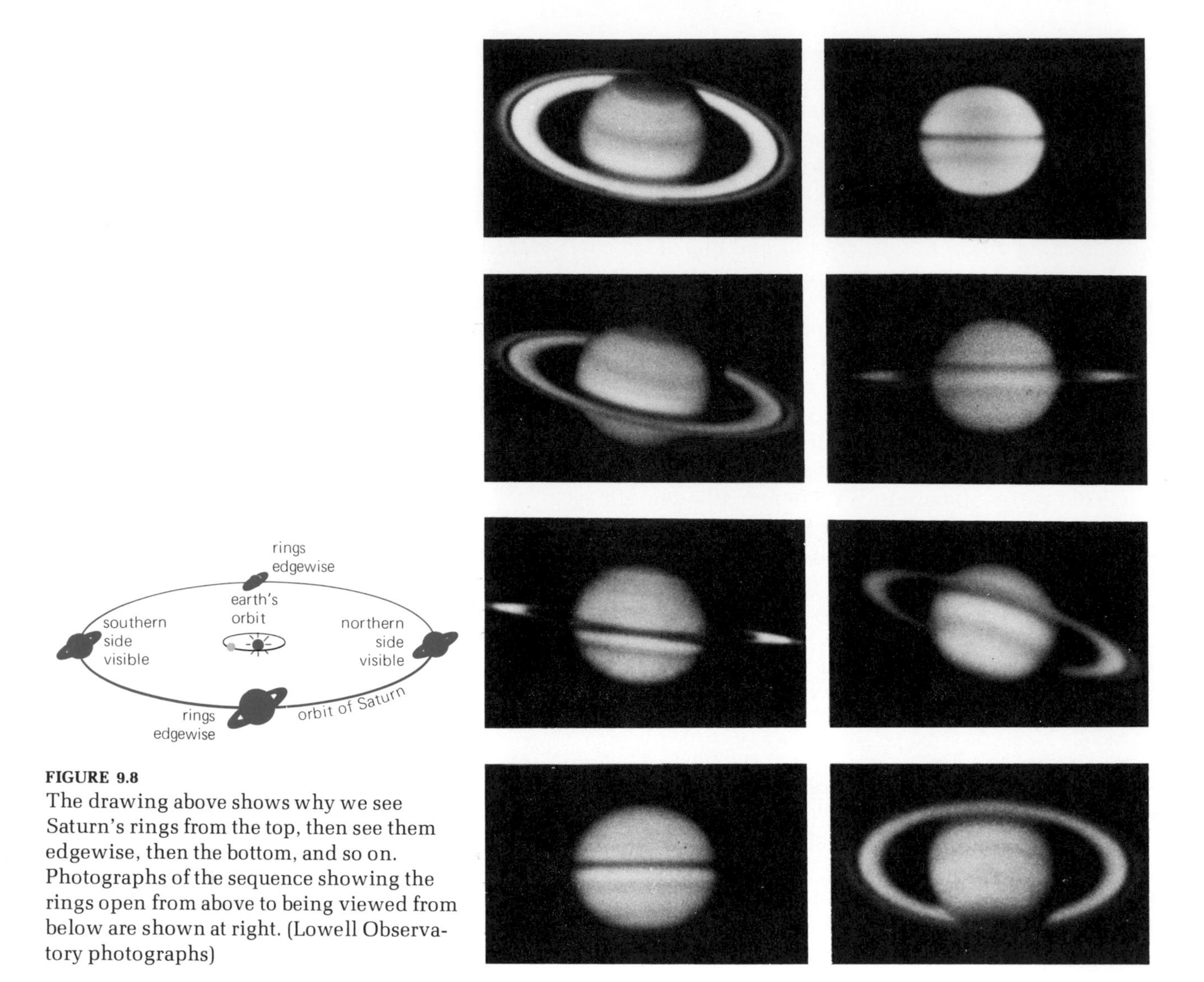


**FIGURE 9.8**
The drawing above shows why we see Saturn's rings from the top, then see them edgewise, then the bottom, and so on. Photographs of the sequence showing the rings open from above to being viewed from below are shown at right. (Lowell Observatory photographs)

*Neptune* Neptune has two moons, one of which, **Triton,** with a diameter of 6000 km, is probably the largest moon in the solar system. **Nereid,** Neptune's other moon, is over ten times smaller than Triton. Of all the major moons in the solar system, Triton is the only one that revolves east to west and has a very inclined orbit compared to the equatorial plane of its planet. One theory to explain the tilt of Triton's orbit invokes the ejection of Pluto as a satellite of Neptune. Little is known about the moons of Neptune simply because Neptune is so remote.

### Planets without Moons

Three of the terrestrial planets, Mercury, Venus, and Pluto, have no known moons. Pluto, as we have indicated above, has been suspected of originally being a moon of either Uranus or Neptune. It may be that none of the terrestrial planets formed with moons. Capture theories have recently gained favor and have been proposed for explaining the moons of Mars. Some astronomers also believe that our moon was captured. It has been difficult to determine the masses of planets without moons. In the case of Mercury and Venus, space probes have been used to calculate their masses, but Pluto's mass continues to be a guess at best.

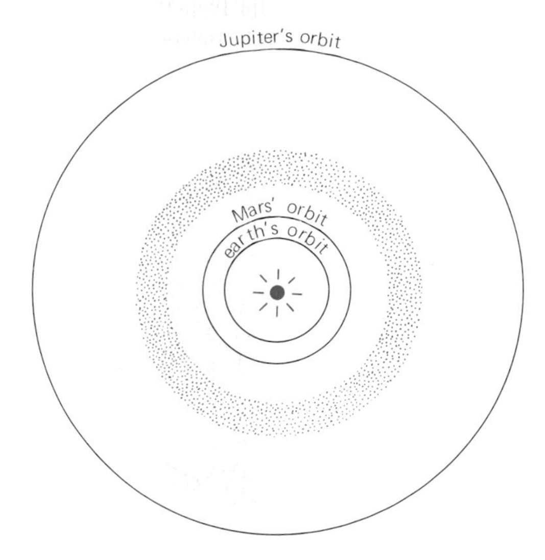


**FIGURE 9.9**
The orbits of earth, Mars, and Jupiter showing the main region of the asteroids between Mars and Jupiter.

## SMALLER BODIES IN THE SOLAR SYSTEM

### Asteroids

In Chapter 8, we mentioned that bursts of energy from the young sun cleared out the gases and small particles in the solar nebula. Much of what was left was collected by the planets through collisions. Today, 4.5 billion years after the formation of the solar system, there are still significant amounts of material between the planets. Most of the small bodies in the solar system are in stable orbits, and there are relatively few collisions.

The **asteroids** make up one class of such objects in the solar system. Indeed, we may have encountered members of this group as captured moons of Mars and Jupiter. Our earlier discussion of the discovery of the asteroids (Chapter 3) gives the impression that all of them are to be found very close to 2.8 AU from the sun. In reality, they are distributed more or less evenly in a belt between 2.1 and 3.5 AU—still closer to Mars than to Jupiter (Figure 9..9). There are distinct gaps or sparse regions in the asteroid belt analogous to the gaps in the rings of Saturn. They are called **Kirkwood's Gaps,** after the astronomer who first noted the effect. Kirkwood's Gaps are the result of Jupiter's influence on the asteroids. The periodic gravitational pull by giant Jupiter slowly clears out certain regions in the belt.

Asteroids also occur outside the asteroid belt. Two groups, called collectively the **Trojan asteroids,** are each at 5.2 AU from the sun and hence on the same orbit as Jupiter. This is shown in Figure 9.10. Both groups form an equilateral triangle with Jupiter and the sun. That is, they are 5.2 AU from the sun and 5.2 AU from Jupiter. One group leads Jupiter and one group follows. They are located at two relatively stable points in the dynamics of the sun-Jupiter system. Individual Trojan as-

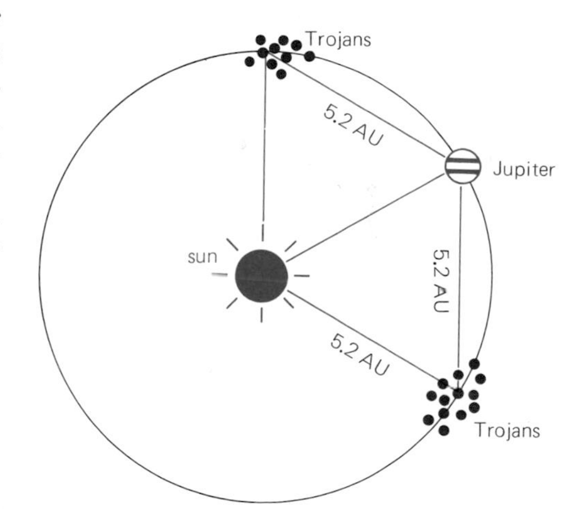


**FIGURE 9.10**
A drawing of Jupiter's orbit showing the location of the Trojan asteroids leading and trailing Jupiter.

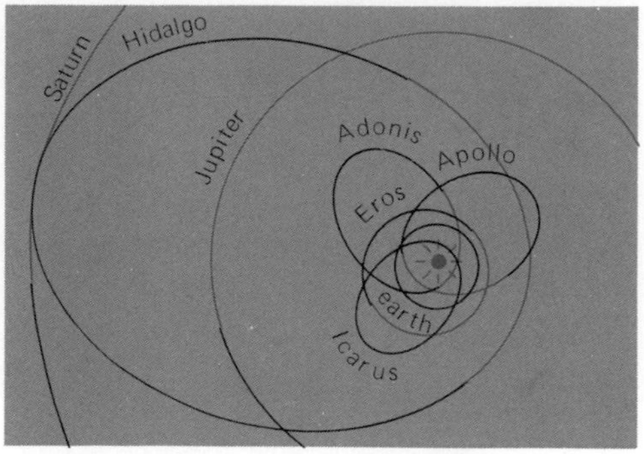

FIGURE 9.11
The orbits of some of the well known asteroids that pass relatively close to the earth's orbit.

teroids have been named after the Homeric heroes of the Trojan War. Those in the leading group are called the **Greek Planets** and have Greek names except for one, the Trojan hero Hector (presumably a spy). The group trailing Jupiter is called the **Trojan Planets** and its members are named after Trojans except for Petroclus, a Greek (and also presumably a spy to even things up).

Some asteroids are on relatively circular orbits about the sun; some deviate wildly and have highly eccentric orbits. The orbits of a few are shown in Figure 9.11. Hidalgo, for example, gets almost out to the orbit of Saturn. Many asteroids have orbits that cross the earth's orbit. These are called the **Apollo asteroids** after one member of this group. The asteroid that comes closest to the sun gets twice as close to the sun as Mercury. It is appropriately named Icarus after a mythical Greek who escaped prison by fashioning wings of feathers and wax but crashed to his death when he flew too close to the sun and the wax melted.

Orbits have been determined for nearly 2,000 asteroids. Many more than that have been observed at one time or another. Due to faintness or an irregular orbit, asteroids are often "lost." When an accurate orbit is found for an asteroid, the discoverer is usually given the right to name it. Asteroids have been named after mythological figures, states, politicians, loved ones, and so on.

**Ceres,** the largest asteroid, is 780 km in diameter. There are only two other asteroids larger than 300 km. About 220 asteroids are larger than 100 km. Statistics indicate that there are thousands of these small objects. Even at that, the total mass of all the asteroids together is less than half of our moon's mass.

Many of the asteroids, especially the smaller ones, have irregular shapes. If an asteroid is elongated in shape and tumbles end over end, we would expect to observe a change in its brightness as it moves through the sky. This is exactly what is observed with many asteroids. A rare radar observation of an occultation by Eros recently gave a better value for the size and shape of that asteroid. It turns out to be almost slab shaped measuring about 7 by 19 by 30 km.

The asteroids are just large chunks of rock or stony-iron material. Their surfaces are probably cratered. The largest asteroids are more spherical and resemble our moon in appearance. The best idea of what a smaller, irregular shaped asteroid would look like is that it would be similar to Phobos and Deimos, Mars' two small moons. On photographs, asteroids appear starlike, hence their name, and they are detectable only by their day-to-day motions (Figure 9.12). None has a perceptible disk when viewed through the telescope.

The actual origin of the asteroids is still a controversy. One school of thought says that the asteroids are the result of two or more small planets forming at about 2.8 AU from the sun and then colliding.

Increasing numbers of collisions result in smaller and smaller fragments. Another school of thought holds that in the presence of Jupiter, no planet of significant size could form in the region of the asteroids in the first place. Jupiter's gravitational attraction would tidally destroy any large planet located where the asteroids are.

Some of the meteors that strike the earth's atmosphere are definitely asteroids or debris from them. Studies of the percent of reflected sunlight of many asteroids agree with remarkable precision with those of various meteors.

## Comets

A relatively minor solar system object of great popular interest and awe is a **comet.** Searching for comets has always been an active pursuit for professional and amateur astronomers. Part of the excitement of discovering a comet is that comets are named after their discoverers. Part must also be the challenge of being the first to see these unusual objects. Unlike the planets, comets usually have very large, elongated orbits. A few of them are periodic and travel once around the sun in less than a hundred years. Only about a hundred comets are known to take less than a thousand years to complete one orbit. The majority of comets are on open-ended, almost parabolic orbits and are regarded as unique visitors to our skies.

**FIGURE 9.12**
The discovery of the very faint asteroid Copernicus. The motion of the asteroid is followed in order for its image to build up. This causes the star images to trail. (Goethe Link Observatory photograph by F. K. Edmondson)

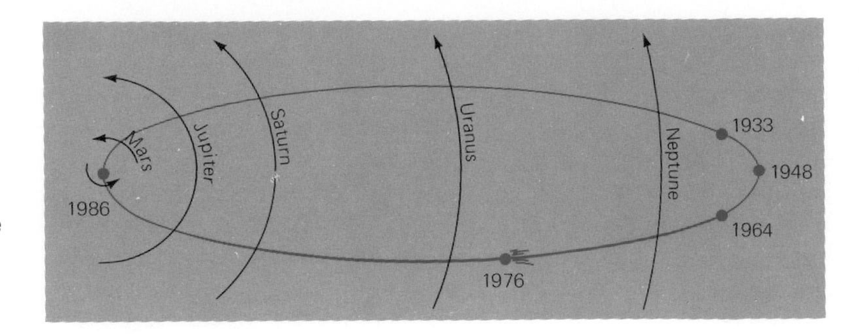


**FIGURE 9.13**
The orbit of Halley's Comet. Note how little the comet moved between 1933 and 1964 and how far it will move between 1964 and 1986.

Indications are that comets on parabolic orbits come from the outer reaches of the solar system, often from as far away as 1 lt-yr. Hence it is proposed that there is a supply of the original frozen solar nebula material around the outer edge of the solar system. We might think of this material as a swarm of comets in a spherical cloud at about 1 or 2 lt-yr from the sun. Being so far from the sun, the comets are moving quite slowly. They are so distant from the sun that the movements of the nearby stars can disturb them and alter their motion. Occasionally the orbit of a comet is disturbed enough that it starts its long journey into and back out from the inner portion of the solar system. If an incoming comet passes very near one of the giant planets, its orbit can be changed by the gravitational action of the planet. Jupiter probably is the dominant influence on incoming comets. Under the proper conditions of encounter, comets can be "captured" and become **periodic comets,** with orbits located entirely in the inner part of the solar system. Comets that have been captured by Jupiter, in the sense just stated, are referred to as Jupiter's family of comets. Neptune has a family of comets as well, the most famous being Halley's Comet (Figure 9.13).

The spectacular and awesome appearance of comets sets them aside from most other celestial objects. Five or six new comets appear in our skies each year. Most are visible only in telescopes. Several periodic comets reappear every few years. We must wait until 1986 to see the famous Halley's Comet, which last appeared in 1910. The scarcity of bright comets, coupled with the general unpredictability of their occurrence, gives them an aura of suspense. It is no wonder that comets were once regarded as evil omens. Enterprising individuals as recently as 1910 sold "comet pills" to ward off the influence of Halley's Comet.

One astronomer has aptly described comets as "more nothing than anything known that can still be called something." For their size, comets consist of extremely little material. The main parts of a comet are its **nucleus,** its head or **coma,** and its **tail** (Figure 9.14). The tail forms only when a comet is near the sun. It begins to be detectable when the comet is at about 2AU from the sun and extends as the comet draws closer to the sun. Before that, the comet is just nucleus and head. The word comet comes from a Greek word literally meaning "hairy star"—a fitting description for these strange lights in the sky.

Nuclei of comets have been described as dirty snowballs. They are icy chunks of material consisting of frozen hydrocarbons with trapped dust and rocks. At most, they are several hundred kilometers in diameter. As the nuclei approach the sun, some of the material warms and evaporates, forming the coma, which may extend 100,000 km or more into space. As the comet continues to approach the sun, the sun's radiation pushes out material from the nucleus and coma to form the tail. Comet tails always point away from the sun in space. They can grow to be millions of kilometers in length (Figure 9.15). The air in an average room contains more matter than an entire tail of a typical comet. The densities of comet tails are lower than that of a good vacuum on earth—almost nothing. The earth can pass easily through a comet tail with no effect at all, as happened in 1910 when we passed through the tail of Halley's Comet.

Comet tails can grow to be the most extended things in the solar system. They may be straight, have a straight and curved component,

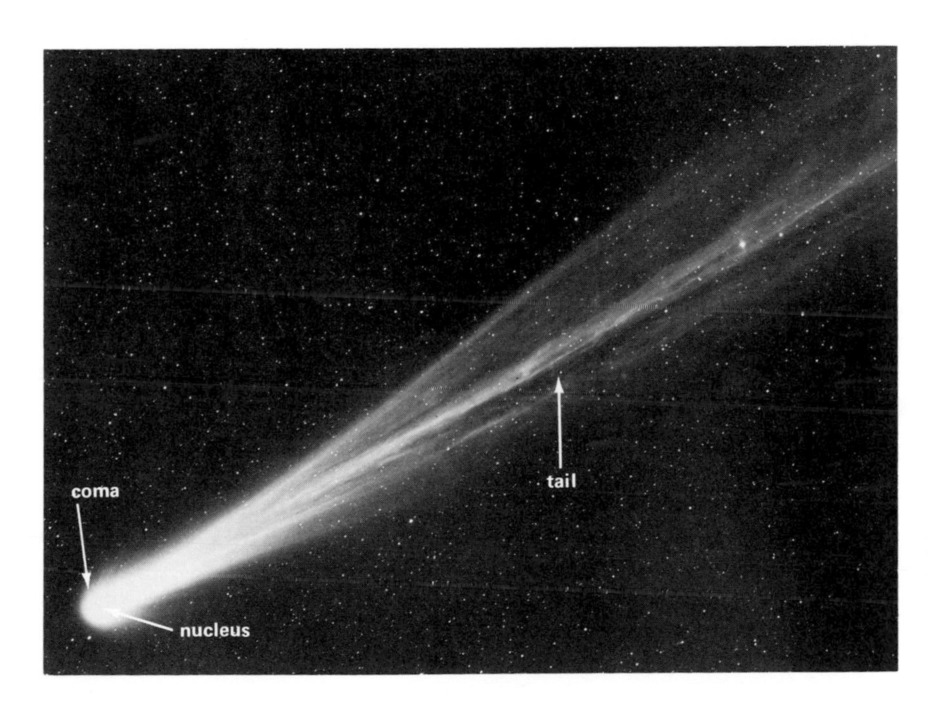


**FIGURE 9.14**
Comet Tago-Sato-Kosaka. The three principle features of a comet are labeled. (Cerro Tololo Inter-American Observatory photograph)

**FIGURE 9.15**
Halley's Comet. Note the motion of the comet between the left (12 May) and right (15 May) pictures as well as the growth in the length of the tail. (Hale Observatories photograph)

or have only a curved component (Figure 9.16). The straight portion, called the **gaseous tail,** is composed of ionized molecular material driven out of the nucleus. As the ices of the comet melt, the dust and rocks are released and pushed away into a tail, giving a curved appearance. This is called the **dust tail.** Often the tail will develop twists and kinks, indicating that it is being acted upon by fast-moving particles or a magnetic field.

The light from comets is a combination of reflected sunlight and fluorescent light (ultraviolet light from the sun is absorbed by the gases and then emitted as visible light) produced by the gases of the coma.

Most comets survive their close approaches to the sun. As they recede from the sun, the matter that has spread out into the tail cools down and the entire comet fades back into a cold, barely visible dirty snowball in space. Some of the icy particles and dust and rock from its tail are left in its wake along the orbit. Periodic comets gradually lose their luster with each successive passage near the sun. Some have been known to break up into several parts or to disappear completely near the sun. Although very spectacular, Halley's Comet in 1910 did not compare to Halley's Comet of 240 B.C., which totally dominated the night sky. With each passage near the sun, Halley's Comet loses some of its gases. Its brightness will depend in part on how much material it has lost and in part on how far away from earth it is. At some point in the future, Halley's Comet will become a collection of dust along its orbit. When the earth crosses the orbit, the dust will produce a meteor shower. An obvious but not very spectacular comet suffered such a fate in 1862. This comet is the origin of the Perseid meteor shower observed in mid-August each year.

Although unlikely, it is possible for a comet nucleus to hit the earth. It is widely believed that a collision did occur in Siberia on June 30, 1908. The object or comet nucleus obviously exploded in the air before hitting the ground, since trees were seared and leveled radially outward in an area with a radius of about 15 km. No crater and no ordinary meteorite fragments were found at the site of the explosion. Some small spherules were found that were thought to be part of the object or comet nucleus. Presumably most of the icy nucleus vaporized in the explosion, and dust and rocks imbedded in the ice melted into spherules and fell to the ground.

The main scientific interest in comets is that they may be samples of the oldest unchanged material in the solar system. Studying them may give us valuable clues to the origin of the solar system.

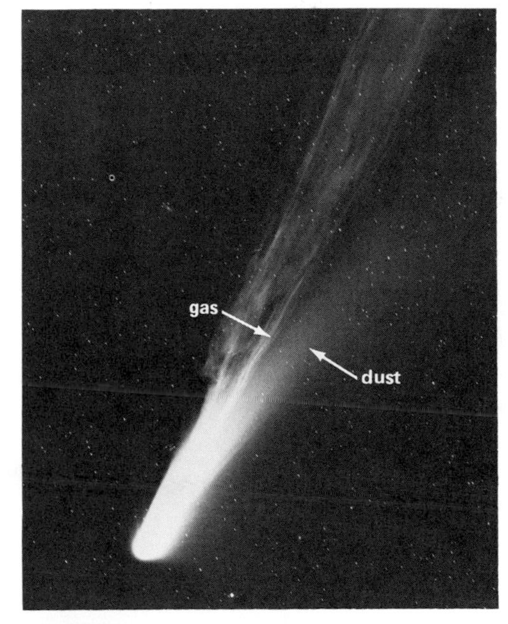


**FIGURE 9.16**
Comet Mrkos in 1957 clearly displayed a gaseous tail and a dust tail. (Hale Observatories photograph)

## Meteors

**Meteoroids** are irregular stony and metallic particles orbiting around the sun. If they happen to be on a collision course with the earth, or if they get close enough to be attracted by the earth's gravitational pull, they plunge into the earth's atmosphere and become visible as **meteors,** or the so-called shooting stars. Meteoroids are melted and vaporized when they strike the air molecules about 100 km above the earth. They produce luminous trails across the night sky (Figure 9.17). The average meteoroid is about the size of a grain of sand. Friction with the air due to their speed of some 40 km/sec causes them to glow brightly. Unusually bright meteors are called **fireballs** and are about the

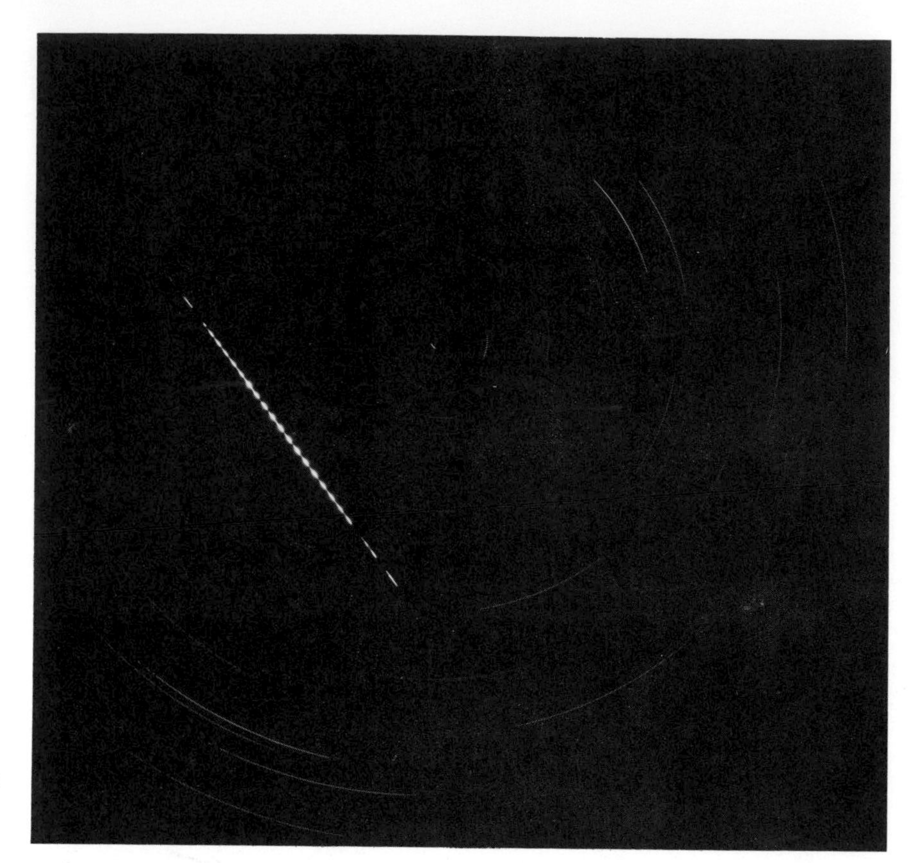

**FIGURE 9.17**
A bright meteor passing near the pole star. The meteor was photographed by a special camera equipped with a chopper. A chopper interrupts the photograph periodically and allows us to obtain the angular motion of the meteor. A similar photograph with another camera would allow us to calculate the orbit, speed, and so on. (Harvard College Observatory photograph)

size of a walnut. On any clear, dark night, about five or six meteors can be seen per hour. These meteors seem to come from random directions.

Meteor showers occur when the earth encounters a swarm of meteoroids in space. Shower meteors always look as though they are radiating from the same direction in space (Figure 9.18). They are usually named after the constellation that is in the direction of their apparent origin. The **Perseids,** for example, seem to radiate from the direction of Perseus. Meteor showers happen about the same time every year. The reason they can be predicted so well is that swarms of meteoroids in space are usually debris from old comets. The dates when this debris encounters the earth are just the dates when the earth's orbit crosses the old comet orbit. Table 9.2 lists the dates of some meteor showers. On the other hand, random, unpredictable meteors are generally due to debris from asteroids.

**Table 9.2 A List of Well-Determined Meteor Showers**

| Shower | Maximum Display (Date) | Associated Comet (and Notes) |
|---|---|---|
| Quadrantids | 3 Jan. | |
| Lyrids | 21 Apr. | 1861 I |
| Eta Aquarids | 4 May | Halley (?) |
| Arietids* | 8 June | (= delta Aquarids) |
| Zeta Perseids* | 9 June | |
| Beta Taurids* | 30 June | Encke |
| Delta Aquarids | 30 July | (two streams) |
| Alpha Capricornids | 1 Aug. | 1948n |
| Perseids | 12 Aug. | 1862 III |
| Draconids | 10 Oct. | Giacobini-Zinner |
| Orionids | 22 Oct. | Halley (?) |
| Taurids | 1 Nov. | Encke (two streams) |
| Andromedids | 14 Nov. | Biela |
| Leonids | 17 Nov. | Temple |
| Geminids | 14 Dec. | |
| Ursids (Ursa Minor) | 22 Dec. | Tuttle |

*Shower in daytime.

**FIGURE 9.18**
A time exposure during the Leonid Meteor shower of 1966. The meteor trails trace back to the radiant. A non-Leonid meteor appears on the left. (Photograph by D. McLean)

**FIGURE 9.19**
The Barringer Meteorite Crater near Winslow, Arizona. The diameter of the crater is 1.3 km. (Photographed by Meteor Crater Society, Winslow, Arizona)

Very rarely, a meteoroid will be massive enough to withstand burning up in the atmosphere, and a portion of it will crash to earth. These are called **meteorites.** Most meteorites are less than a centimeter in size, but there are a few very large ones weighing several thousands of tons known to have blasted out craters on the surface of the earth. Barringer Crater near Winslow, Arizona, a circular depression of 1280m across, is the best preserved of about a dozen earth craters that have been produced by meteorite impacts (Figure 9.19). This crater is only about 30,000 years old and is in arid country. There must have been many more meteorite craters in the earth's early history, but these have been obliterated by millions of years of erosion.

If a small meteorite struck the ground right in front of you, you would be able to pick it up immediately without any danger of burning yourself. Many people are surprised to learn this. We can explain this by thinking of the heat shields on reentering space vehicles. The meteoroid as is the case for the heat shield, heats up and burns away a small portion of its surface, but the interior is unaffected. In the case of the meteorite, the interior is still at the nearly absolute zero temperature of outer space. Even though the outside may melt and burn somewhat, the whole meteorite will immediately reach an equilibrium temperature as soon as it reaches the ground. The meteorite will hardly even be warm to the touch.

FIGURE 9.20
The zodiacal light and Comet Ikeya-Seki. The zodiacal light is the triangular glow that extends from the horizon up to the bright star, Regulus. (Photograph by H. Gordon Solberg, Jr.)

## THE INTERPLANETARY MEDIUM

### Dust

Besides the meteorites, the earth collects several thousand tons of dust each day from the **interplanetary medium.** The dust particles just settle gently to the surface, and while "tons" sounds large, the total mass is inconsequential compared to the mass of the earth. Some of the interplanetary dust is the result of asteroidal material that has been ground up over millions of years of collisions. Some of it is comet debris.

The interplanetary dust is concentrated rather strongly in the ecliptic plane defined by the orbits of the planets. The dust reflects sunlight, and under the right conditions, we can see it. It is the cause of both the **zodiacal light** and the **gegenschein** or counterglow. The zodiacal light is the faint glow of reflected light extending along the ecliptic or band of the zodiac on either side of the sun's position in the sky. It can best be seen in middle northern latitudes after sunset in the western sky in the early spring or in the eastern sky just before sunrise in the early fall (Figure 9.20). The gegenschein is a faint glow in the night sky directly opposite the position of the sun. It is best seen in northern latitudes on a clear, moonless night in December (Figure 9.21).

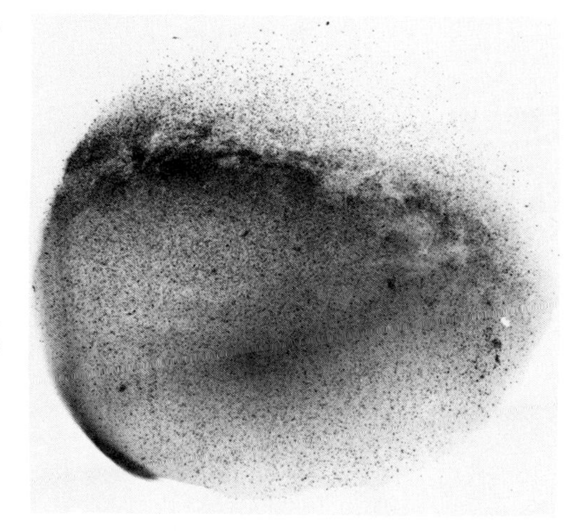

FIGURE 9.21
The gegenschein can just be detected in the center of this negative print. The Milky Way goes across the top. (Photograph by Shohei Suyama)

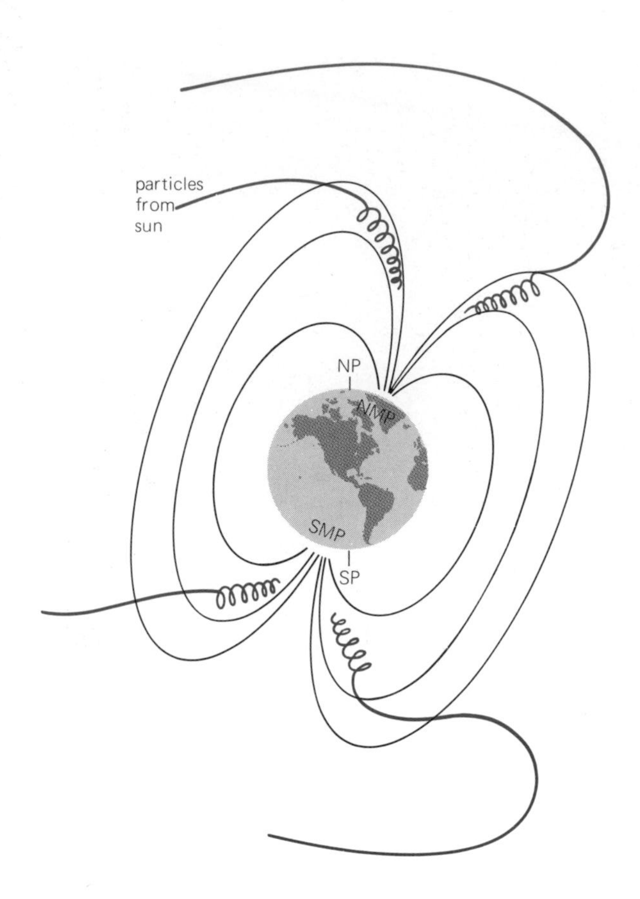


**FIGURE 9.22**
A schematic drawing of the earth and its magnetic field. Trapped particles interact with the atmosphere in a ring above the magnetic poles.

## Solar Wind

Interplanetary space is also filled with subatomic particles—protons and electrons—moving away from the sun. The sun has a magnetic field, and these particles move outward along the magnetic lines of force. These outward moving particles are called the **solar wind.** The solar wind interacts with the earth's magnetic field. When the solar particles are especially energetic and numerous, they can interfere with radio communication on the earth and also produce **auroras** (Figure 9.22). The light of an aurora is produced when the particles in the solar wind get trapped in the earth's magnetic field. The particles are pulled down around the earth's north and south magnetic poles, where they interact with the atmosphere to produce arches and bands of glowing light, in colors of white, green, red, and blue.

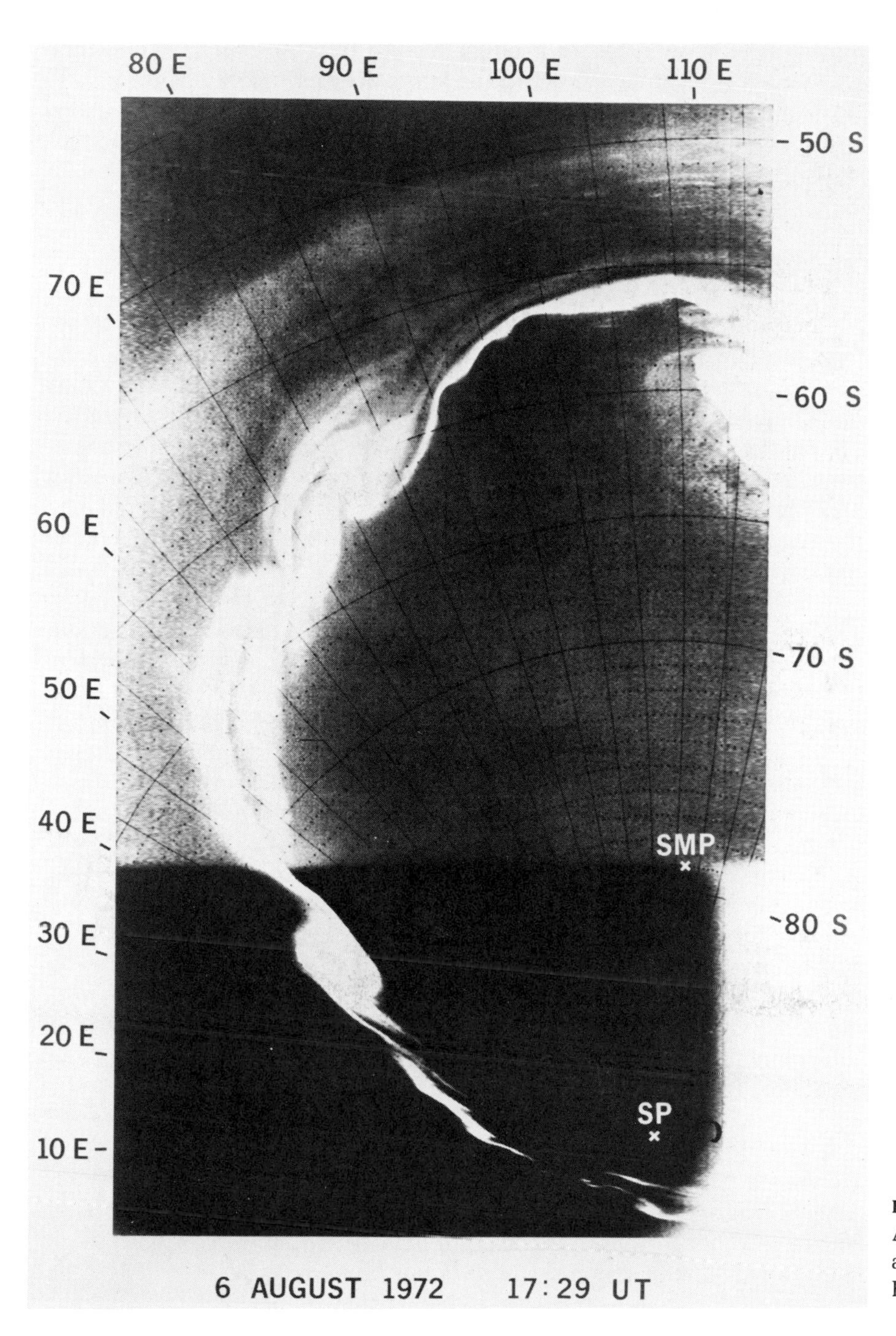


**FIGURE 9.23**
An aurora around the south magnetic pole as seen by a very high satellite. (U.S. Air Force Weather Service satellite)

Evidence for the solar wind and the extent of the sun's magnetic field was first seen by their effect on the tails of comets. Today the existence of the solar wind and the sun's magnetic field has been fully confirmed by direct sampling and measurements by numerous spacecraft. Generally, the earth's magnetic field shields us and protects us from all but the most energetic solar particles and cosmic rays.

## SUMMARY

Several of the solar system's thirty-five known moons are larger than Mercury. Some of the moons must have formed with the planets, like miniature solar systems. Other moons resemble asteroids and may have been captured by the planets. Moons give us a way to determine the masses of the planets they orbit. The earth and its moon are sometimes regarded as a double planet. The moon affects our ocean tides and also stabilizes the tilt of the earth's axis. The particles that make up the rings of Saturn and Uranus behave as billions of tiny moons orbiting those planets.

Asteroids have more irregular orbits and shapes than the major planets. All of them together would not make up a planet the size of our moon. Their orbits are influenced by gravitational effects of the giant planets. Comets may be remnant material from the original solar nebula. In general, they travel on highly elongated orbits around the sun. Sometimes a comet will be captured by one of the giant planets and become a periodic comet. Comets develop their characteristic tails only when they are near the sun. The sun's magnetic field often produces twists and turns in comet tails.

Some meteroids are the debris left over along old comet orbits. Meteors are visible on any clear night. Heavy bombardment of the planets by meteorites ended early in the development of the solar system, over 3.5 billion years ago. Occasionally a meteorite will strike the earth. Tons of micrometeorites (dust) settle on the earth daily. Both the zodiacal light and the gegenschein are due to sunlight reflected off interplanetary dust particles. The solar wind interacts with earth's atmosphere to produce auroras.

### *FALLACIES AND FANTASIES*

*A "shooting star" is a star.*
*Halley's Comet gets more spectacular at every appearance.*
*Our moon is the largest moon in the solar system.*

## REVIEW QUESTIONS

______ 1. What are the three components of the structure of comets near the sun? Give their approximate sizes.

______ 2. Why is the composition of meteorites of particular interest to astronomers?

______ 3. What observations of the rings of Saturn show that they are not solid?

______ 4. Is there a way to tell a comet from a star before the comet develops a tail? Explain.

______ 5. What specific information would you need about the moons of Mars in order to determine the mass of Mars?

______ 6. Describe one relationship between comets and meteors.

______ 7. What are auroras? Why do we have them?

______ 8. What makes the Trojan asteroids special?

______ 9. If we had a permanent observatory on the moon, could we observe meteors? Would they look like meteors on the earth? Explain.

______ 10. How do we know interplanetary dust exists?

# 10 The Sun: From Eclipses to Energy Production



## SOLAR ECLIPSES

### The Spectacle

The sun has drawn the attention of observers since before recorded history. Now, as in the past, the disappearance of the sun from a blue, cloudless sky demands attention and provides the most awe-inspiring spectacle in nature. From the moment the moon first occults the rim of

**FIGURE 10.1**
The 1970 total solar eclipse. The partial phases procede slowly and are not very interesting (left). Then, seemingly instantaneously, the sun is covered up and the corona bursts into view (right).

the sun, it is clear that something unusual is happening. But seeing more and more of the sun covered up does nothing to prepare you for the moment when the solar disk is totally covered, and the sun's pearl white corona bursts into full view (Figure 10.1).

The event of a **solar eclipse,** that is, when the moon passes between the earth and the sun and hence covers up the sun, demands silence and respect, and it receives just that—except from astronomers who, because of their studies, cannot be silent. They are counting photographic exposure times, shouting orders to assistants, and only occasionally glancing at the spectacle above them. However, when a large crowd of people gather to watch an eclipse, with no astronomical team present, the response is dramatic (Figure 10.2). In 1970, with 7,000 people packed at one person in each 4 square feet, not a word was heard. No one cared to speak—except one seventy-year-old gentleman who said quietly, "Ah, at last, at last I have seen one!" And then when the moon moved off the solar disk and the corona was gone, a young girl shouted "That was neat, Daddy; have them run that again." Thus is the power of a solar eclipse and, alas, of television.

If an eclipse in our modern era can silence multitudes, imagine the effect of solar eclipses before the advent of writing. What was happening to the sun? Various tribes in the Far East thought that a dragon was eating the sun, and the dragon had to be driven away. Even today, some

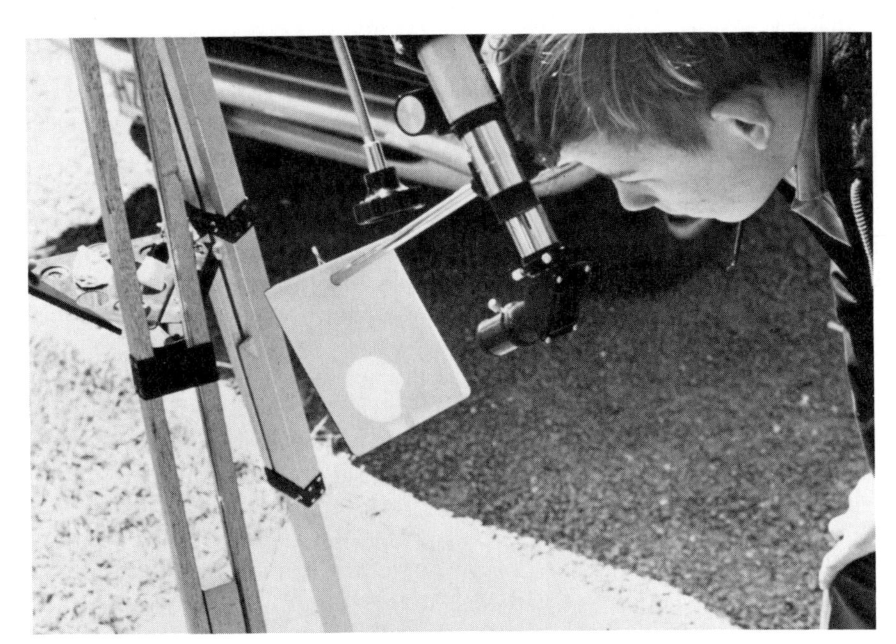

**FIGURE 10.2**
A portion of the crowd gathering for the 1970 eclipse. By eclipse time the entire field in the background was filled with cars and people. In the bottom photograph, an observer looks at a projected image of the sun during the early phases of the eclipse.

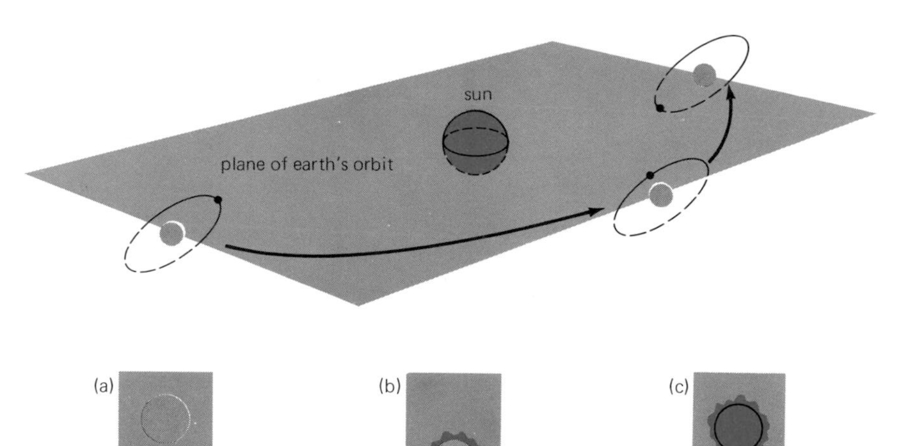


**FIGURE 10.3**
Because of the tilt of the moon's orbit to the plane of the earth's orbit, eclipses do not occur every time the moon is directly between the earth and the sun. The moon may be above (a) or below (c) the plane. Only when the moon is in the plane (b) will an eclipse occur.

of these people set up bamboo pipes and beat a strange, melodic cadence until the sun is free again. When questioned as to why they go through this ritual, their leader responds, "It worked, didn't it?"

One of the earliest tasks of astronomy was to explain the disappearance of the sun. Painstakingly, observations were put together and patterns began to emerge. Several early cultures independently came to the conclusion that the patterns meant that eclipses could be predicted.

## The Conditions

Every month when the moon is in its new phase, it is almost in the same direction in the sky as the sun. We do not, however, have an eclipse of the sun every month because the moon's orbit is tipped compared to a line connecting the sun and earth. At most new phases, the moon passes either above or below the sun's position in the sky. This can be seen in Figure 10.3. Each year there are two **eclipse seasons** during which it is possible for the moon to be new and at the same time be directly on a line between the earth and the sun. These are the conditions necessary to produce eclipses. During solar eclipses, the moon's shadow touches the earth (Figure 10.4) and anyone located inside that shadow sees the sun eclipsed by the moon.

Since the moon is moving during an eclipse, the moon's shadow is also moving. On the earth, the shadow traces out a narrow band called the **path of totality.** Figure 10.5 shows the path of totality for the 1991

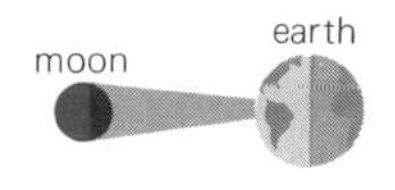


**FIGURE 10.4**
When the apex of the moon's shadow falls inside the earth a total solar eclipse will be seen by an observer inside the shadow cone. In the diagram sizes and distance are not to scale.

**FIGURE 10.5**
The ground track (path of totality) of the 1991 total solar eclipse. An observer located between the boundry lines will observe the total eclipse.

solar eclipse. Since the paths of solar eclipses are so narrow, around 200 km across, seldom do two solar eclipses occur at the same place close in time. Today, the orbits of the earth and moon are understood well enough to determine both future and past eclipses with great accuracy. Future eclipses are of scientific interest; past eclipses are often of historic interest in dating events. For example, the eclipse predicted by the biblical prophet Amos, and used to tell when Amos lived, occurred on 15 June 763 B.C. It was the only eclipse visible in Samaria during a period of five hundred years.

Depending on the moon's position and distance, three types of solar eclipses are possible: partial, annular, and total (Figure 10.6). During a **partial eclipse,** the moon hides only a fraction of the sun's disk. An **annular eclipse** occurs when the moon is directly between the earth and sun, and also at the most distant point of its orbit from the earth. When this is the case, the moon appears slightly smaller than the sun and does not completely cover it. At the middle of the annular eclipse, a bright ring (or *annulus*) of the sun remains uneclipsed. During a **total eclipse,** the moon completely covers the visible disk of the sun. This allows astronomers to see the faint atmosphere of the sun and to learn more about the atmospheric structure of our nearest star.

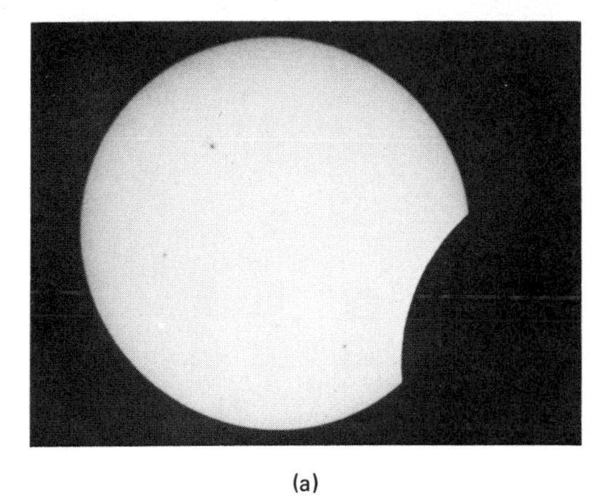

(a)

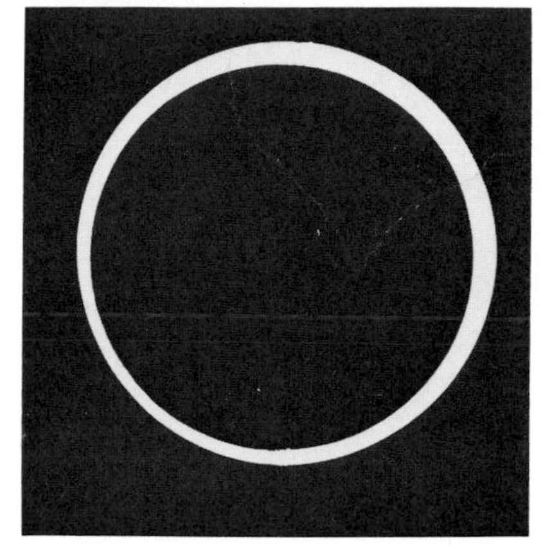

(b)

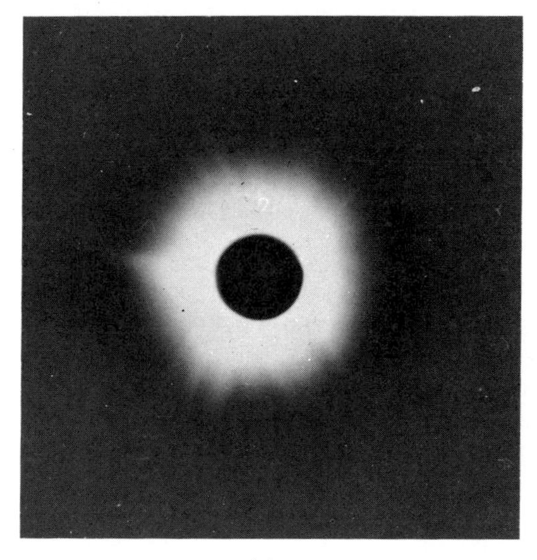

(c)

**FIGURE 10.6**
(a) A very partial eclipse. The observer was so far from the ground track that the moon only covered a few percent of the sun's surface. (b) An annular eclipse. Here the observer is located in the ground track but the moon is so far from the earth that the apex of its shadow cone does not reach the surface. (c) A total eclipse. The shadow cone reaches the earth's surface.

# THE PHYSICAL NATURE OF THE SUN

## Structure

The sun is a globe of very hot gas 1,390,000 km in diameter, or 109 times the earth's diameter. If we hollowed out the sun, it would take 1.3 million earths to fill it up again. The sun's mass is one-third of a million times greater than the earth's. Its density is one-fourth the earth's mean density, or 1.4 times the density of water.

The interior of the sun, below the visible surface, is known to us only indirectly by means of theory. Its temperature increases from 6000°K at the visible surface to 13,000,000°K at the center. The density and pressure of the sun's gases also increase toward the center. Since the sun is gaseous throughout, these changes occur gradually. For convenience in discussing the sun, we divide it into concentric layers like the layers of an onion.

The most familiar layer of the sun is its visible surface, called the **photosphere.** This is the shallow layer from which the continuous background radiation of the solar spectrum is emitted. The gases of the photosphere are so hot (6000°K) they are opaque. The gases above the photosphere constitute the sun's atmosphere. The atmosphere is composed of the chromosphere and above that the tenuous corona of the sun. These two atmospheric layers are too faint to be seen except during a total eclipse. We will examine each component of the solar atmosphere in more detail after we take a look at the sun's energy production in its core (Figure 10.7).

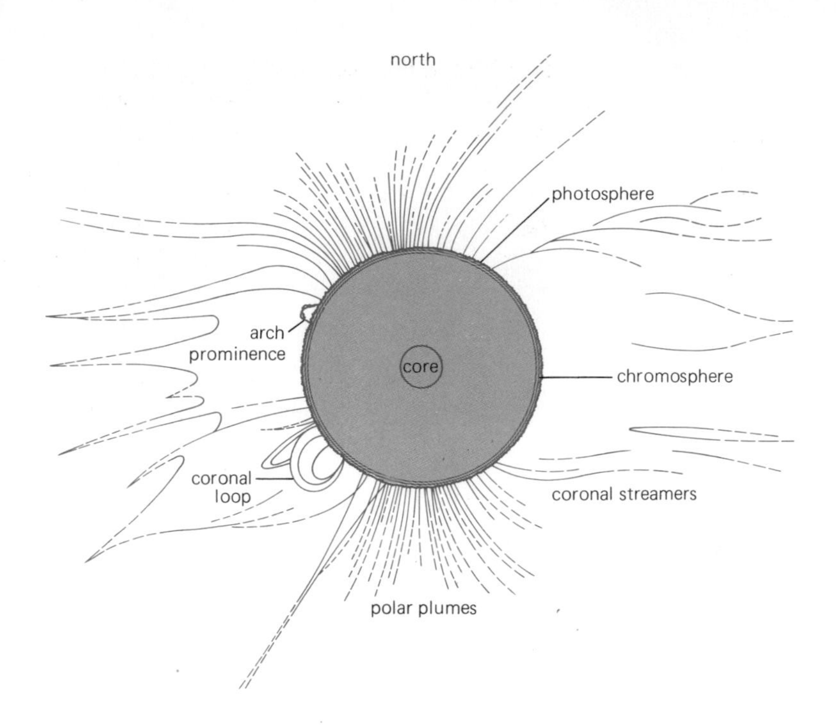


**FIGURE 10.7**
A schematic drawing of the sun showing features discussed in the text pp. 210 *ff*.

Analysis of the solar spectrum tells us that the sun is composed of roughly 71% hydrogen, 26% helium, and 3% all other elements by mass. The vast quantity of hydrogen is the fuel that keeps the sun shining steadily. The sun's energy is produced deep in its interior, where hydrogen is converted into helium.

## Energy Production

The sun emits an immense amount of energy. The total amount of energy that leaves the sun every second is usually measured in units called **ergs.** One erg can be thought to equal the energy of impact of a slow-flying mosquito. The sun produces 4 billion trillion trillion ($4 \times 10^{33}$) ergs every second. That is equivalent to 500,000 billion billion ($5 \times 10^{23}$) horsepower. We know how much energy the sun sends out by measuring the amount we receive at the distance of the earth and then working backwards to find out how much left the sun. All the energy passing through a sphere whose radius is equal to the earth's orbital radius left the sun. When we add up the energy of the sun in all directions, we get the large values given above.

The energy is produced deep in the central core of the sun, where temperatures are extreme. One of the milestones of 20th-century astronomy has been the discovery that nuclear reactions in the sun are converting mass into energy. One of Einstein's most famous equations states the equivalence of mass and energy. That is, mass and energy are interchangeable. In the sun, nearly 5 million tons of mass are changed into the equivalent amount of energy every second. That is a terrific amount of mass by earthly standards, but compared to the entire mass of the sun it is very little. Every second, for the last several billion years at least, the sun has been annihilating that much mass and converting it to energy. There is enough mass in the sun for it to keep producing energy this way for at least 5 or more billion years.

The sun is made up primarily of hydrogen, and this is the material that furnishes the mass to be converted into energy. There are two basic nuclear reactions that can convert hydrogen into something else and release energy. Actually, the reactions convert, or fuse, hydrogen atoms into helium atoms and in the process release some energy. One is called the **proton-proton reaction,** and the other is called the **carbon,** or **CNO, cycle.** These kinds of reactions, which convert lighter elements into heavier ones, are called **nuclear fusion reactions.**

The proton-proton reaction proceeds in three specific steps, which are shown schematically in Figure 10.8. Step 1: Two protons combine to form **deuterium.** This combination releases a **positron** and a **neutrino.** Protons are the nuclei of hydrogen atoms. At the center of the sun, the density and temperature are so high that all hydrogen atoms have been stripped of their electrons. Hydrogen nuclei (protons) are abundant. When two of them combine, they form deuterium, which is sometimes called heavy hydrogen. The positron that is released is similar to an electron except that it has a positive instead of a negative charge. The positron quickly finds an electron. The collision results in the annihilation of the particles and releases some energy. The neutrino that is released is a small, massless, chargeless particle that carries a little bit of the energy of the reaction away. Neutrinos are interesting because they can travel at the speed of light right through the layers of the sun. Neutrinos from the sun reach the earth in a little over eight minutes.

Step 2. The deuterium combines with another proton to form an **isotope of helium.** Normal helium has two protons and two neutrons in its nucleus. This helium has two protons and one neutron in its nucleus and is referred to as an isotope of helium. A gamma ray is released in this process. A **gamma ray** is an extremely high-energy, very short-wavelength photon. Gamma rays are absorbed and reemitted millions of times by atoms in the sun. It can easily take a million years for the energy from the gamma rays to reach the surface of the sun. By that time, the gamma rays have become many visible light rays. This is the light we see coming from the sun. Step 3. Finally, two helium

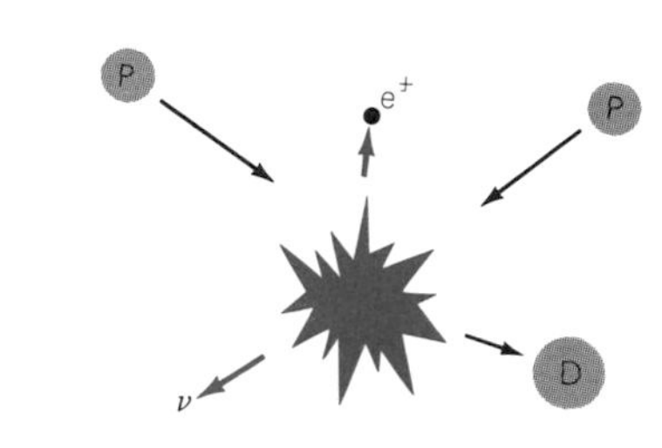


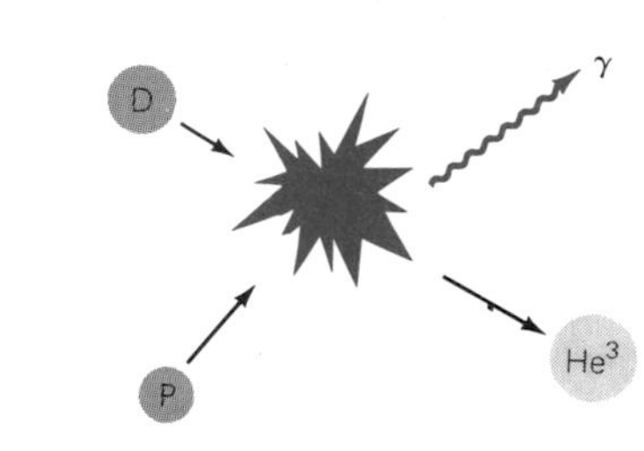


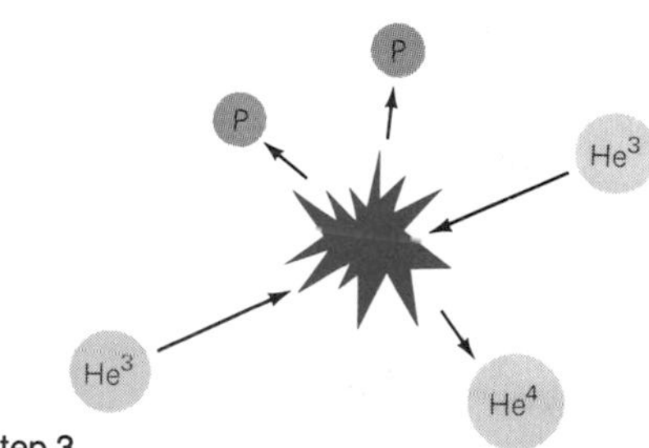


**FIGURE 10.8**
A descriptive rendition of the three steps in the proton–proton reaction. The symbols are: P-proton, $e^+$-positron, $\nu$-neutrino, D-deuterium, $\gamma$-gamma ray, $He^3$-light isotope of helium, and $He^4$-normal isotope of helium.

isotopes combine to form a normal helium atom and two protons. As these three steps are repeated over and over again, the hydrogen in the center of the sun is converted to helium and energy is released.

The other possible hydrogen conversion method, the carbon cycle, is more complicated. Repeated fusions of protons convert carbon into nitrogen and then into oxygen. Hence the alternate name, CNO cycle, for carbon-nitrogen-oxygen cycle. In the end, the oxygen decays into carbon and helium, accomplishing the conversion and leaving the carbon essentially as it was originally. The carbon is a catalyst in this reaction. That is, it helps make the cycle run but is not used up. During the cycle, two positrons are released, as are two neutrinos and three gamma rays. The neutrinos and gamma rays released in this reaction have different energies than those released in the proton-proton reaction.

We do not know for certain which of the processes described above powers the sun. We suspect that it is the proton-proton reaction. In principle, it would be possible to decide which process is working if we could detect the energy of the neutrinos being released.

## Looking for Neutrinos

In 1968, a clever experiment was undertaken to detect neutrinos from the sun. Since neutrinos pass right through the earth, collecting them is out of the question. Instead, we must detect them indirectly by their effect on certain substances. Chemists point out that neutrinos interact with carbon tetrachloride, tetrachloroethylene, perchloroethylene, and other similar common cleaning fluids. A neutrino passing through these fluids will form a radioactive isotope of argon at some statistical rate. A **radioactive isotope** is an isotope that will decay spontaneously into another form of atom. With a large enough tank of fluid, we ought to be able to determine the amount of radioactive argon produced over a period of time and thus tell how many neutrinos interacted with it.

In Lead, South Dakota, a tank containing 380,000 liters of perchloroethylene (ten railroad carloads) is surrounded by water and buried in a gold mine 1.5 km underground (Figure 10.9). Perchloroethylene is used because it is less toxic than the other fluids. The water prevents neutrons from the earth's own radioactivity from getting to the tank. The idea has been to let the tank sit for a month and then filter the fluid, extracting the argon atoms which, as we noted above, are isotopes of the more normal argon atom. A few atoms of stable argon are put into the tank as a check on the filtering process. The stable argon has always been recovered, but much to the surprise of all, very few of the predicted radioactive isotopes of argon have been found.

The experiment has been checked numerous times and improved in many ways. The results are still discouraging. The rate of recovery of radioactive argon indicates that the sun is emitting at best only about 25% as many neutrinos as were expected. This boils down to the conclusion that we may not really understand our sun as well as we thought we did.

Whether or not we understand the sun and the reactions in its deep interior, the sun seems to be a very stable star. Over the past three to

**FIGURE 10.9**
An interesting view of the tank used in the solar neutrino experiment. The technician at the far end gives an idea of the scale. (Photograph by Raymond Davis, Jr., Brookhaven National Laboratory)

four billion years the sun has not changed very much in brightness or color. The sun has enough hydrogen to continue shining steadily for about 5 billion years more. After that, it will experience an energy crisis that will change its structure. Our sun will then become a red giant star. In the process, the earth's oceans will boil away, and life will cease on earth. Then the sun will gradually shrink, becoming hotter and bluer and much smaller. It will become a white dwarf and then, over billions of years, fade into oblivion. (The aging of the sun and other stars is discussed in Chapter 11.)

### Solar Energy

Every gram of hydrogen that is converted into helium in the sun releases enough energy to keep a 100-watt lightbulb burning for a million years. Hydrogen fusion is such an efficient energy source that we might ask why we do not use it on the earth to generate power. Supplying enough hydrogen is no problem. But there are two basic problems with this energy source. First of all, it takes tremendous temperatures, like the 13 million degrees in the sun's interior, even to begin the reaction by fusing two protons. Second, once the reaction has begun, it must be controlled. We have produced some energy by hydrogen fusion in the hydrogen bomb, but we have not found a way to harness that energy. Experiments are under way right now to try to trigger fusion reactions using laser beams aimed at a pellet of hydrogen. If this energy source is tapped and can be controlled, it may certainly be the most efficient and limitless source of energy ever used.

The next best thing to reproducing the energy processes at work in the sun is to let the sun produce the energy and then harness what we can. Solar heating and cooling systems are already in use in many parts of the country. Solar units of some form or another are placed on roofs to collect the sun's heat. Circulating water is generally used to soak up the heat and carry it to where it can be stored or used immediately. Experimental solar houses have been in operation for some twenty years. One such house is shown in Figure 10.10.

## VISIBLE FEATURES OF THE SUN

### The Photosphere and Sunspots

The photosphere of the sun, when studied in detail, has a mottled appearance resembling the skin of an orange. This mottled appearance is called **solar granulation** (Figure 10.11). Granules vary from 240 to 1400 km in diameter and are separated by narrow dark spaces. The

**FIGURE 10.10**
A working solar energy house at Queche Lake, Vermont. (Photograph courtesy of Grumman Energy Systems)

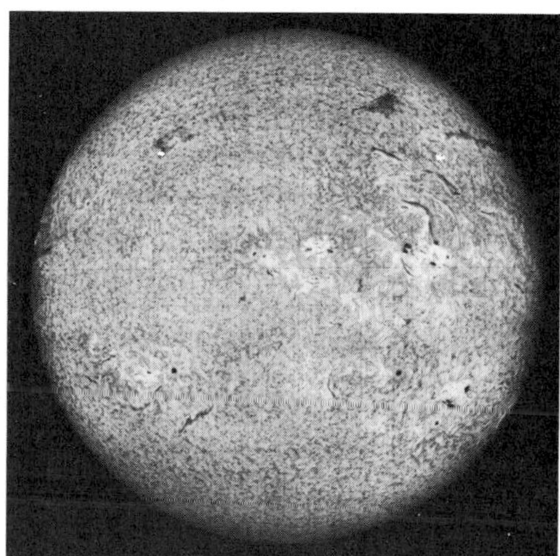

**FIGURE 10.11**
Large scale granulation on the sun. This effect is often called the orange peel effect and is enhanced here by photographing the sun and filtering out all but a few specially selected wavelengths of light. (Hale Observatories photograph)

formation and disappearance of the granules is a continuous process. Each granule lasts only a few minutes. What we see is the hotter gaseous material from below the photosphere welling up in the center of the granules. As this gas cools, it appears darker because a cooler gas is not as bright as a hot gas. We see the cool gases sinking back down around the edges of the granules. The boiling motion of the gases forming the granules is very similar to that of lava in a caldera or, on a smaller scale, to oatmeal cooking in a pot.

**Sunspots** are dark spots on the sun's disk or photosphere (Figure 10.12). They are composed of the same gases as the rest of the sun and its atmosphere. The reason that sunspots appear darker is that they are a thousand to fifteen hundred degrees cooler than the gases surrounding them. Sunspots range in size from a few hundred kilometers to more than 100,000 kilometers in diameter. The largest spots are many times larger than the earth and can be detected by the naked eye, but it is not advisable to look at the sun unless you know how and when. Visual sightings of large sunspots form a valuable historical record of sunspot activity dating back more than four thousand years.

The safest way to look for sunspots is to project the image of the

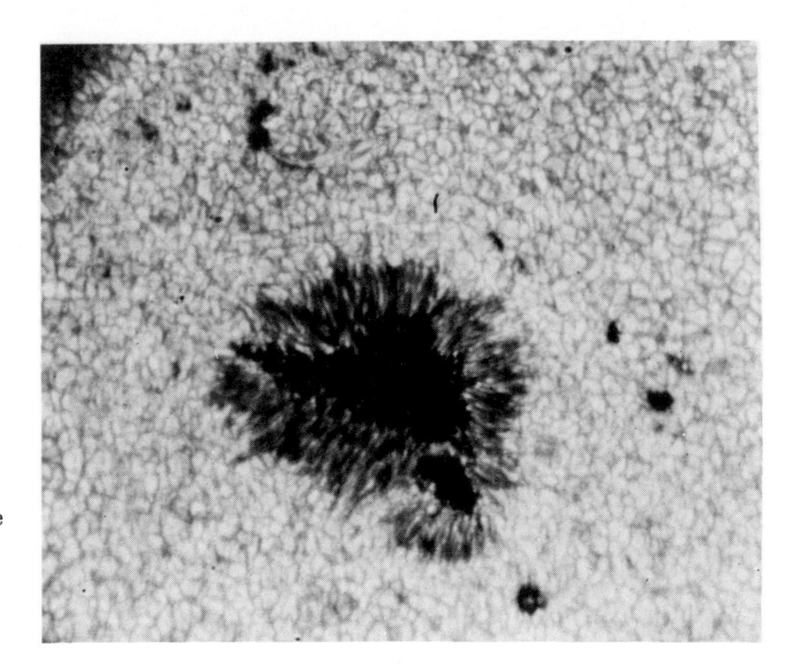

**FIGURE 10.12**
An excellent photograph of the small-scale solar granulation and a sunspot. This photograph was taken by a telescope on a balloon 24,000 m above the surface of the earth. (Princeton University photograph)

sun through a pinhole or binoculars or a telescope onto a white paper, as shown in Figure 10.13. Special filters for viewing the sun directly are available, but they must be used with extreme caution and the directions that come with them must be followed. Polaroid sunglasses, or any other kind, are not suitable for viewing the sun directly.

Sunspots tend to occur in pairs and groups (Figure 10.14). The total number of sunspots on the sun varies in a cycle of about eleven years (Figure 10.15). When a sunspot cycle begins, there are only a few spots on the sun, and they are located about midway between the sun's equator and its poles. Individual spots may last anywhere from a few days to a few months before they disappear. As the eleven-year cycle progresses, the total number of spots on the sun increases, and the spots appear nearer and nearer to the sun's equator. After five or six years of increasing numbers of sunspots, a maximum is reached; the number of spots then begins to decrease. The location of the spots moves closer and closer to the sun's equator. Finally the cycle fades away at the equator, and the first spots of the next eleven-year cycle appear midway between the equator and the poles. The most recent cycle began in 1976.

Sunspots move with the rotation of the sun (Figure 10.16). Tracking sunspots across the disk of the sun was one of the first ways to establish the rotation period of the sun. Since the sun is gaseous, it does not rotate at a single rate the way the earth does. The earth, being solid, rotates with a single period. The rotation period of the sun ranges from about 40 days near the sun's poles to 25 days at its equator. Many other stars rotate faster than the sun.

Strong magnetic fields are associated with the sunspots. When pairs of sunspots are observed, one spot will act like the north end of a magnet and the other spot like the south end. The area around sunspots is always active. The great prominences and flares described below occur in the sun's atmosphere above the sunspots.

## The Chromosphere and Corona

The **chromosphere** extends to a height of several thousand kilometers above the photosphere. It received its name because of its red color, which is quite apparent during total solar eclipses. The red hue is due to strong hydrogen emission in this layer of the sun. Many bright, narrow protuberances, called **spicules,** keep emerging at the upper surface of the chromosphere, giving it a grasslike appearance. It is

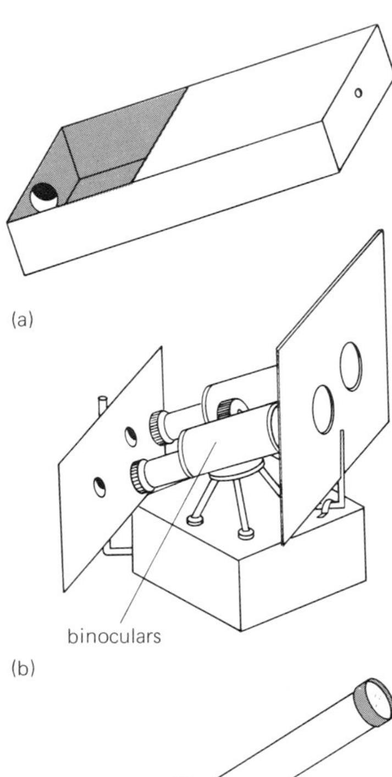


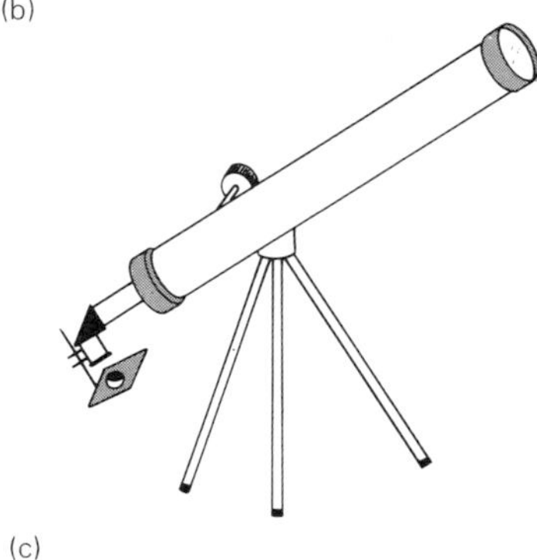


**FIGURE 10.13**
Three ways to observe the sun either before or during a solar eclipse. (a) A long box with a pinhole on one end and a white screen on the inside of the other end. The sun is viewed from above through a small cutaway in the box. (b) Binoculars projecting the eclipse on a white screen. The front screen is to cut off the direct light. (c) A telescope projecting the eclipse onto a screen. At the moment of totality it is safe to view the sun directly.

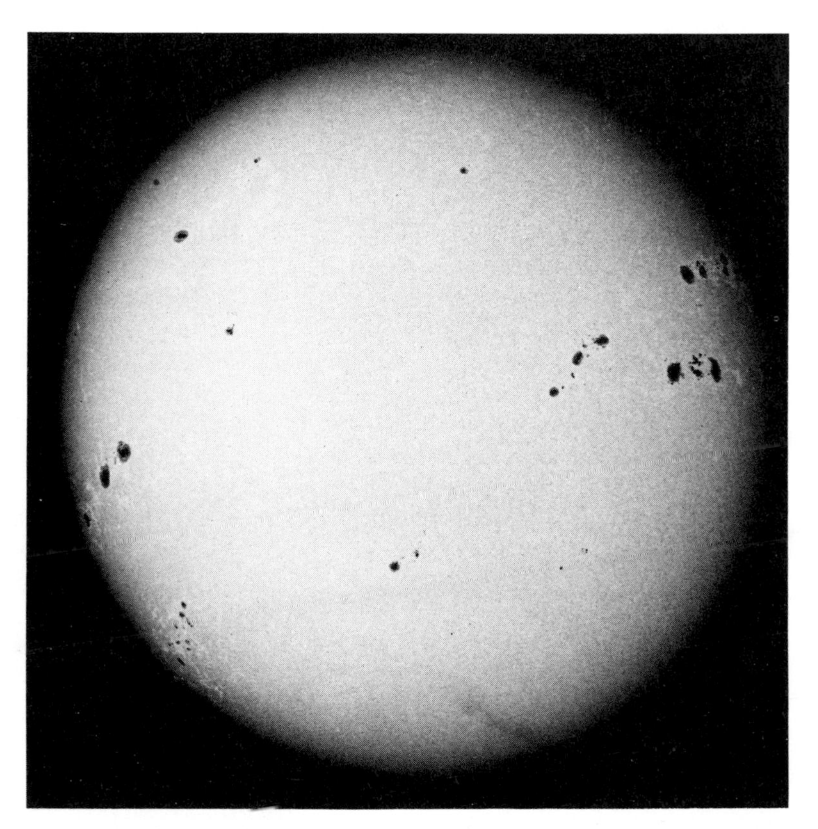

**FIGURE 10.14**
Sunspots occur in pairs and groups. (Hale Observatories photograph)

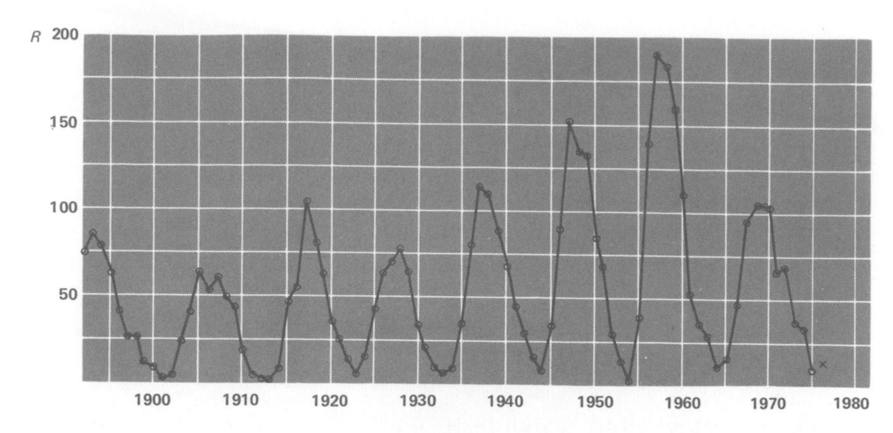


**FIGURE 10.15**
The relative number of sunspots (R) as a function of time. The number goes through clear short-term maxima and minima with a period of about 11 years. There is a longer-term cycle of about 100 years—see Figure 10.22.

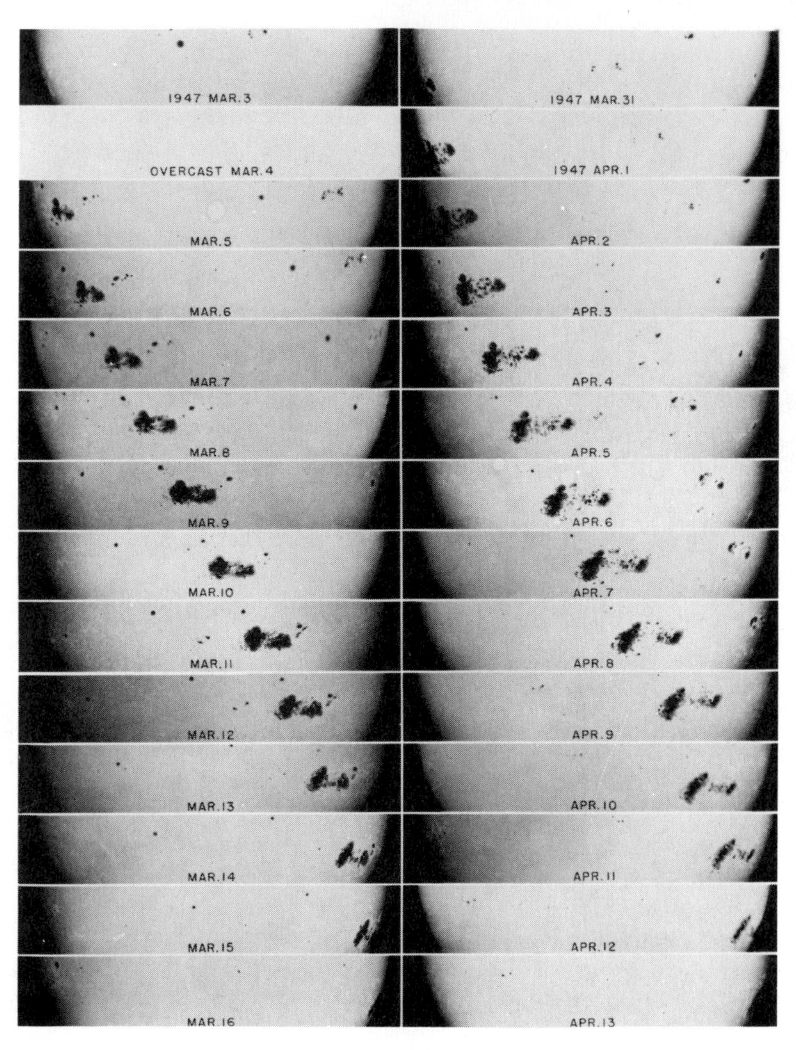


**FIGURE 10.16**
A long enduring group of sunspots shows the solar rotation quite clearly. (Hale Observatories photograph)

thought that these spicules may be associated with the granules of the photosphere.

Flamelike pillars of gas best observed during eclipses around the edge of the sun beyond the chromosphere are called **solar prominences** (see color plates). Some prominences rise to great heights above the chromosphere, as shown in Figure 10.17. Other prominences originate high in the sun's atmosphere and fall downward into the chromosphere. Still other prominences look like great arches. The prominences often occur above groups of sunspots.

Occasionally hot gases will emerge explosively from below the photosphere of the sun. These gases burst into space and are called **solar flares** (Figure 10.18). They may take from a few minutes to fifteen minutes to reach their maximum brightness. Very intense x-rays and radio waves are always emitted from the flares. These radiations reach the earth at the speed of light, in only about 8 minutes. Flares also eject energetic particles, mostly protons and electrons, into interplanetary space. These particles travel slower than the speed of light and reach the earth from several hours to several days after the flare. Flares produce some of the effects of the sun on the earth which we discuss in the next section.

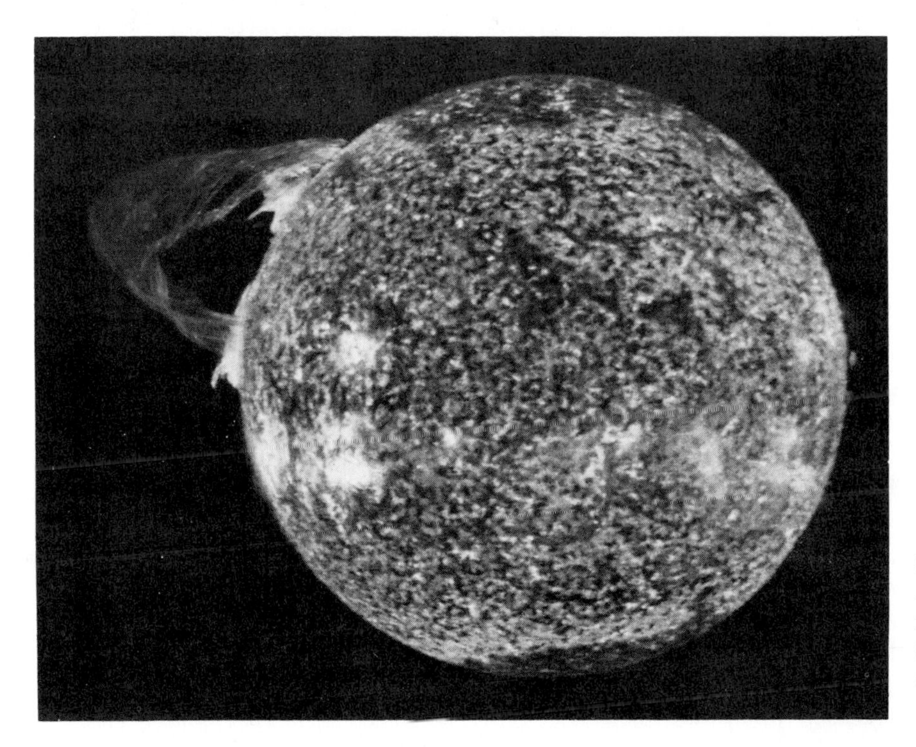

**FIGURE 10.17**
A magnificent arch prominance as observed from Skylab. This photograph was made in red light and shows the cool polar regions as dark. (NASA photograph)

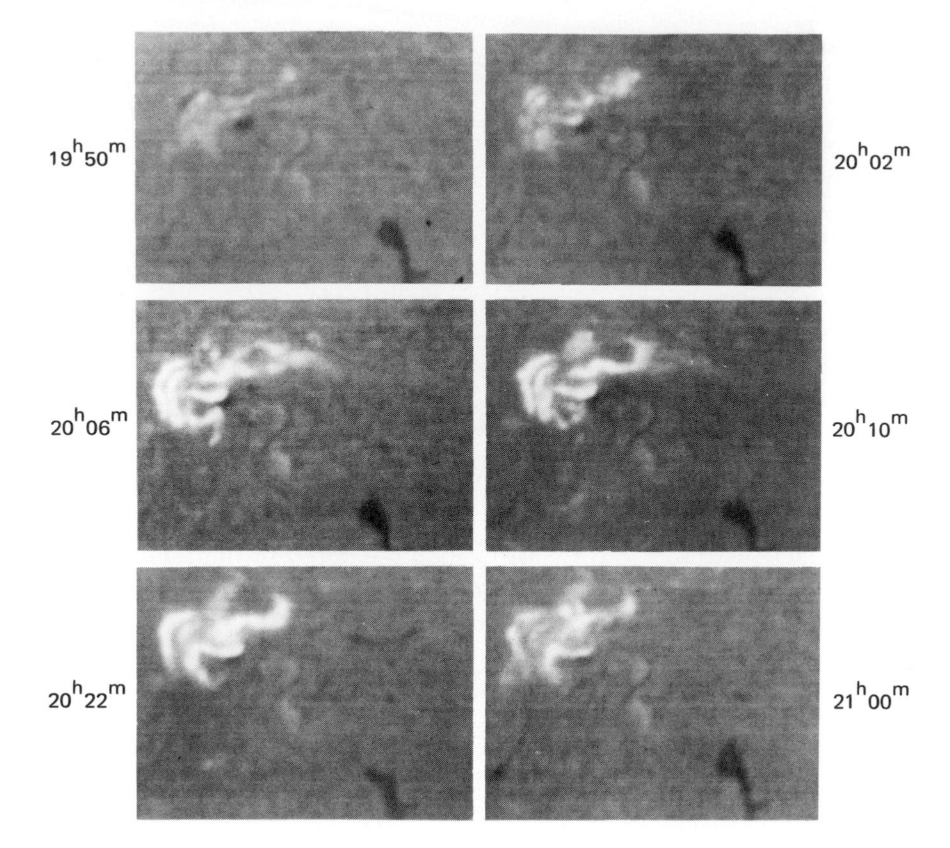


**FIGURE 10.18**
A sequence of pictures of a solar flare. The flare is moving up from the center of the sun's disk directly at the earth. The whole event took less than two hours. (Hale Observatories photograph)

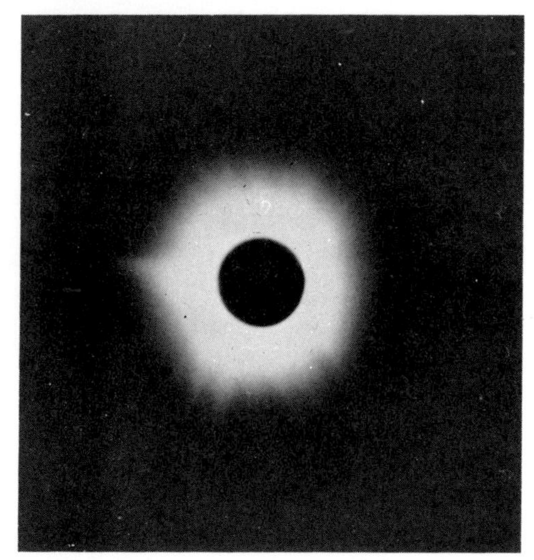

**FIGURE 10.19**
A solar eclipse at sunspot maximum. Note that the corona is quite large and round.

The **corona** is the pearl-white, extensive atmosphere of the sun seen during solar eclipses. Delicate streamers radiate outward through the corona. The shape of the corona varies with the activity and number of sunspots on the sun. Near sunspot maximum, that is, when sunspots are most numerous, the corona is roughly circular; petallike streamers give the appearance of a dahlia (Figure 10.19). Near sunspot minimum, the corona is flattened in the polar regions (Figure 10.20). Long streamers in the equatorial regions may reach out more than a million kilometers.

The total brightness of the corona is only half that of the full moon and hence far too faint to be seen in daylight from the earth. Until recently, the faint corona had to be studied during the brief minutes of the total phase of a solar eclipse or with a few special instruments called coronagraphs. Astronomers prepared for months and worked hard to observe the corona during eclipses. Today, with telescopes in space, we can study the corona by artificially producing our own eclipse.

**FIGURE 10.20**
A solar eclipse at sunspot minimum. The long delicate equatorial streamers are typical of the solar corona during sunspot minimum. (High Altitude Observatory photograph)

The Skylab astronauts observed the sun's corona over a 30-day period. During that time, they saw several great loops in the corona (Figure 10.21). These loops were previously thought to be very rare, since they were seen only twice in three hundred years. Observing the corona only during solar eclipses did not give us a large enough sample of the activity of the corona. Great coronal loops seem to be a common occurrence.

Material consisting mainly of electrons and protons from the corona continually streams outward into space. This is called the solar wind, mentioned in Chapter 9. The electrons and protons move along the sun's magnetic field lines and eventually are lost to the sun.

**FIGURE 10.21**
An immense coronal loop observed from Skylab. Several such loops were observed during the one month mission. (NASA and High Altitude Observatory photograph)

## EFFECTS OF THE SUN

### On the Earth

In addition to providing us with a stable amount of heat and light so that life can exist, the sun interacts in more direct ways with the earth. When energetic particles ejected from the sun during flares reach the earth, they can disrupt radio communications and cause magnetic compasses to oscillate wildly. The more constant bombardment of solar wind particles produces the auroras described in Chapter 9.

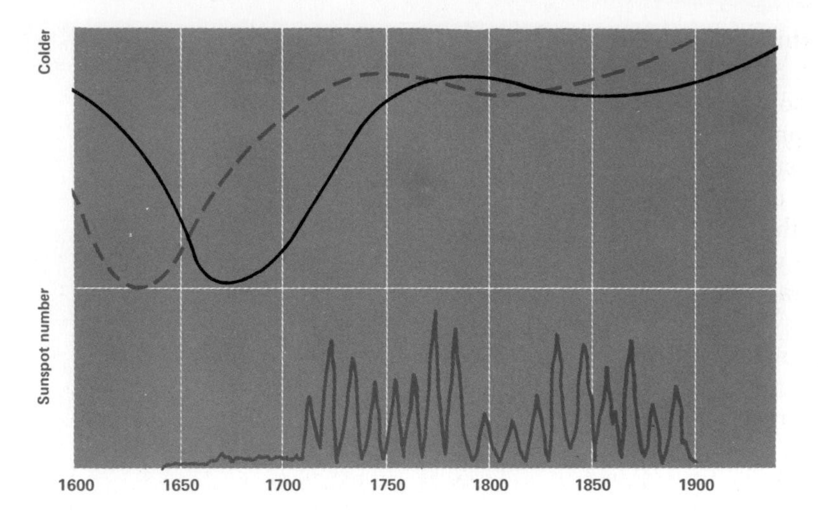


**FIGURE 10.22**
A plot of the relative sunspot number from 1600 to 1900 (bottom) and relative average temperature (top). The dashed temperature line is the observed line. This diagram follows a suggestion of correlation by John Eddy. Note the almost total lack of sunspots between 1650 and 1700.

Another example of the interaction of the sun and earth has to do with the earth's climate. Here we are referring to long-term heating and cooling cycles (Figure 10.22), not the short-term local weather patterns. Occasionally the eleven-year sunspot cycle disappears completely, and this disappearance somehow causes a cooling trend to set in on the earth. This was the situation from 1645 to 1715, when few or no sunspots were observed. Similar events have occurred over the last four thousand years. Cooling periods and warming periods are well documented in ancient history, coral growth records, tree ring records, and, until the mid-1800s, in radioactive carbon deposits. When sunspots were visible to the naked eye, the earth was invariably warmer than normal; when they were absent the earth was cooler than normal. Since the 19th century few records of naked eye sunspots have been kept because other records have superseded this method of observing. However, the records of the ancients, especially the Chinese, are invaluable in testing for long-term effects.

## On the Light from Background Stars

Relativity theory predicts that a photon emitted by a massive body like a star will lose some of its energy while escaping from that body. This is called the **Einstein,** or **gravitational, redshift.** The Einstein redshift should not be confused with the Doppler shift, which may be a red shift or a blue shift, presented in Chapter 4. Although the sun is a massive body by our standards, the Einstein redshift for the sun is very, very small and difficult to measure. Relativity theory also predicts that light which passes close to the sun from distant stars will be deflected or bent. Starlight just grazing the edge of the sun will be deflected by 1.75 seconds of arc, an easily measurable amount.

Experiments over the years have confirmed the predicted deflection of starlight. One technique used was to photograph the star field directly around the sun during a total eclipse. Then, six months later, when the same star field was visible at night, a second photograph was taken (Figure 10.23). When the positions of the stars on the photographs were measured, it could be seen that the sun had deflected the light of the distant stars.

Another technique is to study the positions of radio sources as the sun passes them in the sky. This technique makes basically the same type of measurements but has the advantage that the measurements are made entirely with the sun visible. Hence, there is no six-month wait involved, during which many parts of a telescope and the experiment can change. The radio technique, for this reason, is much more reliable than the photographic technique. It fully confirms the theory of relativity.

Because the sun is 275,000 times closer than the next nearest star, it is the only star we can really study in detail. Because of the sun's direct effect on the earth, it is only natural that it will continue to be studied carefully.

## SUMMARY

Partial, annular, and total solar eclipses can be predicted with great accuracy. Any particular eclipse is visible over only a small portion of the earth. During a total solar eclipse, the chromosphere and corona of the sun are visible.

Hydrogen fusion produces tremendous quantities of energy in the sun's core. Some of the energy is in the form of neutrinos, which reach the earth at the speed of light. The rest of the energy begins as gamma rays and gradually works its way to the photosphere of the sun, where we see it as visible light.

Sunspots are associated with the sun's magnetic field. They can be used to determine the rotation period of the sun. There is an eleven-year sunspot cycle from the maximum number through a minimum to the maximum number visible. Granulation, spicules, prominences, and

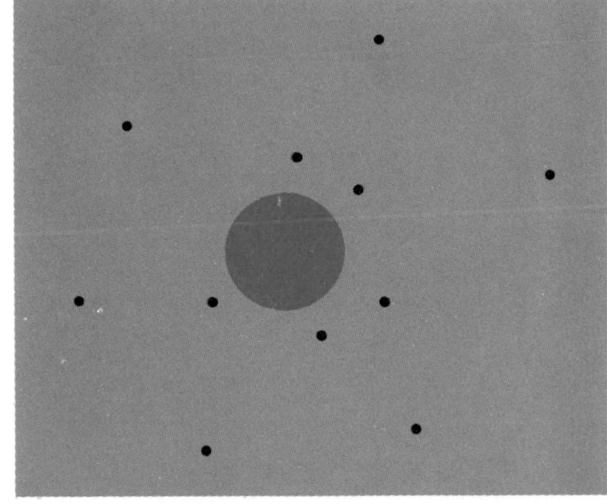

(a)

(b)

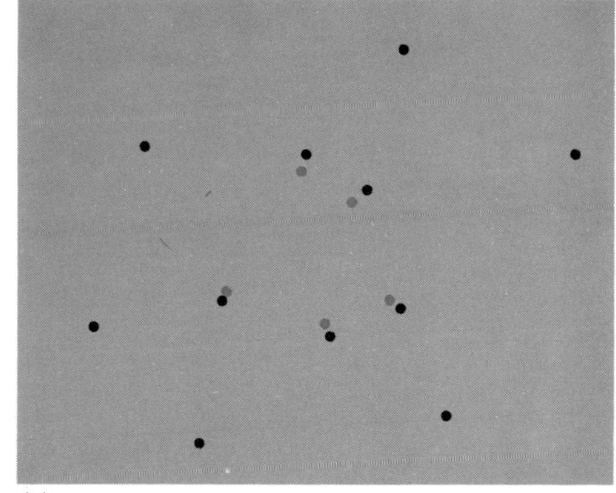

(c)

**FIGURE 10.23** A relativity experiment. The stars are photographed when the sun is in eclipse (a) and again six months later (b). They are then compared (c). Stars near the sun appear farther from the sun during the eclipse because the mass of the sun bends their light.

flares are all evidence of moving gases and activity in the sun.

The sun's radiation affects our climate, while the solar wind and solar flares produce various atmospheric effects on earth. In accordance with Einstein's relativity theory, the sun's strong gravity deflects starlight passing near it.

## *FALLACIES AND FANTASIES*

*The sun is the largest star.*
*The sun is an old star near the end of its life.*
*The sun burns coal to produce its energy.*

## REVIEW QUESTIONS

______ 1. Why do sunspots appear dark?

______ 2. How do we know the sun is gaseous throughout?

______ 3. What are the conditions for a total eclipse of the sun? When is this eclipse an annular eclipse?

______ 4. How do we know the sun rotates faster at its equator than at its poles?

______ 5. What makes the sun shine?

______ 6. How old is the sun?

______ 7. When should we expect the next maximum of the sunspot cycle?

______ 8. Describe the photosphere.

______ 9. What elements (principally) compose the sun?

______ 10. Describe a relation between the corona and the sunspot cycle.

# 11 Stellar Evolution: From Main Sequence Stars to Stellar Corpses



## LIFE CYCLES OF STARS

### Review and Perspective

To begin the story of stellar evolution with the sun as it is today would be to begin with middle age. First we must learn how stars are born and which stars are young ones and which are old. Since we cannot hope to study a star over its whole lifetime, we must draw conclusions in other

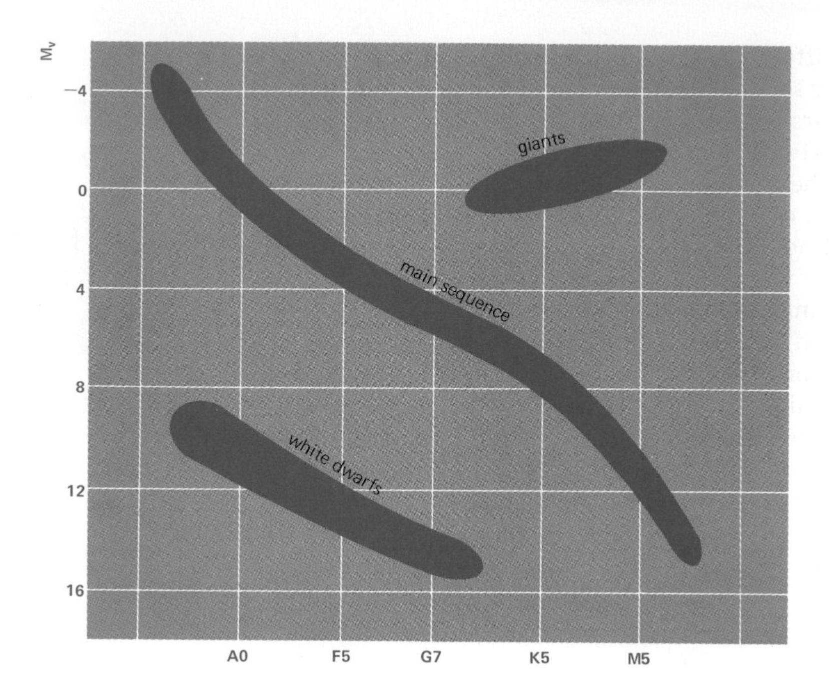


**FIGURE 11.1**
The standard H-R diagram showing the regions of the main sequence, giants, and white dwarfs. The main sequence is well defined because stars spend 90% of their lifetime as stable stars.

ways. Astronomers collect data as census takers on earth do. To find out about human life cycles, we do not have to wait seventy years. We merely need to examine a large number of people at a single point in time. Imagine looking at the population of a city. Count the number of babies, children, and adults. Weigh them, measure them, find their ages, and determine that age corresponds with time. Those statistics would provide a nice summary of the human life cycle. People start off young and small, and they grow rapidly to a certain size, where they remain for about fifty years. Then they die. The majority of people at any one time are adults in a stable weight-height condition. So it is with stars.

The sun has reached the long, stable part of its life. It is an "adult," or main sequence, star. It is with great difficulty that astronomers have been able to measure and weigh other stars and determine their temperatures, colors, and ages. In Chapter 4 we organized some of this information on the H-R diagram (Figure 11.1). Most of the stars are main sequence stars. Red giants and white dwarfs also appear on the H-R diagram. In this chapter, we will place some of the more unusual stars on this diagram. Basically, the H-R diagram can be used as a chart of the life cycle of stars. Young stars have certain spectral types and magnitudes that place them in one section of the diagram; old stars have different spectral type-magnitude relationships that place them in another section.

Stars spend about 90% of their lives as main sequence stars. How they arrive at this relatively stable stage of their lives and what happens when they die are the most intriguing parts of the story of the stars. In the case of people, heredity determines where we will fit into the pattern of adult characteristics such as height and weight. Heredity plus the way we live will determine how we die. In the case of stars, the initial amount of material—the mass— of the star will determine where it will fit into the main sequence and spend most of its life. Now that we understand how energy is produced in the stars (Chapter 10), we can predict what will happen as the nuclear fuel is used up. Our story of the life cycles of stars is a story of the delicate balance between nuclear energy and gravitational forces. Gravitational forces are forever, but when the nuclear fuel is used up, stars die.

## Young and Middle-Aged Stars

Earlier we noted that it takes billions of years to create a protostar. During its next million or so years, the protostar rapidly shrinks and glows by converting the energy of its inward-falling material to radiant energy. The energy comes from the collisions of the atoms, molecules, and dust making up the protostar. The central region of the protostar packs in tighter and tighter, and the material becomes hotter and hotter as it tries to resist the fall of the material above it. Finally, the temperature becomes sufficiently high to cause the proton-proton cycle to start. The ignition of hydrogen fusion sends a burst of high-energy radiation outward. It distends the protostar so much that the proton-proton cycle is stopped. Again the material presses downward, and again ignition happens. After several thousand years of youthful changing back and forth, the hydrogen fusion becomes self-sustaining, and the star is born. If the star is sufficiently massive, something like the sun or heavier, the carbon-nitrogen-oxygen cycle ignites and takes over in place of the proton-proton cycle.

Once the nuclear processes begin, the star searches for **equilibrium**—the right balance between the outward pressure of the hot gas and radiation and the inward pressure of the material above it caused by gravitational attraction. In finding this critical balance, the star flickers and flashes irregularly. We see this as irregular changes in the star's brightness. These are the same kinds of flashes that blew away the solar nebula when the sun and solar system were forming. Stars with this type of variation are called **T Tauri stars**, named after the the first such star observed. Now we find T Tauri variable stars in every region of space where we know stars are being born. On the H-R dia-

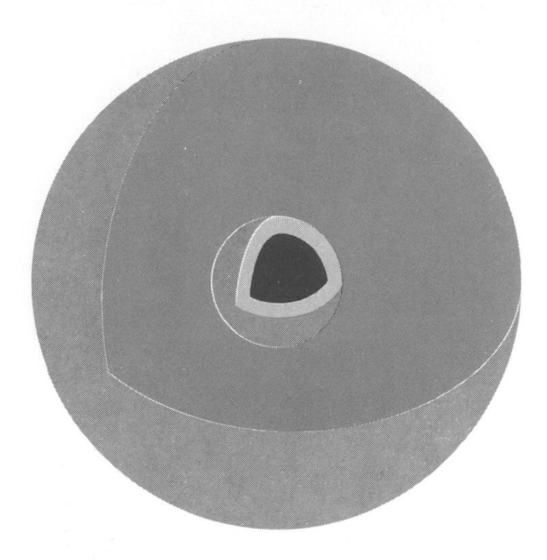

**FIGURE 11.2**
When hydrogen is used up in the core of a star, the hydrogen "burning" continues in a shell around the core. The core is primarily helium. If the star is massive enough, the helium eventually ignites and sustains the star.

gram, T Tauri variable stars are located slightly above the main sequence.

Finally, the star establishes equilibrium. It is no longer flickering as it did in the T Tauri stage. It is now a main sequence star. Where it is on the main sequence depends on its original mass. The most massive stars will be at the upper end—very hot, blue, and bright. The least massive stars will be at the lower end—very cool, red, and dim. The sun ended up about in the middle of the main sequence. The energy output of main sequence stars is only relatively constant. All stars are variable to some degree. Even the sun varies, but never more than 1% in its energy output.

The more massive a star is, the more quickly it will burn its fuel and reach the end of its life. A hot, blue star may remain in the main sequence stage for only a few million years. A yellow star like the sun will remain a main sequence star for about 10 billion years. A red star may remain a main sequence star for 100 billion years.

## Old Age for the Sun and Less Massive Stars

Ultimately, a star will use up all the hydrogen in its core. When this happens, the nuclear "burning" moves outward toward other layers in the star (Figure 11.2). The outer layers begin to expand and cool. The star becomes larger faster than it cools, so it becomes brighter. As the process of expanding and cooling continues, the star becomes a red giant. Becoming a red giant can take hundreds of millions of years. Stars in this phase are found in the upper right portion of the H-R diagram. When the sun becomes a red giant, it will probably be 150 times its current size (Figure 11.3).

After the red giant phase, a star less massive than the sun simply begins to contract until it becomes a white dwarf. In these stars, all nuclear energy has been used up, and the matter is packed tightly together. In order to picture a white dwarf, imagine packing all the sun's material into a star the size of the earth. The electrons in a white dwarf are stripped from their atoms and move among the atoms as a gas. The electron gas keeps the matter from collapsing even more. Since these stars cannot collapse any more, and since all the nuclear fuel is used up, they will gradually cool and will remain the same size. After 50 billion years, white dwarfs will be invisible cold cinders in space, called **black dwarfs.** Actually, the term white dwarf is a bit of a misnomer because such stars can also be yellow, orange, or red. On the H-R diagram, they are located to the lower left, well below the main sequence.

One of the most famous white dwarfs is a companion to Sirius, the bright star in Canis Major. Since Sirius is sometimes called the Dog

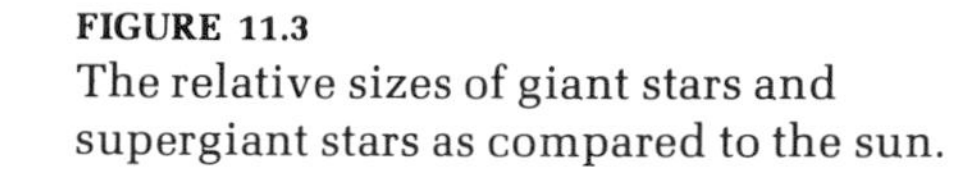


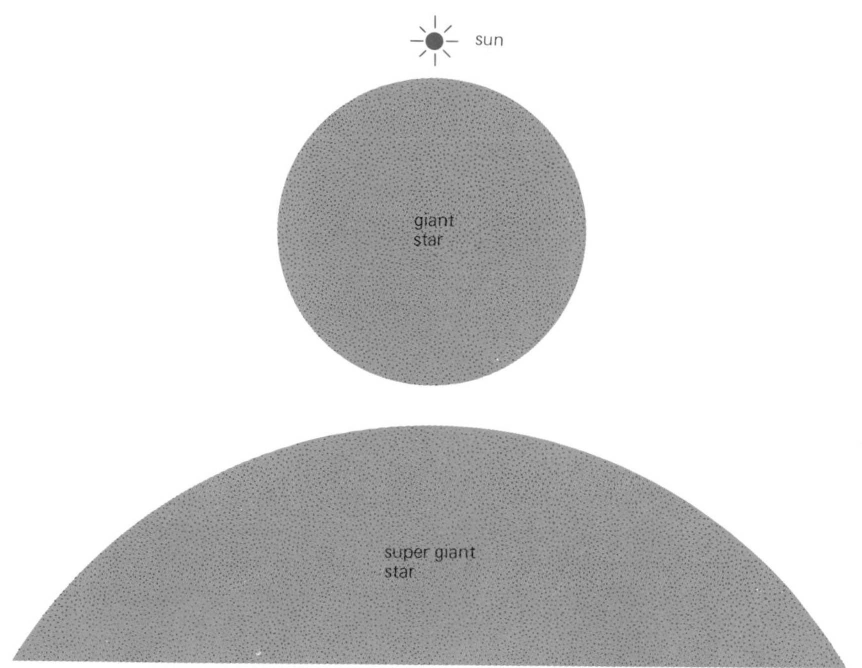


FIGURE 11.3
The relative sizes of giant stars and supergiant stars as compared to the sun.

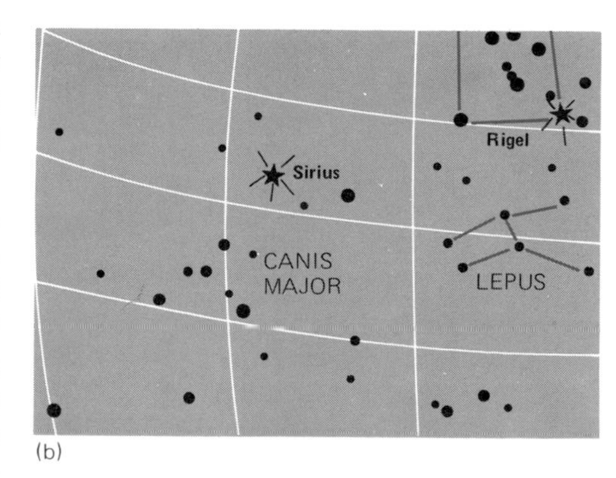


FIGURE 11.4
Sirius and its faint white dwarf companion (above). A special masking technique allows seeing the companion to the east of Sirius. (Sproul Observatory photograph) The chart below shows the position of Sirius in Canis Major.

Star, the white dwarf companion is known as the Pup. Ever since 1844, studies of the motion of Sirius indicated that it had a companion star in orbit around it. The Pup was first observed in a telescope in 1862 by Alvin Clark. At the time, Clark was using the image of bright Sirius to test the lens of a telescope he had just constructed. Figure 11.4 shows Sirius and its faint white dwarf companion.

Stars like the sun follow the same life-cycle stages up to red giants. If there is sufficient mass present as the outer layers of the stars expand, their cores will contract and the central temperatures may become hot enough so that the nuclear process fusing helium into carbon may begin. Successive nuclear reactions in these stars may ignite producing energy and at the same time building heavier and heavier elements from the lighter ones. The stars would expand rapidly as each new reaction sets in. Finally their outer atmospheres are pushed away from them, leaving the core of the star as a hot white dwarf. The outer atmosphere is now an expanding shell of gas. Stars in this phase of their life are called **planetary nebulae** (Figure 11.5 and color plate). Their name is derived from their appearance. In the earliest telescopes, which were not capable of high magnification, they looked like planets. We do not observe very many planetary nebulae in space, so this period of a star's life must not last very long. The shell of gas spreads out into space and dissipates. The white dwarf star gradually fades to a black dwarf.

### Old Age for Massive Stars

Stars more massive than the sun go through their life cycles much more quickly. In only a few million years, they become red giants. At this point, they will not collapse to a white dwarf; they follow a different course of development. The central temperature and pressure in these stars will be great enough for the helium to start fusing into carbon.

As this happens, the star becomes hotter and bluer. It enters an unstable period of its life as a pulsating giant star, alternately swelling and shrinking and changing its brightness by a noticeable amount. These pulsating variable stars are the cepheids we discussed in Chapter 5. Their periods are very regular, from several days to fifty days, depending directly on their average brightness.

As the core of a massive star is converted to carbon, the core contracts and heats up. The star becomes brighter, the outer layers expand and become cooler, and the star is a red giant again. Now the next nuclear reaction, carbon fusion, sets in. This begins a cycle of successive nuclear reactions and pulsations, ending only when the massive star is a red supergiant with a core made up of iron. Before these stars die, they must shed most of their mass. The next section describes some of the spectacular ways such stars lose mass.

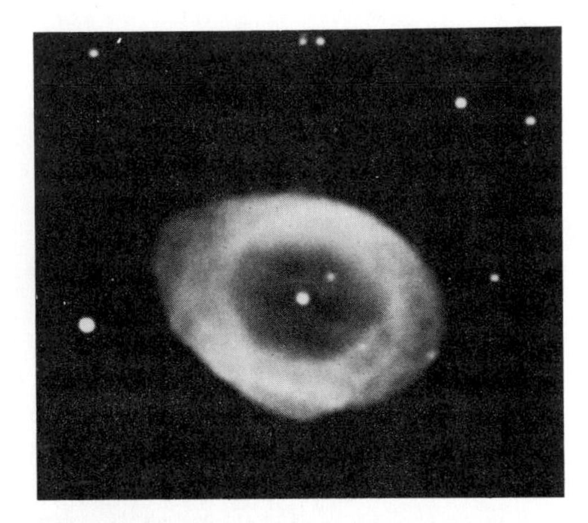

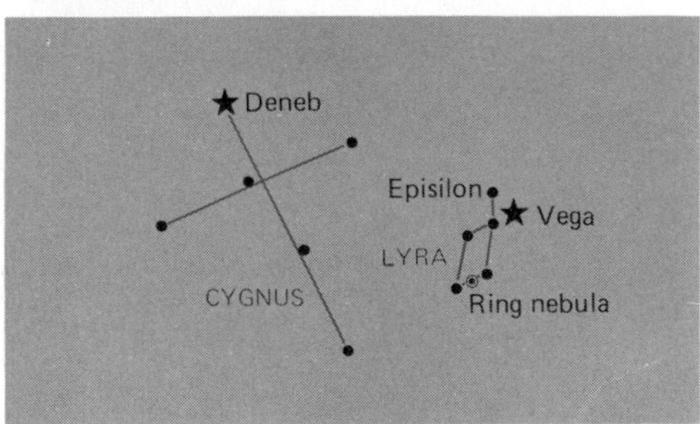


**FIGURE 11.5**
The Ring Nebula in Lyra. This is a beautiful object when viewed with a telescope. The chart below helps to locate this nebula in the sky. (Dominion Astrophysical Observatory photograph)

## WAYS OF SHEDDING MASS

### A Little Bit at a Time

Stars are always losing a little of their mass. When we described the sun, we indicated that the solar wind and solar flares send charged particles out into space. Other stars experience similar winds and flares. These events account for minute but constant mass losses. Another kind of mass loss we have described is when a star sheds a shell of gas forming a planetary nebula. In these ways a medium-sized star may lose just enough mass to become a normal white dwarf star. (A normal white dwarf can never have more than about 1.5 times the sun's mass.) But even this is not enough loss for the heaviest stars.

Certain massive stars make more violent adjustments in their old age. These stars are called **novae,** or new stars. The word "new" is a little misleading. Novae are only new in the sense that they brighten so much that often they become visible where no star was seen before. A star that becomes a nova undergoes an explosion. When an excessive amount of energy is produced below the surface of the star, an outer layer can be ejected. The gas that is ejected is sent out into space at tremendous speeds, over 1000 km/sec (Figure 11.6). The gas disperses into space and is no longer visible after several years. The high speeds and quick dispersal of gases make the nova events different from those of planetary nebulae. A star undergoing a nova event loses less than 1% of its mass.

Novae occur quite frequently. There are probably forty or so such events in the Milky Way galaxy each year, although we actually observe only a few of them. The remainder are hidden from our view by the great clouds of gas and dust in the galaxy. The stars that become novae eventually become white dwarfs and gradually fade to black dwarfs.

Some stars have been observed to erupt more than once during their lifetimes. These recurrent novae are almost always part of a binary star system.

## Everything at Once: Supernova

Great stellar suicides, called **supernovae,** occur when a star explodes and sends almost all its material out into space. The explosion probably comes about in the following way. The life cycles of the most massive stars proceed until the core is made up of iron. The temperature in such a core is several billions of degrees. Electron gas pressure in the core supports the weight of the upper layers of the star. The next nuclear reaction to occur uses up energy rather than producing it. This tips the balance so that the downward pressure becomes dominant. When the downward pressure in the core becomes too great, the electrons penetrate the iron atoms and no longer support the weight of the star. The core of the star collapses, and the outer layers collapse on top of that. The lighter atoms from above—carbon, oxygen, helium, and hydrogen—undergo furious nuclear reactions when they encounter the high temperatures near the core. The event is so violent that the outer regions of the star are completely blown away, as are most of the heavier elements in its interior. The star material shoots out into space at speeds up to 7000 km/sec (Figure 11.7). The entire supernova explosion takes only minutes.

**FIGURE 11.6**
Material ejected by Nova Persei 1901 can be seen in this photograph. (Hale Observatories photograph)

When a supernova explosion occurs, the star becomes many millions of times more luminous than the sun. Because of this tremendous brightening, it can be seen over great distances. Supernova events are not as frequent as novae. In the Milky Way galaxy, they occur once every twenty-five to fifty years. Six or seven outbursts have been seen

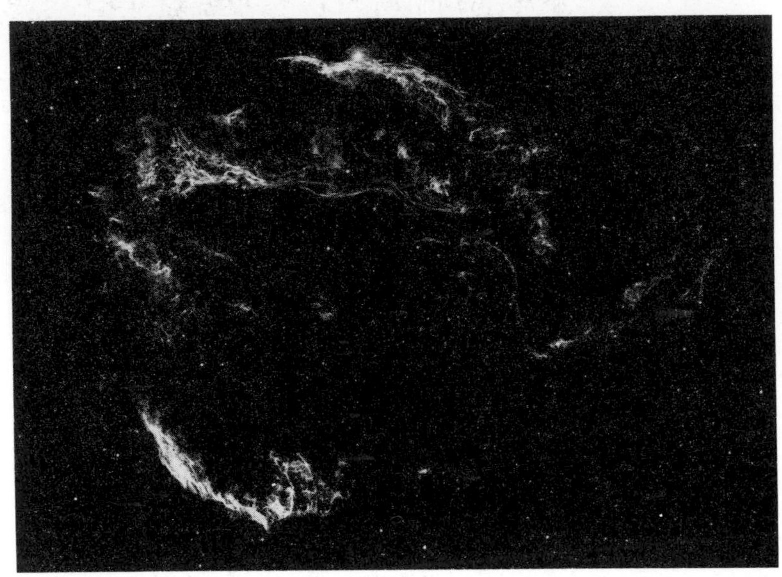

**FIGURE 11.7**
Filamentary nebula in Cygnus referred to as the Loop nebula. This is the remnant of a supernova event that occurred 70,000 years ago. (Hale Observatories photograph)

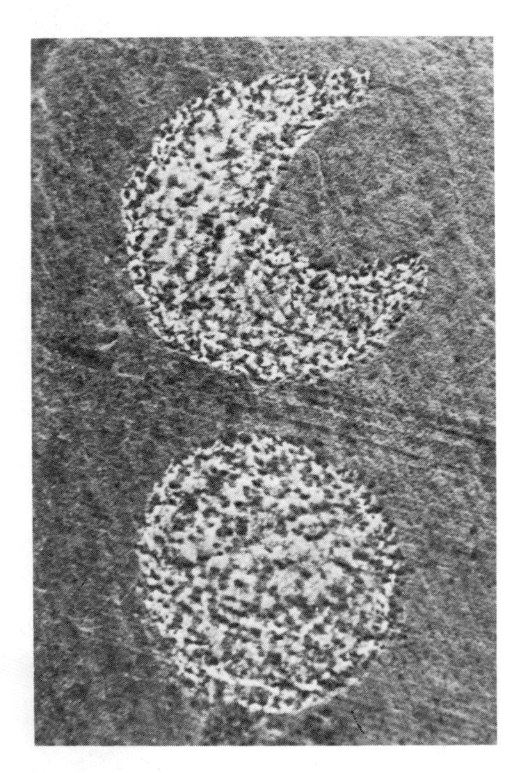

**FIGURE 11.8**
A pictograph carved into the north wall of Navajo Canyon. The pictograph reproduces the configuration of the moon and the Crab supernova event as seen on 5 July 1054 quite accurately. (Museum of Northern Arizona photograph, courtsey of W. C. Miller)

during the past two thousand years. The most famous supernova is probably the one that occurred in Taurus in 1054. It is recorded in Chinese records and on cave walls by North American Indians (Figure 11.8). This supernova became as bright as Jupiter and remained visible for two years. European observers either missed this supernova or left no records of it.

The explosion left a large nebula of gas and dust in space called the Crab Nebula (Figure 11.9 and color plate). The Crab Nebula is expanding at the rate of 1100 km/sec away from its center. Almost every kind of radiation imaginable—visible light, radio waves, x-rays, gamma rays, and so on—is being emitted from the Crab Nebula. It is an excellent astrophysical laboratory for studying high-energy radiation.

When supernovae occur in other galaxies, they are often as bright as the entire galaxy they occur in (Figure 11.10). In other galaxies, we see three or four supernovae per century per galaxy. Curiously enough, supernovae all achieve about the same absolute brightness at maximum light. If all supernovae have nearly the same absolute magnitude at maximum, we can measure their apparent magnitudes and use those values to calculate distances to the galaxies.

Although a supernova is the end of the life of one star, it may represent the beginning of other stars. As the expanding remnant nebula of a supernova spreads out and mixes with other material in space, it may become the trigger for protostars to form. A supernova is also a most efficient way of recycling star material. The sun and planets, with their heavy elements, are certainly offspring of earlier generations of stars. Every heavy element and molecule in our bodies must have been produced in the interiors of much earlier massive stars.

After a supernova explosion, there is something left besides a nebula. The very core of the star that exploded remains. During the explosion, the core suffers a fate that changes it into a far more unusual star than a white dwarf.

## REMAINS OF SUPERNOVAE

### Neutron Stars

When the core of a massive star begins to collapse, triggering the supernova event, the violence of the collapse is so great that electrons combine with protons inside the iron nuclei and turn into neutrons.

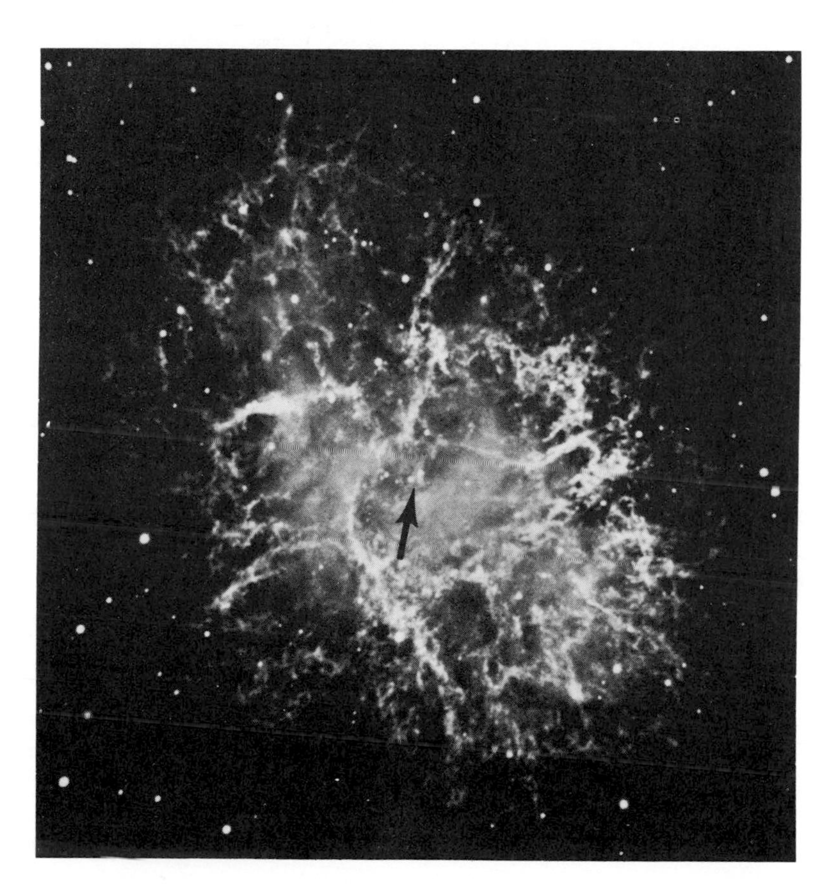

**FIGURE 11.9**
The Crab Nebula in Taurus. The remnants of the 1054 supernova is now this strange filamentary structure with a pulsar (marked by the arrow) at the center. Eventually it may look like the Vela Nebula—Figure 11.13. (Hale Observatories photograph)

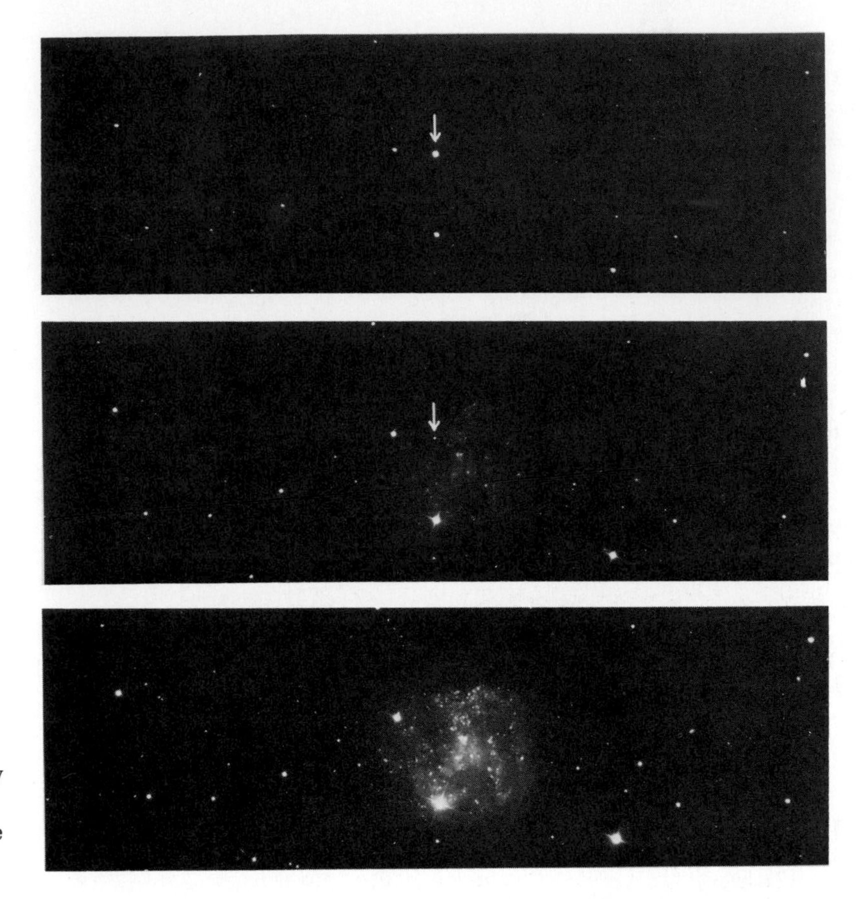

**FIGURE 11.10**
A supernova in a galaxy known only as 1C4182. The supernova was brighter than the whole galaxy—top. The galaxy is only visible in a very long exposure—bottom. By the time the bottom photograph was made the supernova had faded to obscurity. (Hale Observatories photograph)

Soon the entire core is composed of neutrons and is incompressable. The entire mass of the core, perhaps twice the mass of the sun, is now a sphere of neutrons only 10 or 15 km in diameter. One teaspoon of this matter brought to earth would weigh some 100 million metric tons (a metric ton is slightly heavier than 2000 pounds). Such a **neutron star** is at the center of the Crab Nebula.

Because visible radiation from neutron stars is very feeble, they are hard to see. There are, however, two characteristics of neutron stars that allow them to be detected. First, when the core collapses it carries the star's magnetic field with it. A weak magnetic field around a normal star of some 2 million km in diameter becomes very compressed and intense when the star is shrunk down to 10 km. The magnetic field is frozen into the star and turns with it. Second, the spin of the star is preserved. As the star collapses, the rate of spin increases. This is like the ice skater who spins faster when the arms are drawn in toward the body. A star that shrinks to a 10 km diameter must spin rapidly indeed. Depending on the initial conditions, a neutron star might spin 300 times per second.

These two characteristics, an intense magnetic field and rapid spinning, combine to produce radiation. The radiation is produced by electrons trapped in the magnetic field and moving near the speed of light. This kind of radiation is called **synchrotron radiation** because it was first seen in atomic accelerators called synchrotron machines.

The radiation region around the neutron star is a disk, and the energy is emitted only in the plane of that disk. Suppose a neutron star has a spin axis in the plane of the sky and has its radiation disk fixed to it at an angle (Figure 11.11). The edge of the disk from which the energy is being radiated will cross our line of sight twice during each spin of the star. Whenever it crosses our line of sight, we will see a flash of radiation. This model of a neutron star is referred to as an **oblique rotator model.** A neutron star could be oriented so that the radiating disk will never cross our line of sight, and we will not see the star. But in cases where the disk does cross our line of sight, rapid pulses of radiation should reveal the presence of a neutron star.

Although astronomers had concluded that neutron stars should exist, they did not expect to find one. The discovery of neutron stars came quite by accident. In 1968, while investigating radio waves from distant galaxies, a rapidly pulsating source was discovered in our own galaxy. Shortly afterward, another pulsating source was discovered, this time in the Crab Nebula. This second one was pulsing 30 times each second with incredible accuracy (Figure 11.12). The very distinct, sharp pulses of radiation observed from the spinning neutron stars suggested their other name—**pulsars.**

Several hundred pulsars are known, but only a few are near enough to be observed at all wavelengths of the spectrum. Most are beyond the nearby gas and dust and hence invisible at visible wavelengths. Only two pulsars have been seen in visible light—the Crab Nebula pulsar and the Vela pulsar, each of which is at the center of a supernova remnant (Figure 11.13). The Vela pulsar, however, is pulsing about 12 times a second, which is slower than the Crab pulsar. There is some indication that pulsars slow down as they age. Since the Crab pulsar is spinning so rapidly, it is one of the most recently formed pulsars. We observed its creation in 1054, which completely supports the theory.

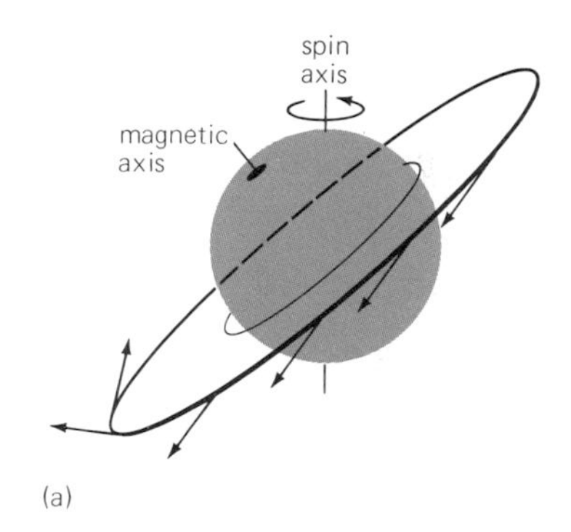


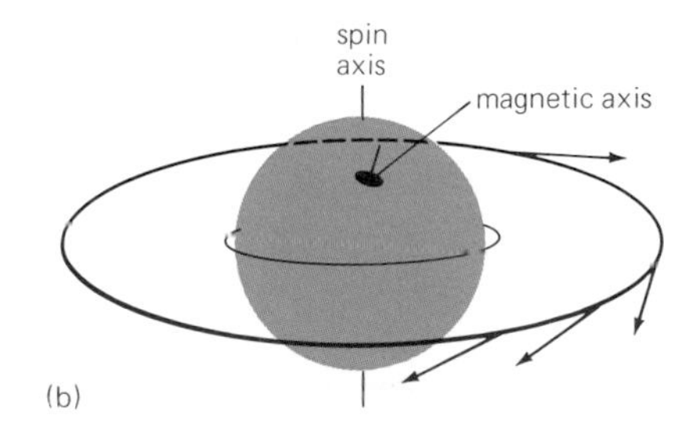


**FIGURE 11.11**
Radiation from electrons in the disk around a pulsar is emitted in the plane of the disk. The tilted magnetic axis causes us to see the radiation twice in each rotation.

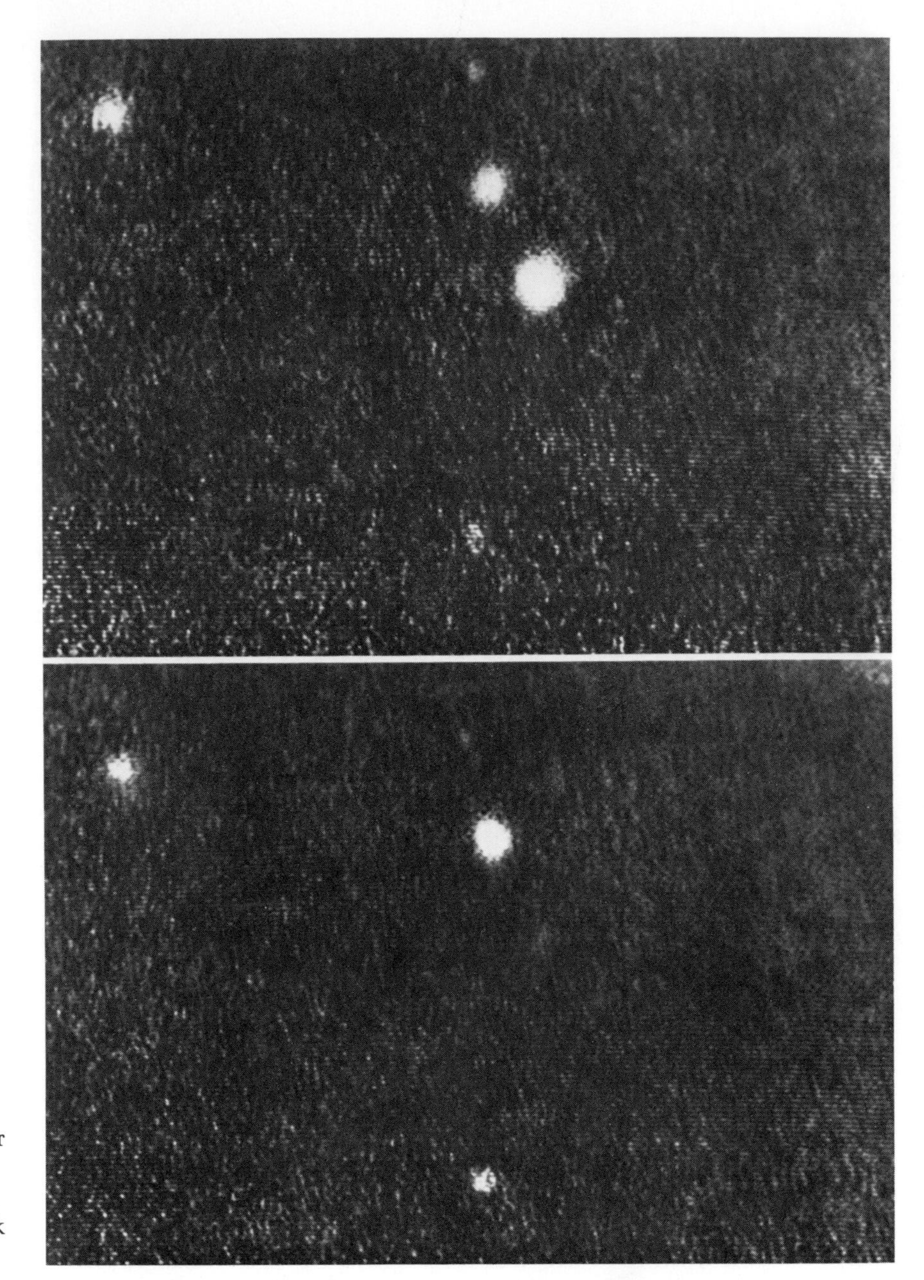

**FIGURE 11.12**
Two photographs of the Crab Nebula pulsar taken using a special technique. In the top frame the pulsar is on. In the bottom frame, one-thirtieth of a second later, it is off. (Lick Observatory photograph)

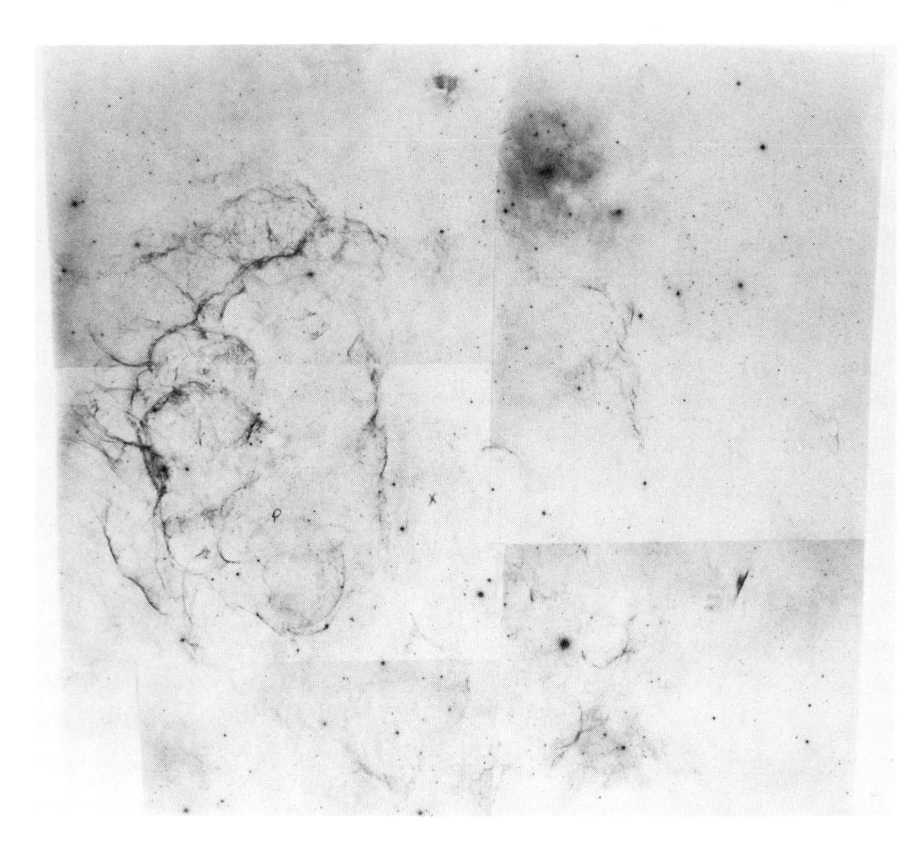

**FIGURE 11.13**
This negative print shows the supernova remnant in Vela. The filamentary structure covers 5° of the sky. The pulsar is marked with a P. (Cerro Tololo Inter American Observatory photograph)

## Black Holes

Scientific theory predicts even more exotic objects than neutron stars. Suppose a massive star evolves, and the core implodes with such violence that it collapses beyond the point where a neutron star is stable. The matter crushes down completely. The stellar corpse collapses until it occupies a space too small to imagine and has a density too great to imagine. Mathematicians would call the object a singularity. The star's gravity becomes so strong that anything too near will be captured and will never escape. Even photons of light traveling at the speed of light could not escape from such a star. The area of space around this bizarre object from which there is no returning is called a **black hole**. The name comes from the fact that no light will ever leave a black hole, and hence we cannot hope to see one directly.

The only way we can detect a black hole is by watching for its influence on other objects in space. For example, suppose some material

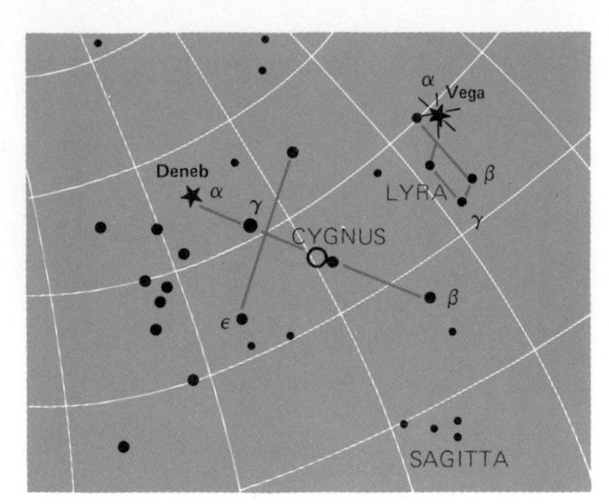


**FIGURE 11.14**
The location of Cygnus X-1 on a star map (marked by an O).

ventures near a black hole, is captured, and disappears. While the material is falling toward the black hole it will be heated to high temperatures, and collisions of atoms will occur. The high-energy, violent collisions in the material will produce x-rays which we should be able to detect. Earth-orbiting satellites have discovered some strong sources of x-rays from certain points in space. Many of these are suspected black holes. One x-ray source, called Cygnus X-1, is thought to be the first confirmed black hole, although many astronomers are cautious on this point (Figure 11.14).

Cygnus X-1 is not only an x-ray source; it is also part of a binary star system. The x-rays appear and disappear with great regularity. This indicates that the two stars (or the star and the black hole) are eclipsing one another. Analyzing the light from the eclipsing binary system tells us that Cygnus X-1 is extremely small and extraordinarily assive. It has all the characteristics that a black hole would have ac-ırding to the theory.

To see how eclipsing binary star systems can give us so much valuable information, look at Figure 11.15, where we have drawn the situation for two stars. One star is much larger than the other. The orbits are aligned so that when the small star goes behind the large one, the small star is totally eclipsed. As we go from position 1 to position 2, the small star is covered up bit by bit, and the light declines from the combined light of both stars to just the light of the large star. Between positions 3 and 4 the picture reverses, and the total light increases until we see the combined light of both stars again.

Astronomers measure the time it takes to go between positions 1, 2, 3, and 4 by measuring the total brightness of the stars at x-ray wavelengths or radio wavelengths. Knowing how fast the star disappears and how long it takes to cross behind the large star tells us the relative sizes of the two stars. If the stars are nearly the same size, it will take a longer time for one star to disappear than it will take to cross behind the other star. If the stars are very different in size, the smaller star will disappear quickly and remain hidden a long time. From just such an analysis, we learn that Cygnus X-1 is exceedingly small.

Since binary stars obey Kepler's laws, we can also determine the masses of the two stars. The procedure is like the one we described in

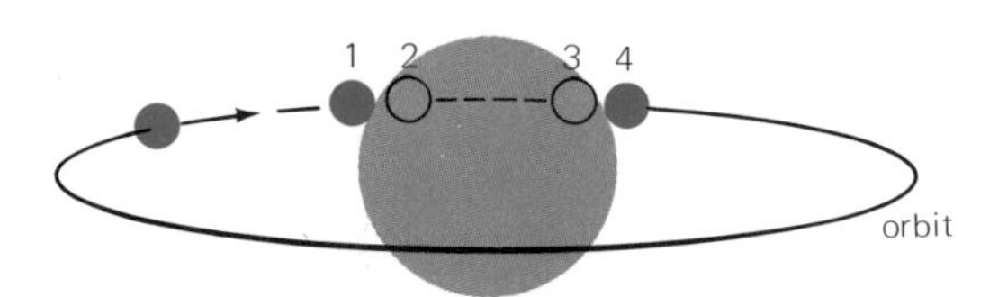


**FIGURE 11.15**
A schematic drawing of a binary star. A star that disappears behind its companion and takes a long time to reappear must be very small compared to its companion.

Chapter 9 to determine the masses of planets with satellites. For Cygnus X-1, the mass is determined to be very great.

The model of Cygnus X-1 that emerges is this (Figure 11.16): one star is a blue supergiant; the other is a black hole. The gravity of the black hole is so strong that it produces a tidal effect which actually distorts the large visible star. Gases from the outer atmosphere of the supergiant are pulled into a swirling disk around the black hole. As this gas spirals in toward the hole, it produces the x-rays we observe. All the while, the two stars are orbiting around each other.

The two questions most often asked about black holes are these: What happens if I get too close to a black hole? Why don't black holes consume everything in the universe? The answer to the first question is that if anything gets too close to a black hole, it will be pulled in and destroyed. In the particular case of a person falling in feet first, the body would be pulled apart lengthwise since the pull on the feet would be greater than the pull on the head. Eventually the strong gravity would pull limb from limb and atom from atom until the person was an unrecognizable mass of minute particles. Actually, everything falling into a black hole loses its identity. It is possible, however, to pass by a black hole if we keep a safe distance.

Presumably, black holes get more massive over time as they pull in material around them. Their gravitational attraction will increase, and they will grow larger. This leads to the next question. Will they eventually consume everything? The answer is no. Distances in space are so vast, and black holes are so small, that rarely would any large amount of material suffer the fate of getting too near a black hole.

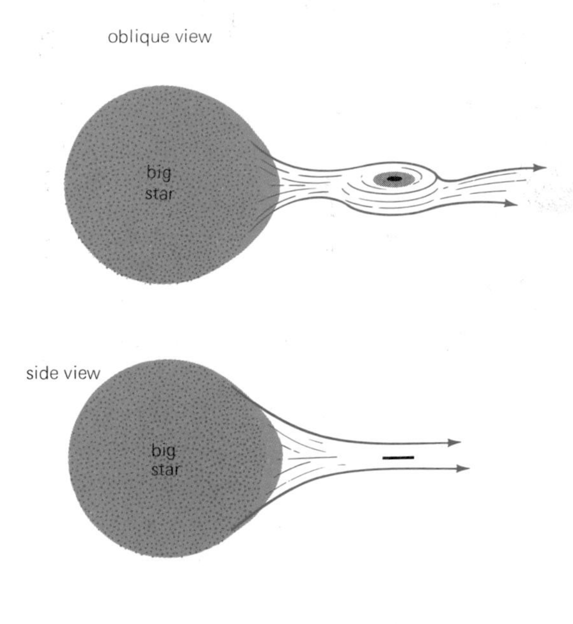


**FIGURE 11.16**
A schematic drawing of a black hole and its companion star. X-rays emitted from the hot disk surrounding the black hole reveal the presence of the black hole.

## OTHER VARIABLE STARS

### RR Lyrae and A Type Variables

So far in this chapter we have mentioned only four types of variable stars—T Tauri stars, cepheid variables, nova and supernova variables, and pulsars. There are some thirty classes of variable stars that we have not mentioned. We do not understand where many of these variable stars fit into the life cycles of stars. Many of these classes are shown on the H-R diagram in Figure 11.17.

Most notable among these other variable stars is a large group of spectral type A stars called RR Lyrae variables. Their distinguishing feature is a periodic light variation ranging from nine hours to one day. No matter what the period of the light variation, all RR Lyrae stars have the same average absolute magnitude. This makes them excellent distance indicators, as we noted in Chapter 5.

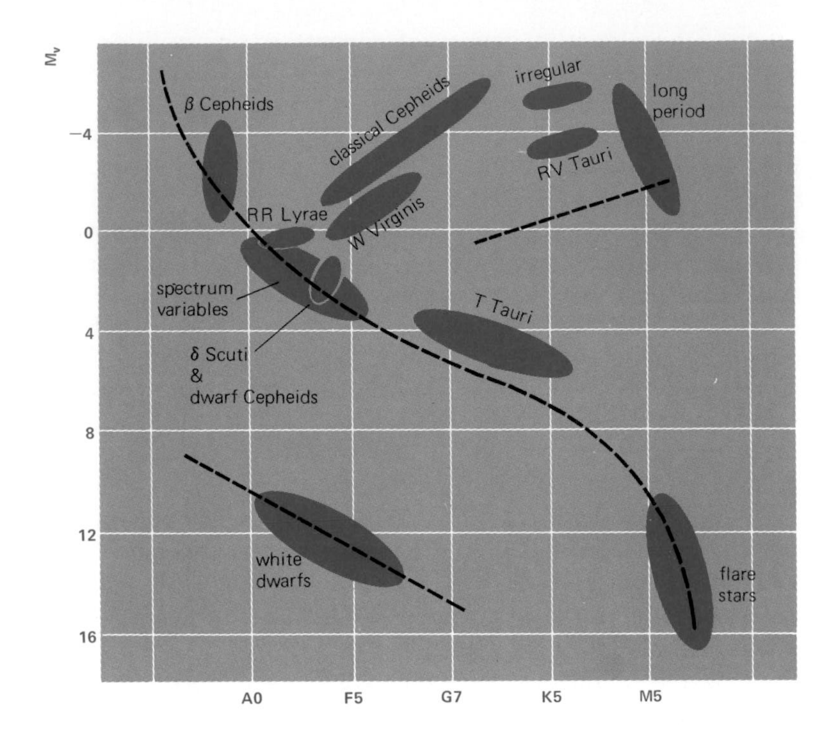


**FIGURE 11.17**
Various major classes of variable stars are located on the H-R diagram. Most types of variable stars lie off of the main sequence indicating that they have undergone significant evolution.

Many other stars of spectral type A show variations in one way or another. Some A stars have unusually strong lines of heavy elements appearing in their spectra. These lines vary in strength even though the overall light of the star does not vary. We call them **spectrum variable stars.** Other A stars have strong magnetic fields that show variations. These are the **magnetic A stars.** Other A stars, called **dwarf cepheids,** have light variations like the cepheids but with periods as short as 55 minutes.

## Dwarf Novae and Long-Period Variables

There is a large group of stars called **dwarf novae.** These stars have novalike outbursts several times each year. The outbursts are not nearly so violent as a normal nova event, but in other respects they look like a nova. As with many variable stars, we do not understand which stars become dwarf novae or where dwarf novae fit into the life cycle of stars.

The **long-period variable stars** are red giant stars with periods ranging from 100 to 500 days. Their range in brightness is quite dramatic in visual light. Mira, a long-period variable star in the constellation Cetus, has a brightness range of eight magnitudes. At its brightest, it is a star of the second magnitude; whereas 166 days later, at its faintest, it has dropped completely out of sight for naked eye observing (Figure 11.18).

In general, it appears that the regular variability of stars such as cepheids and RR Lyrae variables is connected with a definite stage in the life cycle of stars. The irregular variations, such as dwarf novae and long-period variables exhibit, appear to be associated with the needs of certain stars to lose mass. It is almost as if stars more massive than the sun know that they must lose mass and try various mechanisms to do so.

## Eclipsing and Spectroscopic Binary Stars

When two stars are orbiting each other, and they are aligned so that we see one star eclipse the other, there is a noticeable change in the spectral lines and in the light level we observe. Such binary stars are variable stars of a different nature. The spectrum or the light of the system changes due to the geometry of the orbit rather than to any property of the stars themselves. We merely mention these orbiting star systems here since they are occasionally referred to as variables. The true importance of binary stars is that careful studies of their orbital motions and light changes yields valuable information about the physical parameters of the individual stars, especially size and mass.

## SUMMARY

The mass of a star determines what type of main sequence star it will become and the speed and course of its life cycle. Stars spend most of their lives in the main sequence stage, when they are producing energy by hydrogen fusion processes. The life story of stars is one of the balance between the force of gravity, which pulls the material downward, and the outward push of radiation.

When stars less massive than the sun have used up the hydrogen fuel in their cores, the balance is tipped, and the next stage is a red giant. They eventually contract and become white dwarfs. After billions of years, a white dwarf will become a black dwarf.

Stars like the sun, or those that are more massive, will become red giants more quickly than less massive stars. These red giants will end their lives as white dwarfs, neutron stars, or black holes. Between the red giant stage and the end, many interesting events can take place. The temperature at their centers will increase, allowing the fusion of helium to carbon, carbon to oxygen, and so on, possibly up to iron. Stars may undergo variations in size and brightness as they adjust to

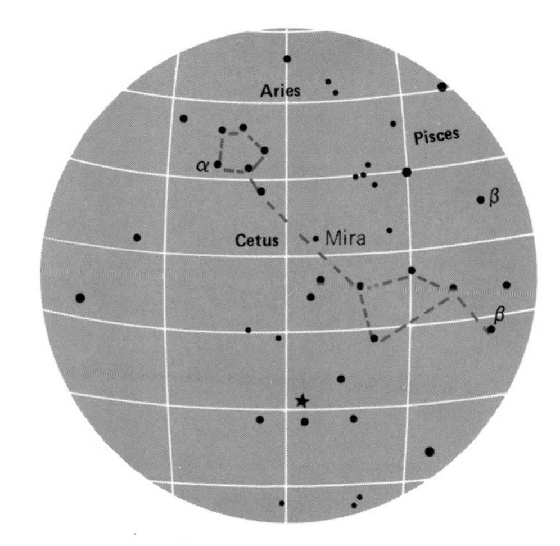


**FIGURE 11.18**
The location of Mira on a star map. Mira is usually easily visible at its maximum and drops out of sight at its minimum.

each new energy process. Many kinds of observed variable stars are thought to be in these transitional stages.

As they age, the most massive stars must also shed some of their mass. Mass loss happens in various ways, from small ejections to catastrophic explosions of matter. Only the most massive stars are thought to become supernovae, returning vast quantities of material to the interstellar medium. The remnant star of a supernova is likely to be a neutron star or possibly a black hole. Black holes can be detected only by their gravitational influence on nearby matter and x-ray radiation emitted by infalling material.

### *FALLACIES AND FANTASIES*

*Stars only live to be about a hundred years old.*
*Red giants are the hottest stars known.*
*White dwarfs are young stars.*
*The sun is a young star.*

## REVIEW QUESTIONS

_____ 1. How does the lifetime of a star depend on its mass?

_____ 2. Distinguish between T Tauri, cepheid, and long-period variable stars.

_____ 3. Describe the next stage in the life of the sun. When will it happen? What will be the consequences for the earth?

_____ 4. Discuss the likelihood of the sun's becoming a supernova.

_____ 5. Why are eclipsing binary stars of great interest to astronomers?

_____ 6. What are pulsars?

_____ 7. What are planetary nebulae?

_____ 8. List the following stars in order of age from youngest to oldest: white dwarf, main sequence, red giant, T Tauri, protostar, nova.

_____ 9. Why are stars main sequence stars for most of their lives?

_____ 10. Are variable stars generally more or less luminous than main sequence stars?

# 12 Galaxies: From the Local Group to the Edge of the Universe



## BACKGROUND

### The Nature of the Galaxies

Scientific discoveries and progress fundamentally alter the way we think and look at our world. Since 1969, children have been growing up knowing that we have the ability to travel to the moon and back, to set up space stations, and to send spacecraft to visit other planets. Most of us have grown up knowing that the Milky Way galaxy is one of billions of other galaxies that make up the Universe. This scientific fact, which we take for granted, has been known for less than sixty years. The study of the galaxies is still in its infancy.

Galaxies were observed with many of the great telescopes of the 18th and 19th centuries. They were called extragalactic nebulae in those times. We have already seen Lord Rosse's sketches of the spiral extragalactic nebulae in Chapter 3, Figure 3.13. As early as the mid-18th century, Immanuel Kant suggested that these nebulae were other "island universes" far beyond the Milky Way. Finally, at the beginning of the 20th century, individual stars in the extragalactic nebulae could be studied in detail, and astronomers were developing new methods to measure vast distances through space. In 1913, using a 61 cm telescope, V. M. Slipher made some marvelous observations of the Doppler shifts of the extragalactic nebulae. He found that most of them were moving away from us.

The closer astronomers came to understanding the extragalactic nebulae, the greater the controversy over their nature. Were they large, distant island universes like the Milky Way galaxy, or were they small, nearby nebulae among the stars? The peak of activity concerning this question occurred during the 1920s. During those years, a debate on the nature of the galaxies received national attention in the United States.

The single observation which was proof that the extragalactic nebulae were really distant island universes like the Milky Way was made by Edwin Hubble. In 1923, Hubble discovered a cepheid variable star in M31, the Andromeda nebula. Using the distance relationship for cepheids, Hubble found that the Andromeda nebula was nearly a million light years away. (The modern value of this distance is 2.2 million lt-yr.) That was it. The Andromeda nebula became the Andromeda galaxy. Henceforth the term "nebula" would be reserved for great clouds of gas and dust; "galaxy" would mean an independent assemblage of billions and billions of stars. Studies of the island universes or galaxies intensified.

## Classification of Galaxies

Any casual study of galaxies shows that they can be divided into three broad classes based on their appearance: (1) elliptical galaxies, (2) spiral galaxies, and (3) irregular galaxies. The classification scheme used most often was developed by Hubble in the 1920s. A sketch of his galaxy types appears in Figure 12.1. The characteristics he used to classify the galaxies are described below.

*Elliptical Galaxies.* Great numbers of stars, usually without gas and dust mixed in, constitute **elliptical galaxies.** The stars are concentrated toward the center of the galaxy and thin out toward the edge. Elliptical galaxies contain very few young stars. The shape of these galaxies may be anything from a spheroid to a highly flattened ellipsoid. Hubble divided the class into seven groups, E0 to E7. The spheroidal galaxies are designated type E0, and the most flattened ellipsoidal galaxies are designated E7. All the other elliptical galaxies are distributed in between, depending on the degree of flattening. Figure 12.2 is a photograph of an elliptical galaxy. Elliptical galaxies include the largest and the smallest galaxies known. A letter designation "g" for giant and "d" for dwarf sometimes precedes the classification of elliptical galaxies. For example, dE0 means a dwarf elliptical galaxy, spheriodal in shape. The class E7 has been dropped because all E7 galaxies have proved to be S0 spiral galaxies.

*Spiral Galaxies.* **Spiral galaxies** generally have two spiral arms containing stars and gas and dust, wrapped around a bright central region of similar makeup. The central, lens-shaped region is called the

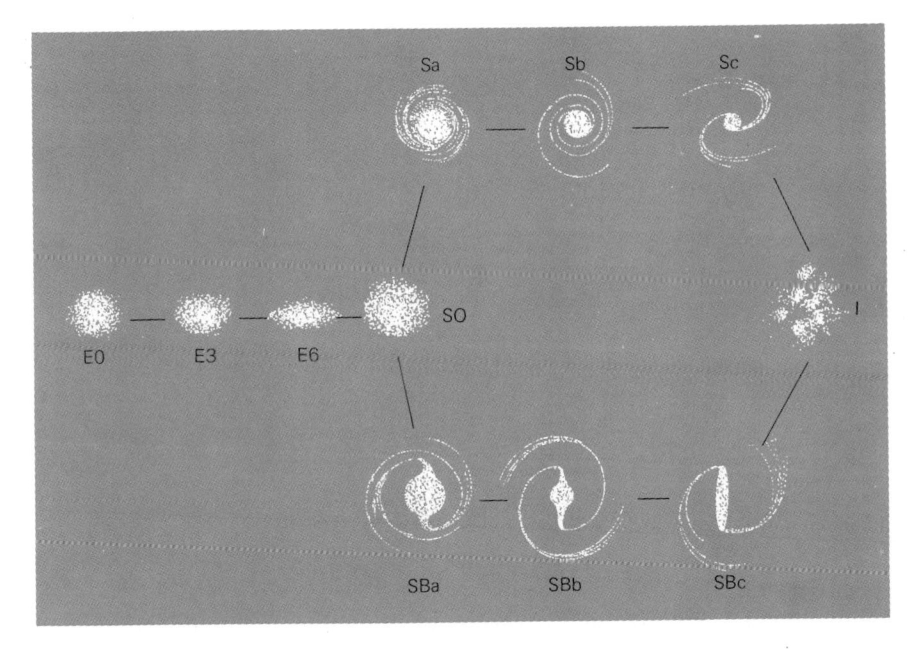


**FIGURE 12.1**
The sequence of types of galaxies following the original classification scheme of E. Hubble.

**FIGURE 12.2**
The elliptical galaxy NGC 147 of type E4. On the original photograph the outer portion of this galaxy is resolved into stars. Individual stars visible in this print are all part of our own galaxy, in front of NGC 147. (Hale Observatories photograph)

nucleus. Stars making up the arms are generally young; those in the nucleus are older.

Since a noticeable bar structure runs through the nucleus of many spiral galaxies, these galaxies are subdivided into two main types, normal spirals designated S, and barred spirals designated SB. The S and SB classes are broken down further depending on the smoothness of the arms and the size of the nucleus. A large nucleus and well-rounded arms is called an Sa or SBa galaxy. A smaller nucleus with less well-defined arms is called an Sb or SBb galaxy. An ill-defined nucleus with open, disjointed arms is called an Sc or SBc galaxy. The Milky Way galaxy belongs to the Sc class. Figure 12.3 shows a sequence of photographs of spiral galaxies with their Hubble classifications. Several spiral galaxies are shown on the color plates.

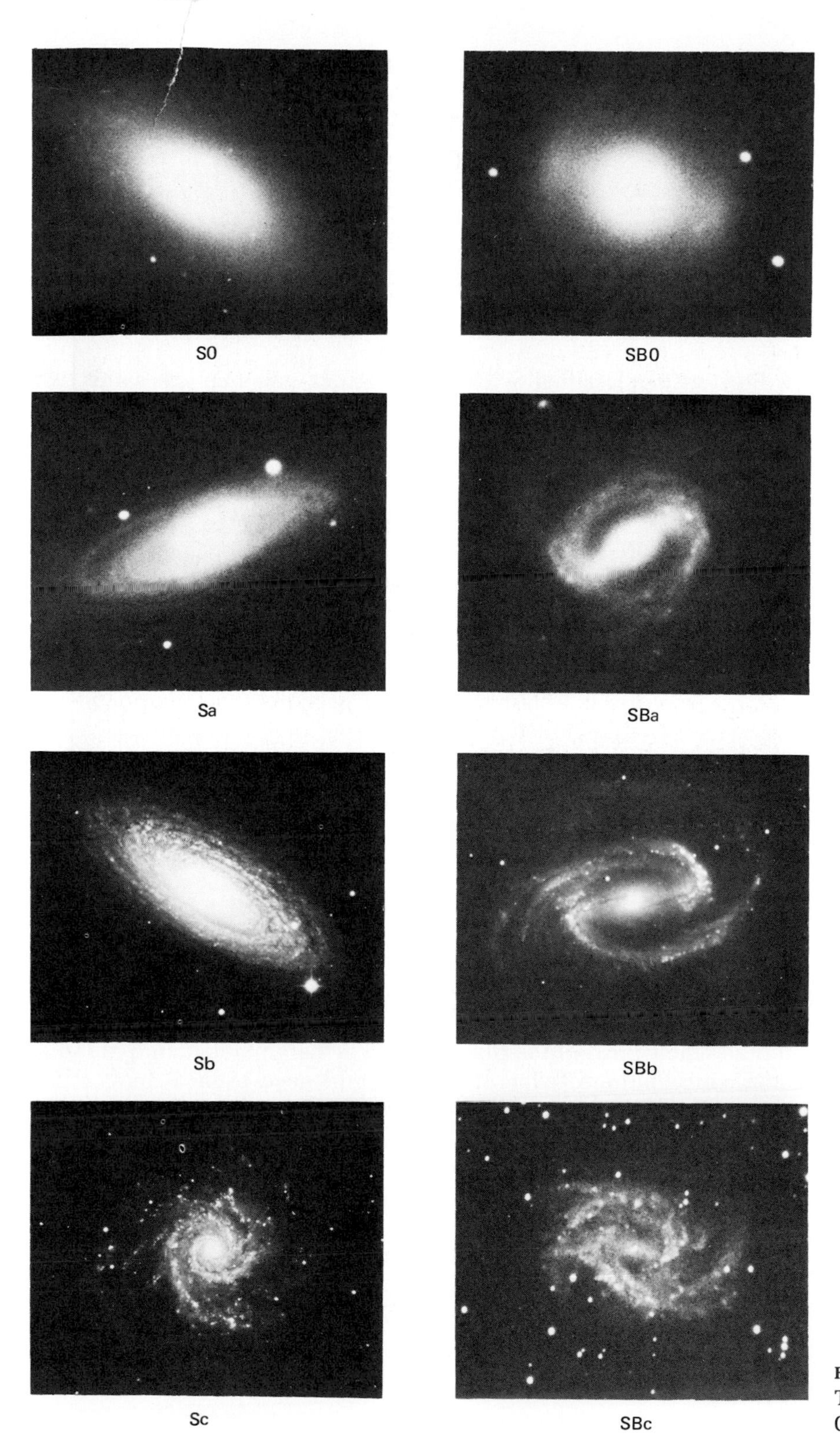


**FIGURE 12.3**
The principal types of spiral galaxies. (Hale Observatories photographs)

One other type of spiral galaxy is indicated in Figure 12.1. Type S0 refers to a galaxy that is much like an elliptical, but that shows evidence of spiral arm structure. As noted above, the S0 galaxies were originally designated as E7 galaxies.

*Irregular Galaxies.* **Irregular galaxies** have no particular forms, except that some appear to be flattened. They are more like spirals than ellipticals in that they contain a lot of gas and dust. Although there are no giant irregular galaxies, there are some dwarf irregulars (Figure 12.4).

Previously classified with the irregular galaxies are some two hundred or more galaxies properly called **peculiar galaxies.** They have strange forms and unusual characteristics. They often show evidence of violent events. We discuss some of these peculiar galaxies later in this chapter.

There have been some attempts to describe an evolutionary sequence for galaxies, saying they evolve from irregulars to spirals to ellipticals or vice versa. Today we know that the Universe is too young for any galaxy to have evolved from one type to another. It is thought that the major types of galaxies have always existed essentially as they are.

More complicated classifications of galaxies than the one above have been devised, but they will not serve our purposes here any better. Regardless of the classification scheme, certain galaxies will be left out—for example, galaxies with a ring of stars and material surrounding their nucleus, which are called ring galaxies. These unusual galaxies will be discussed later.

**FIGURE 12.4**
A typical irregular galaxy, NGC 1313. (European Southern Observatory photograph)

## Galaxies Visible without a Telescope

Only four galaxies are visible to the unaided eye. The nearest of these, and because of this the brightest in appearance, is the Large Magellanic Cloud (Figure 12.5). It fills an area of the sky slightly larger than the whole bowl of the Big Dipper. The Large Magellanic Cloud is in the constellation of Doradus. We would need to be at least as far south as the Caribbean Sea to have a chance to see this galaxy in its full splendor. The Large Magellanic Cloud is usually classed as an irregular galaxy, although some astronomers believe it is a loose, barred spiral. It is about 20,000 lt-yr in diameter and less than 200,000 lt-yr away.

The second brightest and second nearest galaxy is the Small Magellanic Cloud (Figure 12.6). We must be even farther south than the Caribbean to see this galaxy. The Small Magellanic Cloud is in the constellation of Tucana, and it is about half as big as the Large

**FIGURE 12.5**
The Large Magellanic Cloud, which covers 8° of the sky. (Hale Observatory photograph)

Magellanic Cloud. It would fit inside the bowl of the Little Dipper. This galaxy is also classed as an irregular galaxy. Figure 12.7 shows the positions of both Magellanic clouds on a sky map that includes the south celestial pole. The Small and Large Magellanic clouds are considered to be satellite galaxies of the Milky Way.

The next brightest galaxy that we can see with the unaided eye is the great Andromeda galaxy, also known as M31. There are at least nine other galaxies closer to us than Andromeda, but they are all dwarf galaxies. Since we can see Andromeda even at its 2.2 million lt-yr distance, it must be one of the absolutely brightest galaxies. And it is. Andromeda is an Sb galaxy with about 200 billion stars. It is barely visible on a dark, clear night in the constellation of Andromeda, just northeast of the great Square of Pegasus (Figure 12.8). It looks like a tiny smudge in the sky. In binoculars or a small telescope, it stands out more clearly as a hazy patch of light. In a large telescope, it still looks like a hazy patch of light and is disappointing compared to a view of

**FIGURE 12.6**
The Small Magellanic Cloud, an irregular galaxy. The globular cluster 47 Tucanae, part of our own galaxy, is on the right. (Cerro Tololo Inter-American Observatory photograph)

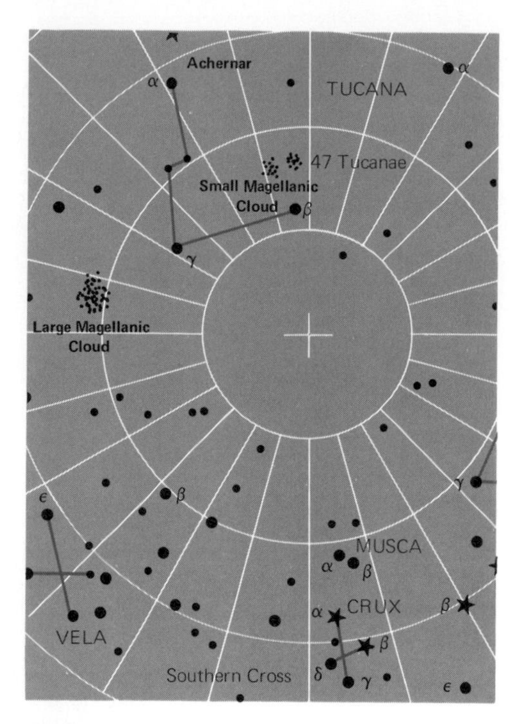


**FIGURE 12.7**
A sky map showing the location of the Large and Small Magellanic Clouds, the globular cluster 47 Tucanae, and the Southern Cross. The south celestial pole is marked with a plus sign.

the galaxy in a long-exposure photograph taken with a large telescope (Figure 12.9 and color plate).

Next in brightness and slightly more distant than Andromeda is the galaxy M33 in the constellation of Triangulum. It also appears in Figure 12.8. In a telescopic view, M33 is observed to be a face-on spiral galaxy of the Sc class (Figure 12.10).

The reason we do not see many galaxies without telescopes is that they are so very far away, not that they are terribly dim. There are only about ten magnitude differences between the absolutely faintest galaxies and the absolutely brightest galaxies. This is far less than the difference in absolute magnitudes between the faintest and brightest stars. If we explain the differences in the absolute magnitude of galaxies in terms of the number of stars in the galaxies, then the faintest galaxies may have several hundred million stars, compared to the brightest galaxies with a thousand billion stars.

*Mass of Galaxies.* How do we know how many stars are in a galaxy or in the Milky Way? Surely no one ever counted them all. In fact, we cannot even see them all. This is really a problem of determining the mass of a galaxy and then converting the mass to a certain number of stars. Once again, we turn to Kepler's Harmonic Law to determine masses. For example, in our own galaxy we use the motion of a star, the sun, around the center of the galaxy to find the mass of the galaxy. Knowing how far we are from the center of the Milky Way, we can calculate the circumference of the orbit. Knowing how fast the sun is moving, we can calculate how long it takes to travel around this orbit. Thus, with the distance from the center and the period, we can apply Newton's version of Kepler's Harmonic Law. From this study we find our galaxy contains a mass equivalent to 100 billion solar type stars.

## DISTRIBUTION OF GALAXIES

### Clusters

If we allow for obscuration by the gas and dust in the Milky Way in our sky, we see that galaxies are found in all directions in space. Many (perhaps all) galaxies occur in clusters that populate the Universe. A cluster of galaxies contains anywhere from 25 to 2,500 galaxies. A few of the most prominent galaxy clusters are commonly known by the names of the constellations in which they appear. An example is the Virgo cluster. At a distance estimated to be 65 million lt-yr, the Virgo cluster is the nearest of the larger clusters containing thousands of galaxies of all types. Spiral galaxies are numerous in this cluster, and their brightest stars can be observed in photographs taken with large telescopes. The Coma Berenices cluster (Figure 12.11), at a distance of over 420 million lt-yr, is reported to have a membership of 1,500 galaxies. The **local group** is a small cluster of about twenty-six galaxies to which the Milky Way belongs (We discuss the local group in more detail below).

Clusters of galaxies are similar to clusters of stars, in that the galaxies of a cluster are gravitationally bound together. Unlike the stars, however, the average separation between galaxies in a cluster is fairly small compared to their size. For example, between the sun and the next nearest star, we could fit at least 25 million other suns. Even in a star cluster, we could fit at least a million stars in the space between two stars. But, between the Milky Way galaxy and the Large Magellanic Cloud, we could not even fit two galaxies like the Milky Way.

There is evidence for material existing between the galaxies in a cluster, but there is nothing to indicate any material between clusters of galaxies.

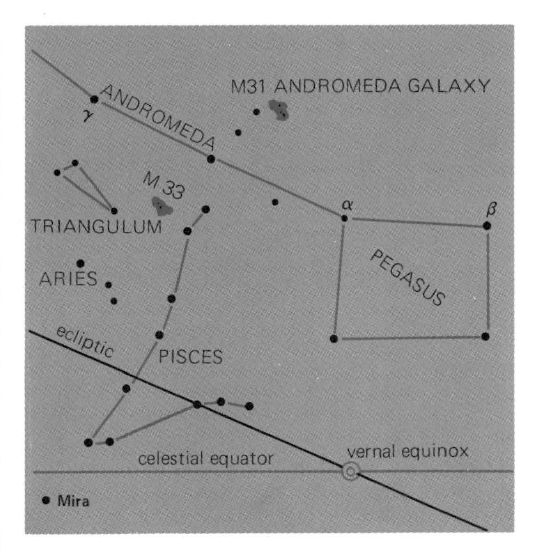


**FIGURE 12.8**
The region of the sky containing M31 (the Andromeda galaxy) and M33.

**FIGURE 12.9**
The Great Andromeda galaxy (M31) and two of its elliptical galaxy companions. (Hale Observatories photograph)

**FIGURE 12.10**
M33, an Sc type galaxy in Triangulum. (Hale Observatories photograph)

**FIGURE 12.11**
A cluster of galaxies in Coma Berencies. This cluster has more than 1000 member galaxies. (Hale Observatories photograph)

## The Local Group

The local group of about twenty-six galaxies occupies an ellipsoidal volume of space over 6 million lt-yr in its longest dimension. The Milky Way galaxy and the Andromeda galaxy (M31) are each near opposite ends of this long diameter. All the known members of the local group are listed in Table 12.1, which contains the galaxy types, their distances from us, and their diameters. There are four normal spirals and five irregular galaxies. The rest are elliptical. Most of the ellipticals are dwarf ellipticals.

**TABLE 12.1 The Local Group**

| *Name* | *Type* | *Distance (millions of lt-yr)* | *Apparent angular diameter* | *Diameter (thousands of lt-yr)* |
|---|---|---|---|---|
| Milky Way | Sc | | | 100 |
| Simonson 1* | dI | 0.07 | 5° | 5 |
| Large Magellanic Cloud | I | 0.16 | 12° | 33 |
| Small Magellanic Cloud | I | 0.20 | 8° | 25 |
| Carina I | dE | 0.20 | 25′ | 1 |
| Ursa Minor system | dE | 0.29 | 55′ | 4 |
| Draco system | dE | 0.33 | 48′ | 4 |
| Sculptor system | dE | 0.33 | 45′ | 4 |
| Fornax system | dE | 0.65 | 50′ | 10 |
| Leo II system | dE | 1.3 | 10′ | 3 |
| NGC 6822 | I | 1.3 | 20′ | 9 |
| NGC 185 | E | 1.6 | 14′ | 7 |
| NGC 147 | E | 1.6 | 14′ | 7 |
| Leo I system | dE | 2.0 | 10′ | 4 |
| IC 1613 | I | 2.3 | 17′ | 10 |
| Andromeda galaxy | Sb | 2.3 | 5° | 170 |
| M32 | E2 | 2.3 | 12′ | 7 |
| NGC 205 | E5 | 2.3 | 16′ | 10 |
| Andromeda I | dE0 | 2.3 | 0.′5 | 2 |
| Andromeda II | dE0 | 2.3 | 0.′7 | 2 |
| Andromeda III | dE0 | 2.3 | 0.′9 | 3 |
| Andromeda IV | dE | 2.3 | 0.′2 | 1 |
| Triangulum galaxy | Sc | 2.4 | 62′ | 50 |
| Maffei 1 | gE | 3.3 | 2° | 114 |
| Maffei 2 | Sa | 3.3 | 23′ | 58 |
| Phoenix 1: | E | — | 2′ | — |

*Provisional.
: Uncertain.

# DISTANCES TO GALAXIES

## Cosmic Distance Scale

Many methods are used to find the distances to objects in the Universe. Each method is appropriate only out to a certain distance. Each depends for its accuracy on measurements of distances to nearer objects. We have described many of these methods in previous chapters. Here we tie them together from the closest measurement to the most distant, and we describe the method of determining distances out to the edge of the visible Universe.

1. We bounce radar off Venus to determine its distance from us in kilometers.
2. Now we can scale the orbits of the earth and Venus to determine the earth's distance in kilometers from the sun. We call this distance one astronomical unit (1 AU). This is like the problem Kepler solved to find the shape and relative size of the orbit of Mars (Chapter 3).
3. We use the method of stellar parallax (Chapter 3) to determine the distance to the nearest stars in astronomical units. Since we know the value of the astronomical unit from step 2, we know the distances to the nearest stars. Once we know that, we can find their absolute magnitudes and develop an H-R diagram for the nearest stars.
4. Using the main sequence of the H-R diagram, we can find the distances to nearby clusters of stars (Chapter 5). Some of these cluster stars are cepheid variables.
5. Knowing the distance to one cepheid, we can use the period-luminosity relationship (Chapter 5) to find the distances to other clusters or galaxies containing cepheids. We can identify cepheids in only a few of the nearest galaxies.
6. Novae and supernovae always reach a certain well-determined brightness when they erupt (Chapter 11). By looking for these events in galaxies, the distances to more distant galaxies can be determined.
7. Finally, by looking at the distances to the galaxies determined from step 5 or 6 above, and by looking at the Doppler redshifts (Chapter 4) of the spectrum lines for those same galaxies, a direct relationship between velocity and distance can be demonstrated. The greater the redshift of the lines, the greater the distance to the galaxy. This method can be extended to any galaxy with a measurable redshift. This is known as the Hubble redshift law.

## The Hubble Redshift Law

As early as 1913, when galaxies were not understood and were still the curious extragalactic nebulae, V. M. Slipher was observing their spectra. He announced that most of the extragalactic nebulae had spectral lines Doppler-shifted toward the red end of the spectrum. Since a redshift meant that the objects were moving away, Slipher had made the first observations indicating an expanding Universe. The amount that the spectral lines are shifted gives a measure of the speed of the object. Not only did Slipher see these objects moving away, but they were moving at speeds of thousands of kilometers per second, indicating a rapid expansion.

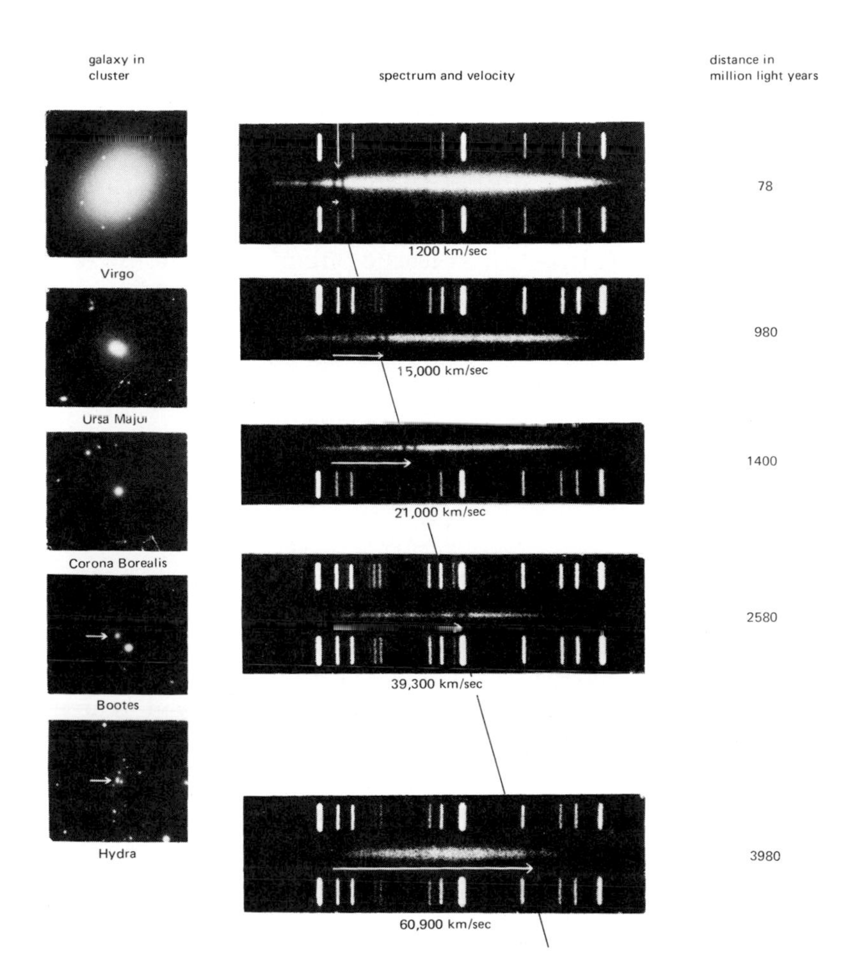


**FIGURE 12.12**
Spectra of galaxies arranged to show the linear relation between redshift and distance. Each of the galaxies is a giant elliptical galaxy located in the cluster of galaxies named in the figure. The arrow under each spectrum shows the amount of redshift of the spectral lines. The diagonal line shows how the amount of redshift increases proportionally with distance. (Hale Observatories photographs)

Hubble pursued Slipher's discovery, aided by the new knowledge that the extragalactic nebulae were distant galaxies. He was able to determine the distances to some of the galaxies using his observations of cepheids. Hubble showed that there is a direct relationship between the distance to a galaxy and its redshift or speed. A galaxy twice as far away as another galaxy is moving away twice as fast, and so on. This relationship is known as **Hubble's Law.** It has been used to show that every galaxy 30 million lt-yr or more away is receding from us, and that we are living in an expanding Universe.

Figure 12.12 (p. 251) shows how the spectral-line redshift increases directly with the distance to the galaxy. Notice how the apparent size of the galaxy gets smaller as it is more distant. The galaxy in Hydra, at a distance of 4 billion lt-yr, is traveling away from us at about 20% of the speed of light. Another galaxy that has been measured is traveling at about 40% of the speed of light. It is twice as far away as the galaxy in Hydra, or about 8 billion lt-yr away. This represents the greatest distance that we have been able to detect galaxies in space using optical telescopes.

The Hubble Law does one more thing for us. We can use it to estimate the age of the Universe. Consider the galaxy that is 8 billion lt-yr away. The light we see from that galaxy left it 8 billion years ago, and we can say that the Universe is at least 8 billion years old. As we will see in the next section, there are objects thought to be beyond the galaxies in space, which lead us to yet older ages for the Universe. As we look farther into space, we are also looking back in time.

## ACTIVE GALAXIES

### Collisions

A natural question arises as to whether or not galaxies collide with each other. The answer is yes, but very rarely. Gravitational interactions, where one galaxy distorts parts of the other galaxy and vice versa, are more likely than collisions. There are many examples of tidally distorted galaxies (Figure 12.13). It is believed that close passages of the Magellanic Clouds have distorted the Milky Way slightly.

What would happen in the case of a collision? If the passage is fast on the time scale of galaxies, meaning something like 100,000 years, almost nothing happens. The gravitational attraction of one star on another does not act over a long enough period of time. Two galaxies could pass through each other without a single star colliding with another star. Recall how vast the spaces are between the stars com-

FIGURE 12.13
A group of four galaxies in Leo. The two spiral galaxies are greatly distorted. (Hale Observatories photograph)

pared to the sizes of the stars. It is unlikely that any stars would collide. If the galaxies involved contain gas and dust, however, the gas and dust do collide. The gas and dust is stripped out of the galaxies and remains behind somewhere between them as they recede from each other. The gas heats up upon collision and radiates energy as it cools later.

Today, high-speed computers are used to simulate collisions of galaxies to see what the results would be. When a small galaxy and a large galaxy collide, where the passage is rather slow, nothing happens to the small galaxy. The large galaxy redistributes some of its stars in a ring around its nucleus. **Ring galaxies** (Figure 12.14) are observational evidence for this type of collision. When we look at photographs of ring galaxies, the undisturbed small galaxy is always seen nearby.

FIGURE 12.14
A ring galaxy designated ESO-034 IG 11. From the locations of the galaxies in the field and the shape of the ring, this galaxy was penetrated by the small elliptical galaxy to the upper left. (European Southern Observatory photograph)

## Explosions

Irregular galaxies classed as peculiar appear to have been normal galaxies where something has gone wrong. An example is the galaxy M82 (Figure 12.15 and color plates). The outer reaches of the disk are almost normal looking, and typically blue. The central region, however, is red and highly distorted, with streamers giving the ap-

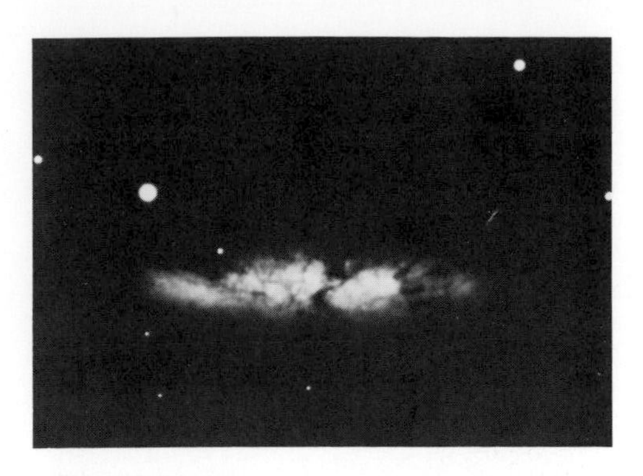

**FIGURE 12.15**
M82, a peculiar galaxy undergoing explosive disintegration. (Hale Observatories photograph)

pearance of ejection. The streamers are moving outward at about 1000 km/sec. Measuring the angular extent of the streamers and knowing the distance to M82, we can figure out that the violent explosion which started the streamers took place 1.5 million years ago. We see similar evidence for explosions in the centers of other galaxies.

When galaxies explode, they do not blow themselves to bits the way a star does during a supernova. Gas is ejected from the nucleus during the explosion, and enormous quantities of energy are liberated. The more violent the event, the more energy released.

The galaxy NGC 1097 is a normal spiral or barred spiral galaxy at first glance (Figure 12.16a). A careful study of the galaxy, however, reveals a discrete break in the northern arm and, in exactly the opposite direction, an absence of material in the arm (Figure 12.16b). A very long exposure photograph shows jets of material exactly in the direction discussed (Figure 12.16c). An energetic but not destructive event has caused this ejection without too much effect on the rest of the galaxy.

If the center of the Milky Way were to explode like M82 it would take about 30,000 years for the radiation to reach us and tell us that something drastic had happened. It would be about 10 million years before the streamers would reach us and material around us would be disrupted.

## Nucleus Variations

Often even normal-appearing galaxies show evidence of events or activity in their nucleus regions. The nucleus is variable in its radiated energy. It has been suggested that some of these events may have been caused by a supernova occurring in the nucleus. On occasion, the brightness of a supernova may exceed the total brightness of the hundred billion stars in the galaxy in which it occurs.

Activity in the nucleus regions of galaxies is much more common than originally believed. For example, many galaxies appear to have small, bright nuclei that vary in brightness (Figure 12.17). In many cases the variation is periodic, with periods of less than a year.

The nucleus of one galaxy is observed to vary in a period of two months. Such a short period light variation in the nucleus region must take place in a very small volume and cannot be due to an assembly of variable stars. A large number of variable stars would not all have the same period and maintain synchronized variations. The only way to keep the periods synchronized would be by signals traveling at the speed of light, and any star more than 2 light months away would not receive its synchronizing signal in the time needed for the variation. Even if all the stars in question did have the same period, we could not crowd enough of them into a volume 2 light months across to get the observed range in brightness. Therefore, some small object or energy

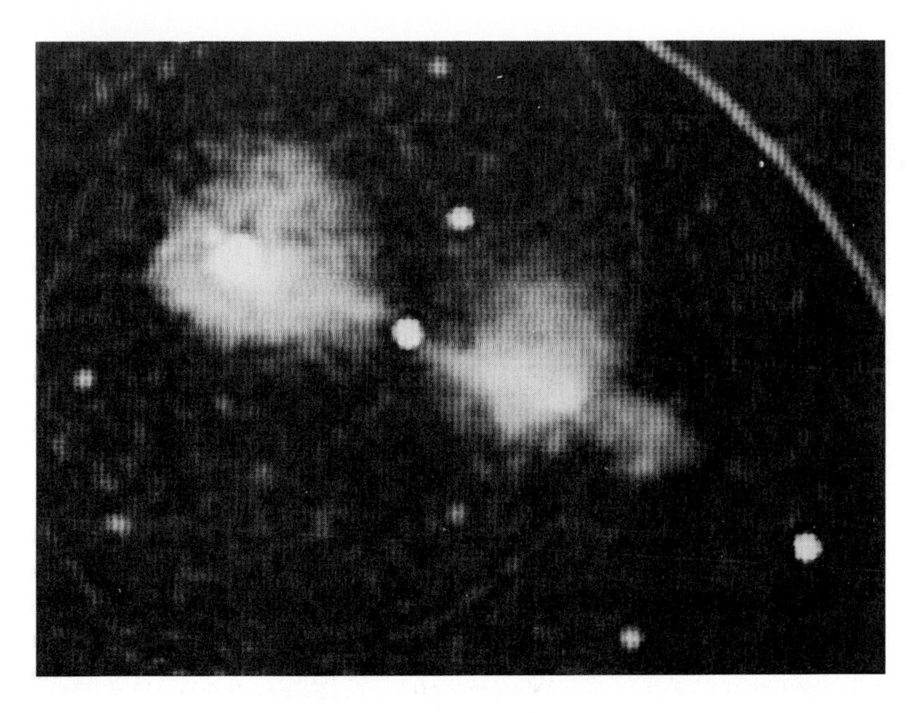

**FIGURE 12.19**
A radiograph of the giant double radio source DA 240. The centrally located object is a normal appearing galaxy. (Courtsey of J. Oort, G. K. Miley, and R. G. Strom, Sterrewacht Leiden)

**FIGURE 12.20**
The bright radio source Centaurus A. This object, known as NGC 5128, is an unusual galaxy as can be seen by studying it on the color plate. (Hale Observatories photograph)

**FIGURE 12.21**
3C273, a quasi-stellar object or QSO. Note the jet extending out from the almost starlike object. (Hale Observatories photograph)

## Nature

The nature of the QSOs is a true puzzle. We have some of the pieces, but so far they do not all fit together. Recall that QSOs show emission lines in their spectra. This is a characteristic of a low-density gaseous object. QSOs have very large redshifts. They are extremely small, perhaps the size of the solar system. Even if their apparent diameters are adjusted for extreme distances, they are nowhere near the size of galaxies (Figure 12.22). Many of them vary in their light emission on a short time scale, further proof that they are small.

In spite of their size, QSOs produce fantastic amounts of energy. The QSO 3C273 outshines any known galaxy by one hundred times. In fact, 3C273 is so small and generates so much energy that no known energy-generating process will explain it. Can that be true? Well, we have made one assumption. To calculate the amount of energy from a QSO, we have assumed that it is at the great distance predicted by its redshift.

If QSOs really are not so far away, then they do not have to produce all that much energy. If they are closer, the amount of radiated energy is less and more easily accepted. But then, how do we explain the great redshift? One possible way is by having a high-density, supermassive

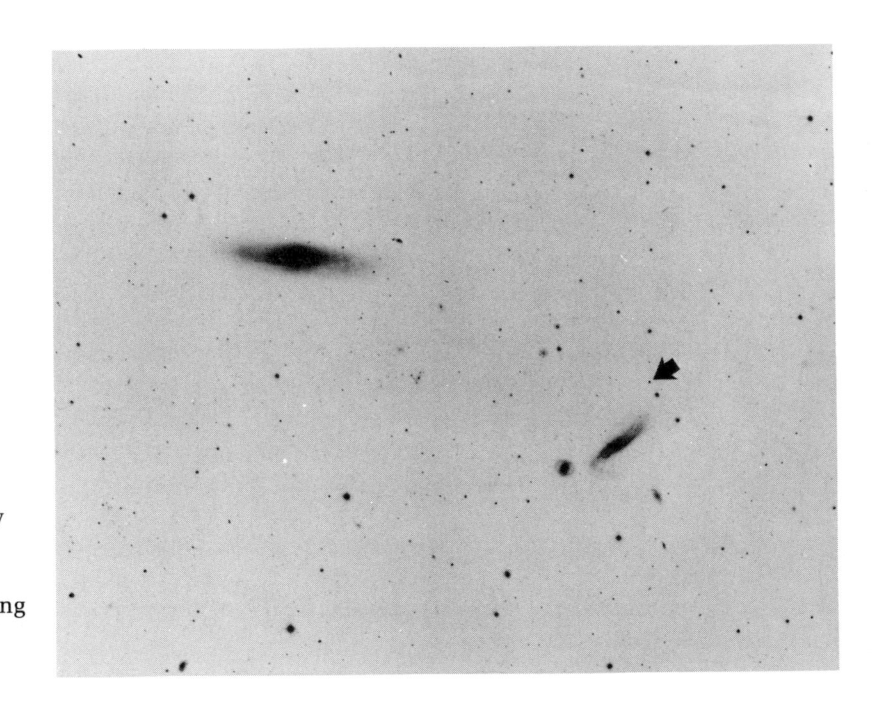

**FIGURE 12.22**
A QSO is marked on this negative print by the arrow and shows a true starlike appearance. The nearby galaxy is quite distorted. Note also the strange-looking ring galaxy. (Hale Observatories photograph, courtsey of H. Arp)

object whose gravity is so large that it "tires out" the light leaving the object. "Tired light" would be redshifted. This is the Einstein redshift mentioned in Chapter 10. This interesting hypothesis seems to be contradicted by the emission spectrum indicating a low-density gaseous object.

Another explanation of QSOs assumes that they are the equivalent of massive stars ejected from some nearby galaxy or even the Milky Way itself. If the event occurred long enough ago, all the objects initially coming toward us will have passed us, and all we see are objects receding from us. Under such conditions, many of the QSOs would have a proper motion component (Chapter 3) of their space motion. No QSOs show proper motions.

Yet another explanation of the QSOs assumes they are galaxies or nuclei of galaxies. In this chapter we have seen the range of peculiarities of galaxies. Perhaps quasars represent one extreme. Whatever the true nature of quasars, most astronomers think that they are at great distances, that their redshifts are Doppler shifts, and that we can use Hubble's Law to estimate their distances. This still leaves the unsolved problem of explaining the great quantity of energy coming from these objects. The QSOs may be the most distant objects we can see. They represent a major frontier of research in astronomy.

## SUMMARY

The true nature of the galaxies, distant island universes, was first revealed in the 1920s when a cepheid variable was observed in the Andromeda nebula. Galaxies are classified into three basic types: elliptical, spiral, and irregular. Each type is subdivided according to finer details. The Milky Way galaxy is an Sc galaxy. Galaxies tend to cluster together in space. Our galaxy is part of the local group.

Determining distances in space is a difficult problem. Astronomers directly measure distances to nearby objects and tie each successive distance determination to those first measures. With galaxies, the Hubble Redshift Law is a means of finding distances. Hubble established a direct relationship between distances to galaxies and their redshifts. With only minor exceptions, all galaxies show Doppler redshifts, indicating that they are moving away from us.

Some of the irregular galaxies are actually peculiar galaxies. These show evidence of collisions and explosions. Other peculiar galaxies have small, bright centers that vary in brightness.

As a group, quasars have the largest redshifts of any objects. Most astronomers use Hubble's Law to say that quasars are at exceedingly great distances. Quasars present many puzzles; astronomers are not certain what these small, energetic objects are.

### *FALLACIES AND FANTASIES*

*The Andromeda galaxy is the nearest galaxy.*
*You can only see a few thousand light years into space without a telescope.*

## REVIEW QUESTIONS

_____ 1. What is a galaxy?

_____ 2. Describe the subtypes of the spiral galaxies.

_____ 3. What differences would there be in an elliptical galaxy type E6 and a spiral galaxy with a large nucleus viewed edge-on?

_____ 4. How many stars are there in an average galaxy? How do we know?

_____ 5. What type of galaxy is most numerous in the local group? Is this type of galaxy liable to appear relatively more or less numerous throughout space? Why?

_____ 6. In 1923, Hubble calculated the distance to the Andromeda galaxy to be about 1 million lt-yr. The modern value is about twice as great. What could cause the change? Could the value change again?

_____ 7. List at least two of the puzzles regarding quasars.

_____ 8. Describe Hubble's Law for determining distances to galaxies. Would the same law work if we were in the Andromeda galaxy?

_____ 9. Describe the distribution of galaxies in space.

_____ 10. What are peculiar galaxies?

# 13 Cosmology: From Cosmic Geometry to Life in the Universe



## THE UNIVERSE

### Observations

Cosmology is the study of the Universe. Cosmogony is the study of the origin and evolution of the Universe. Astronomers have been rather lax in distinguishing between the two studies and now lump all large-scale studies of objects beyond our galaxy as well as their origin and evolution under the term **cosmology.** We will do the same.

In Chapter 1 we presented an overview of the currently accepted picture of cosmology. The picture of cosmology as we have presented it has been developed over the past decade or two. It certainly is not the final view. However, it is based upon a series of observations. These observations are:

1. The farther a galaxy is away from us, the faster it is receding from us. This is a statement of the Hubble Law. It implies that we live in an expanding Universe.
2. The cosmic abundance of the elements has been observed to consist of about 75% hydrogen, 25% helium, and a trace of deuterium. There are local variations to these numbers.
3. There are no QSO's nearby the Milky Way or local group.
4. The density of the Universe is about $10^{-31}$ grams per cubic centimeter according to current estimates.
5. The Universe is filled with radiation at a temperature of about 3°K. Regardless of where we look, we see this radiation.
6. When we count radio sources, we find more and more of them as we go to fainter and fainter intensities.

While we have not mentioned these last three observations previously, they have a direct bearing on the problems of the origin and evolution of the Universe. The last observation, radio source counting, is carried out in a similar way as the star counting techniques described earlier. For the moment, assume that all radio sources outside our galaxy have the same radio intensity. Then, as a source is farther away from us, its intensity will appear fainter. As we go to fainter and fainter levels (farther and farther away) we are taking in a larger volume of space. We must correct our counts to allow for this larger volume. After doing that, we find that there are more radio sources at great distances than nearby; there is a greater crowding of sources at the greater distances. This also means that there were more radio sources in the past than there are today. To understand this, recall that radiation always travels at the speed of light. A flare that we notice on the sun at this moment actually occurred 8 minutes ago. If we observe a flare on Alpha Centauri tonight, it actually occurred 4.3 years ago. A supernova that we observe in the Andromeda galaxy occurred almost 2 million years ago. Thus, the more distant the object, the farther we are looking into the past. When we look at the fainter radio sources, we are looking back into time. By looking to greater distances, we observe the Universe from the present to its early history.

Observation 5 is simply the cooling of the initial radiation to the present value of 3°K. As a gas expands it cools. This is analogous to $CO_2$ gas in a fire extinguisher. When you open the nozzle the gas rushes out and as it expands it cools and becomes $CO_2$ frost. Observation 4 will be discussed beginning on page 268.

To formulate a cosmological theory to explain these observations we must set up an agreed upon rule or set of rules. The rule that astronomers use is called the cosmological principle.

## The Cosmological Principle

In the 17th century Isaac Newton reasoned that the Universe is finite and has a center. His argument was that if the Universe were infinite with matter everywhere it would have infinite gravity and collapse to a single point. On the other hand, G. Leibnitz reasoned that the Universe was infinite and that matter was distributed uniformly throughout space. His argument was that for matter to be distributed otherwise would endow one place as the center of the Universe and that was unacceptable to him. These two early attempts to discuss the nature of the Universe were primarily philosophical. However, Leibnitz was adopting a position which, when slightly reworded, is now a basic starting point for cosmological models, the **cosmological principle.** The cosmological principle states that the Universe must look the same to all observers looking at it at a given time. Since the time can be anytime, the principle must hold as the Universe evolves. Furthermore, it does not assign a center to the Universe.

Our civilization has had a history of overrating its own importance. History began with the local valley as the center of all creation. Then the earth was the center of all the Universe. Then the sun was the center of the Universe. This was upset by showing that the sun is just an average star which, along with 100 billion other stars, orbits the center of the Milky Way galaxy. In our own time we should not make a similar mistake. Just because all distant galaxies appear to be moving away from us should not cause us to imply that our galaxy is at the center of the Universe. Therefore, as a working rule, we adopt the cosmological principle.

We can apply the cosmological principle to our observations of the expansion of the Universe. When we see a distant galaxy receding from us at 100,000 km/sec, observers in that galaxy see us receding from them at 100,000 km/sec. Furthermore, other observers see the Universe expanding, and the expansion is a function of distance. The farther away a galaxy is, the faster it is receding. The Hubble Law looks the same from all galaxies.

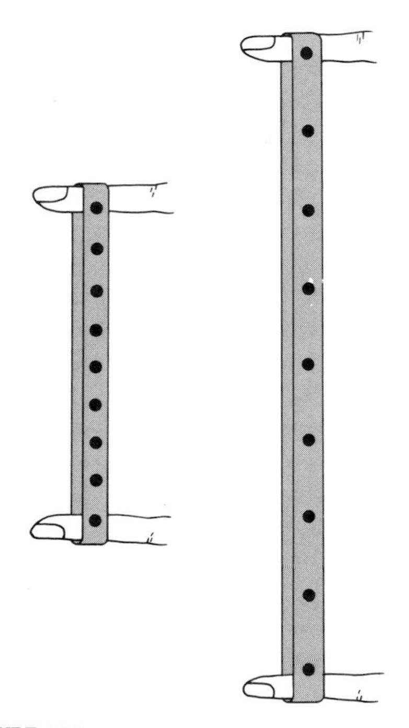

**FIGURE 13.1**
A descriptive explanation of the expanding universe and Hubble's law. The dot separations can be measured and an expansion law derived for the rubber-band universe.

Figure 13.1 shows a large rubber band with dots on it about 0.5 cm apart. Select one of the dots as your galaxy and slowly stretch the rubber band. Stretch it so the dot on either side of your galaxy is now 1 cm away from your galaxy. Note that the next dots out are now 2 cm away, the next dots 3 cm, the next dots 4 cm, and so on. Suppose it took you one second to stretch the rubber band. Evidently the dots nearest to your dot moved 0.5 cm/sec, the next moved 1 cm/sec, the next moved 1.5 cm/sec, and so on. The dots on the rubber band exhibit a Hubble Law: the speed of expansion is directly proportional to distance. The farther away a dot is, the faster it is moving away from your dot. In this experiment, you selected any dot. You probably did not pick the same dot that we did, yet the rule applies to your dot as well as ours. To check this, go back and select a different dot as your galaxy. Repeat the experiment, and you get the same result. From this we may conclude that if the Universe were an expanding rubber band, we could explain the fundamental observation concerning the receding galaxies and fulfill the cosmological principle as well.

We can take our analogy one step further. Take a balloon, blow it up slightly, and put dots all over it. Select one of the dots as your galaxy and slowly blow up the balloon. Regardless of where you are on the balloon, every dot is receding from you. The more distant the dots, the faster they are receding.

## Terminology and Geometry

Astronomers often talk about the extent of the Universe and the expansion of the Universe. In general, the words used to describe the extent of the Universe are *finite* or *infinite*. The expansion of the Universe is described as either *open* or *closed*. An open Universe expands forever; a closed Universe stops expanding.

On a small scale, the geometry of space is Euclidean, that is, the normal three-dimensional space we deal with every day. Newton assumed that the geometry of space was Euclidean and the extent of space was infinite, but he also thought that the material universe in that space was finite. There are many ways to perceive a universe. For example, suppose we have a microscopic creature constrained to live on the surface of a round balloon and another constrained to live on a table top. On a small scale, the microscopic creatures would not be able to tell the difference between the surface of the balloon and that of the table top. However, if the creature on the table top moved far enough, it would reach the edge or boundary. That creature's universe would be bounded and finite. The creature on the surface of the balloon could move in the same direction forever and never come to a boundary. That creature's universe is unbounded, but it is finite since it has a finite surface area. On a small scale the balloon appears flat, but on a large scale the balloon has the curvature of a sphere. In a similar way it is possible that our Universe has a curved geometry on a large scale.

## MODELS OF THE UNIVERSE

### The Big Bang

Albert Einstein's theory of general relativity allows us to build models of the Universe. The models are basically various geometrical expressions that assume the laws of physics , as we know them, hold everywhere and the velocity of light is a constant. Following Einstein, we can try to solve the mathematical equations in his theory and hence make a model of the Universe. After we have our model, we can make some predictions about the Universe which we may then try to verify with observations.

The equations are very difficult to solve so we must make some simplifying assumptions. We assume that on a grand scale the Universe is completely homogeneous and may be treated as a fluid. Such a homogeneous fluid resembles, for example, water, where the individual molecules making up the fluid are clusters of galaxies. The equations for such a fluid can be solved. However, there is actually more than one solution. In the solutions, the Universe turns out to be unbounded and either open or closed.

We will describe some of the different possible solutions. Unfortunately, we cannot picture the real Universe, because we immediately think in terms of Euclidean geometry. We will try, however, to draw some two-dimensional analogies of the Universe. The following solutions or models are part of what is termed the **big bang cosmology.**

Suppose at some time eons ago the Universe was infinitely thin. It contracted to enormous density and temperature and then exploded. All big bang cosmologies begin this way. In one solution, the Universe will expand to infinity. This is called an unbounded, open Universe. On a grand scale, the geometry of this universe has a negative curvature. The best two-dimensional analogy we can picture for this kind of space is a saddle-shaped surface which never closes (Figure 13.2). On such a surface triangles have sides that curve in toward their centers.

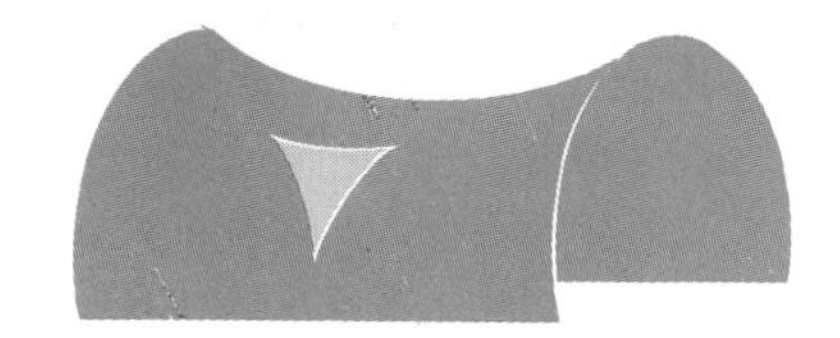

**FIGURE 13.2**
A triangle in a universe of negative curvature has concave (inward bowed) sides.

Another solution to the equations yields an unbounded, open universe that uses Euclidean geometry (Figure 13.3). This universe would expand to infinity at a slower and slower rate. In Euclidean space triangles have straight sides.

A third solution gives an unbounded, closed universe where the geometry shows a positive curvature. This universe would expand to a given size and then collapse. It might do this only once, or it might expand and collapse many times, becoming what is called a **pulsating universe.** The best two-dimensional analogy for this kind of space is a

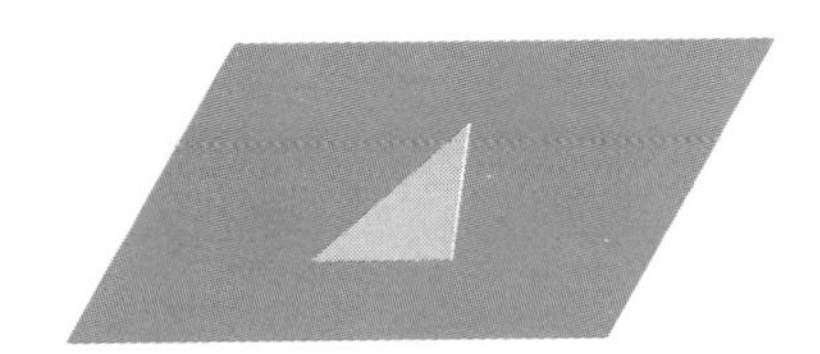

**FIGURE 13.3**
A triangle in a zero curvature universe has straight sides.

spheroidal surface (Figure 13.4), where triangles have sides that curve out from their centers.

All of these solutions involve a time (or times in the case of a pulsating universe) when the universe is at a very dense, high-temperature state. The dense, high-temperature state is called the **primeval atom** and the explosion that occurs is called the **primeval fireball** or big bang.

Big bang cosmology follows the cosmological principle. It also has a unique starting time. It predicts all of the observations we listed earlier: the Universe will be expanding, a trace of deuterium will be formed and be found over eons of time, and there will be background radiation which at this time has a temperature of about 3°K. (The background radiation is simply the cooling radiation from the primeval fireball.) Big bang cosmology also allows an evolutionary universe; thus, there can be more radio sources in the past than at present. The high-intensity radio sources may be one phase in the evolution of galaxies. Indeed, the QSOs may also be a phase in the evolution of galaxies.

## The Steady State

Some cosmologists do not like the idea of a unique beginning for the Universe. Why not make up a universe that is infinite and has a constant density? Although individual components of such a universe evolve, the universe itself is unchanging. This is called the **steady-state cosmology.** Such a cosmology is appealing in its simplicity and follows what is called the **perfect cosmological principle:** The Universe must look the same to all observers regardless of when they observe (Figure 13.5).

Many astronomers objected to the steady state cosmology because it requires the continuous creation of matter. Since the Universe is expanding, new matter must be created to fill in and keep the density constant. The objection to the continuous creation of matter cannot be taken seriously because the matter has to be created sometime in any cosmological theory, either all in one instant or a little bit at a time. The more serious objection to the steady state is that it does not predict observations 2, 3, 5, and 6 stated earlier. It does not produce the observed amount of deuterium. It fails to explain the 3°K background radiation. It cannot explain why there are no QSOs nearby nor the excess of radio sources at a great distance.

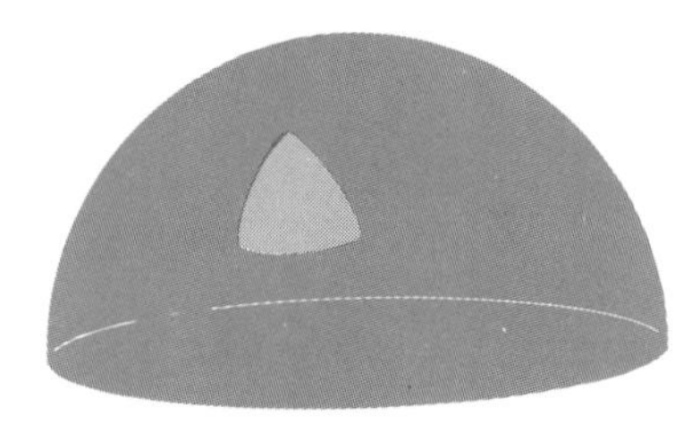

**FIGURE 13.4**
A triangle in a universe with positive curvature has convex (outward bowed) sides.

## Time Varying Constants

We must conclude that the choice between the big bang and the steady state cosmologies lies with the big bang. However, these are not the only choices. If space and time are so intimately related in the theory of general relativity, could it be that space and time interact in such a way that the appearance of an expanding universe is just an illusion?

The basic premise of general relativity is that the laws of physics hold everywhere and that the velocity of light is constant. If you change this premise, or any part of it, you may be able to build a self consistent scheme that will explain the observed Universe. For example, suppose the laws of physics hold, but one or all of the **universal constants** (such as gravity or the charge on the electron) vary with time. To sort out the implications of time-varying constants is mind boggling. We will merely mention the consequences of one variation.

It is possible to imagine that the charge on the electron varies with time. The consequences are not obvious. If the charge on the electron is too small, the chemical reactions that we know could not take place in the same fashion. The same is true if the charge on the electron is too large. Thus, if the charge on the electron slowly changes from being too small to being too large (or vice versa) the Universe might be as we see it only when the charge on the electron passes through the proper range. So far as we can tell from geological records, the charge on the electron has not changed during the past 4 billion years. The splitting of certain spectral lines also depends upon the charge on the electron. Quasars have these lines in their spectra and the lines show the same splitting as in our laboratory. Therefore, if the QSO redshifts are due to the Doppler effect (meaning the QSOs are at great distances)

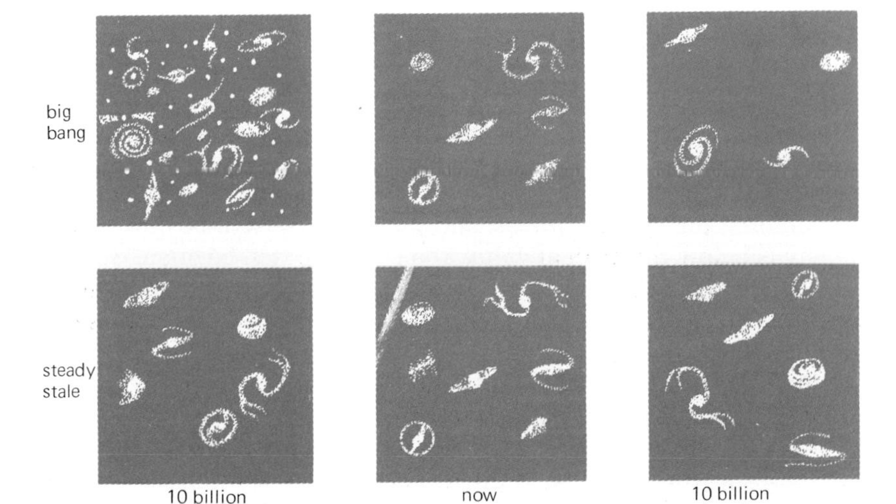


**FIGURE 13.5**
A schematic rendition of the big bang and steady-state cosmologies. The big bang results in fewer and fewer galaxies (on a large scale) per given volume of space. The steady-state universe always looks essentially the same.

then the charge on the electron has not changed much for 15 billion years. Hence, cosmologies based on time variation of the charge on the electron fail. To generalize, we can state that every effort to develop a cosmology involving time-varying constants has been unsuccessful so far. However, this is not to say that such cosmologies are ruled out. We must keep an open mind.

## Age of the Universe

We have already mentioned that the Hubble Law gives us an estimate of the age of the Universe. For example, suppose we measure the Doppler redshift of a galaxy and find that it is traveling away from us at 21,500 km/sec. Using the Hubble Law, the distance to this galaxy is 1400 million lt-yr ($14 \times 10^{21}$ km). Assuming the expansion rate of the Universe is constant and working the expansion backwards tells us that the big bang occurred about 20 billion years ($6.5 \times 10^{17}$ seconds) ago.

Another way to estimate the age of the Universe is to measure the age of objects in the Universe. The oldest stars that we can study in detail are located in the globular clusters surrounding the Milky Way Galaxy. Our studies of the life cycles of stars tell us that the oldest stars are between 8 and 16 billion years old. The age of the Universe can also be estimated by measuring the relative amounts of heavy elements in the Universe. All the elements heavier than iron, including some radioactive elements, are thought to be formed in supernovae that have been occurring since the formation of the Galaxy. The average age of the heavy elements is between 6 and 20 billion years. We conclude that the big bang probably took place between 8 and 20 billion years ago.

## Open and Closed Tests

Given that the best cosmology is the big bang, is the Universe open or closed? That is, will the Universe expand forever or stop expanding and then collapse? One test is to examine the average density of the matter in the Universe. If the density is above a certain critical value, gravitational forces will prevail and the Universe will be closed. If the density is below the critical value, the Universe will be open. Since the Universe is expanding, its density was greater in the past than at present. Thus, density depends directly on the age of the Universe, and age depends on density.

Density is not an easy quantity to measure. One method is simply to count everything (galaxies, stars, etc.) that can be seen in a certain volume of space. Counting visible galaxies gives a value for density that indicates the Universe is open. There is not enough matter in the galaxies to slow the expansion and cause a contraction. But, in addition to galaxies, we must account for other matter in the Universe. The problem becomes very complex because, as we mentioned earlier, there may be matter in the Universe that we cannot see, matter that is not emitting electromagnetic radiation on its own.

Another method is to determine the density of the Universe immediately following the big bang by looking at the abundance of helium and deuterium in the Universe today. Deuterium is thought to have been formed in the first few minutes after the big bang. If the density had been too great, most of the deuterium would have been converted to helium. By measuring the amount of helium and deuterium, we can say that the original density has to be below a certain value. By applying the expansion rate to the original density, we can determine a relationship between density and age.

We have summarized the information about the density of the Universe in Figures 13.6 (a), (b), and (c). In Figure 13.6 (a), the value of density equal to 1 is assumed to be the critical density of the Universe. For values below 1, the Universe is open; for values above 1, the Universe is closed. Which side of the critical density line are we on?

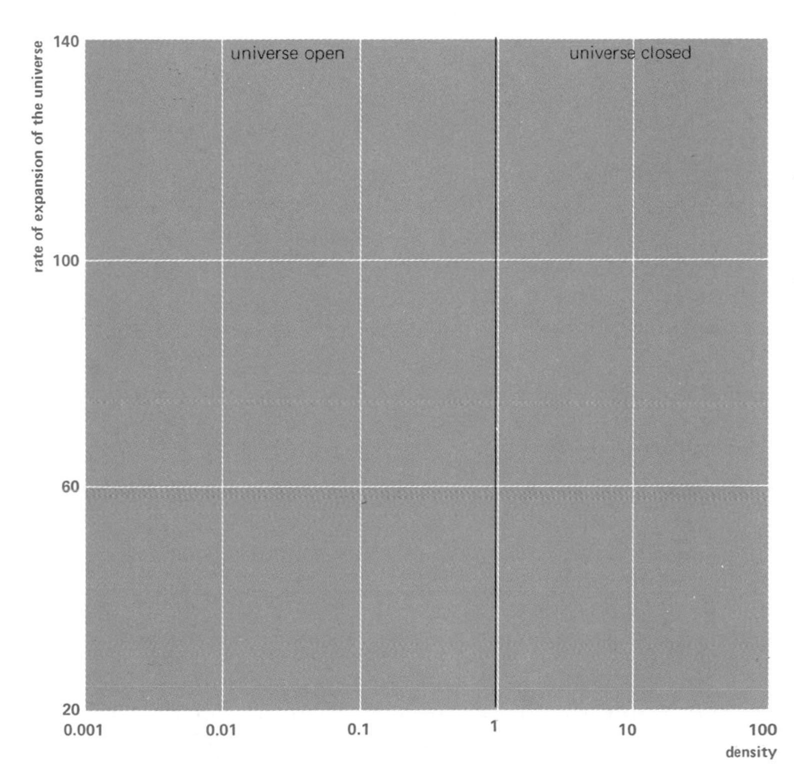


**FIGURE 13.6a**
A plot of the density of the universe against its expansion rate. To the right of the critical density the universe is closed; to the left it is open.

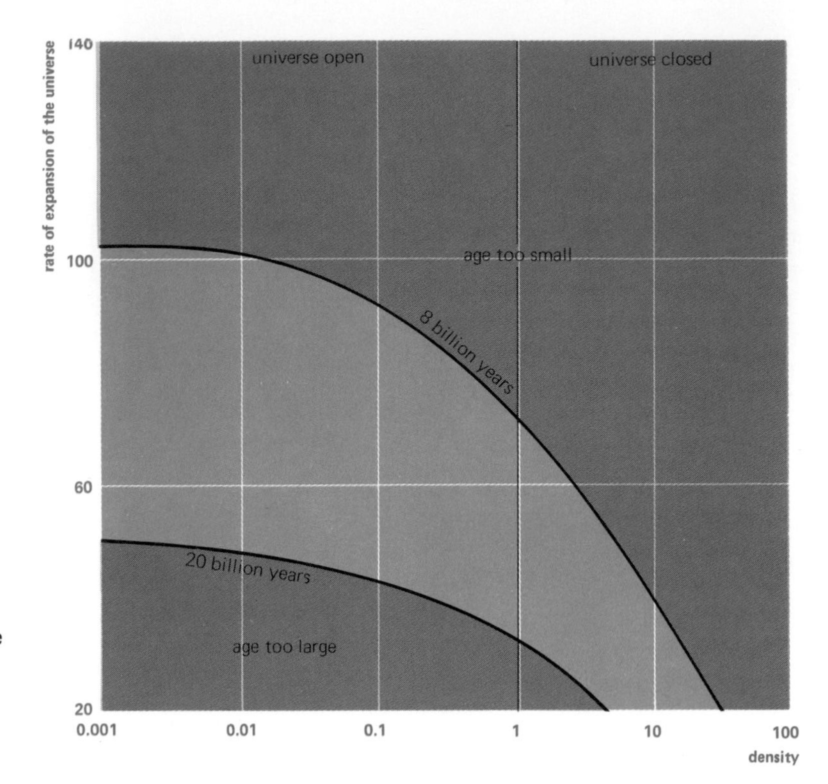


**FIGURE 13.6b**
Here we plot the expansion rates for given densities for the youngest possible universe and the oldest possible universe on the same graph as Figure 13.6(a). This leaves only a narrow curving band as the region where the universe can lie.

If we assume the youngest possible age for the Universe is 8 billion years, we can calculate the required expansion rates for various densities. This is shown by the upper line in Figure 13.6(b) marked 8 billion years. The top portion of Figure 13.6 (b) can be eliminated because the ages are too low. Now, if we assume the oldest possible age for the Universe is 20 billion years, we can also calculate the required expansion rates for various densities. This is shown by the lower line marked 20 billion years. The bottom portion of Figure 13.6 (b) can be eliminated because the ages are too high. The Universe lies between the upper and lower bounds. It still could be open or closed.

The amount of deuterium helps us to narrow down the region even further. In Figure 13.6 (c) we eliminate the region where the abundance of deuterium is too small for the given densities. The conclusion is that for any reasonable age of the Universe, the Universe is open and will expand forever. We must caution ourselves again that this conclusion rests on the validity of our very simple models. The evidence is definitely not all in. At this time there is a strong case for an open Universe, but more complete models may cause this to change. Certainly, the current theories for the origin and evolution of the Universe

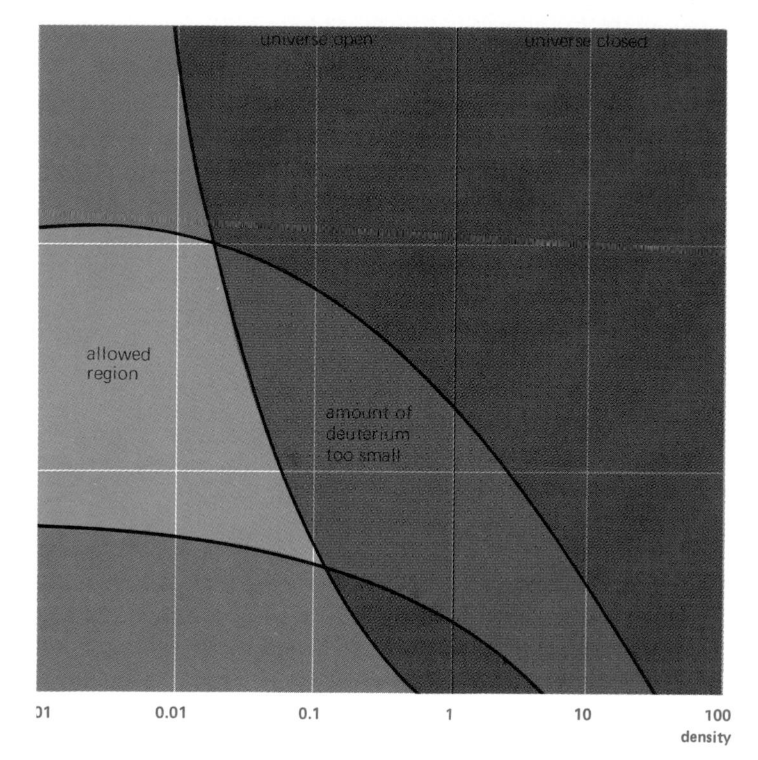


**FIGURE 13.6c**
Here we plot on Figure 13.6(b) the line where the amount (density) of deuterium agrees with observation. To the right of the line the density of deuterium is far too low. This leaves a very restricted region where the universe must lay. The region is entirely on the open side of the critical density.

are quite different from those of 5000 years ago or even those of 200 years ago. Perhaps it is just a matter of our learning and understanding. Given enough time, our civilization may find the correct theory.

## LIFE IN THE UNIVERSE

### Where Is It?

In this immense Universe containing billions of galaxies with hundreds of billions of stars in each galaxy, does life exist? If we define life in such a way that it includes the plants and animals on the earth, then the answer is certainly yes. Yes, because there is such life on the obscure little speck of the Universe that we call earth.

Do planets exist elsewhere in the Universe? The Universe is certainly enormous, and nowhere in the star-making process have we required anything special for making the star we call the sun. Of all the possible stars similar to the sun, it is inconceivable that only the sun has planets. After all, we did not invoke any special conditions to form the planets. In theoretical studies of planet formation, sometimes one of the major planets of a system will have an eccentric orbit. Over a period of time, this causes the disruption of that planetary system. Often enough, all the planet orbits are essentially circular and the planetary system is stable over tens of billions of years.

If it is common for planets to form along with stars, can we see any? Since the distances between the stars are so great, we cannot expect to "see" any of the small, nonluminous planets around other stars. The only way we can hope to detect them is to watch for irregularities in the motion of the parent star. These are caused by gravitational disturbances from the planets. At present, the irregularities can only be observed for the nearest stars with measurable proper motions. Of the dozen or so stars nearest to the sun, about half show unseen companions that are the mass of Jupiter or larger. The most famous star with companions is Barnard's star, about 6 lt-yr away. Barnard's star seems to have two companions about the size of Jupiter in circular orbits around it. The conclusion is that stars other than the sun can and do have planets.

If other stars have planets, then some of the planets must be at just the right distance from the central star for carbon-oxygen chemistry to operate under nearly earthlike conditions. It is not unreasonable to assume that a carbon-oxygen chemistry like ours could develop elsewhere. After all, the interstellar molecules we observe are just those molecules essential to life as we know it.

Does one of the planets outside the solar system have human type life? To answer that question, we will make an assumption. We shall call human type life **humanoids,** and assume that all life forms are convergent toward the most efficient form. This is a restatement of Darwin's principle of the survival of the fittest. A race of porpoises with a cubic body and fan-shaped fins must develop into a race with a torpedo-type body of great agility in water in order to survive. Humanoids will develop so that they walk erect, develop marvelous hands, and develop a brain capable of innovative thinking.

Many astronomers have tried to estimate the number of stars in the Milky Way galaxy with planets, the number of those planets with life, and the number of those with technological civilizations. Some astronomers have said that there could be only one, the earth. A better guess is that there are a million such technological civilizations in our galaxy. If they were spread out randomly, then the next nearest civilization should be about 300 lt-yr away. Information sent to or from that distance would arrive in 300 years traveling at the speed of light. A two-way message would require 600 years.

The design and content of an interstellar message is a complex study in itself. To give you an idea of what is involved, we will take a look at one message (Figure 13.9) that has already been transmitted. In 1974, at the dedication of the Arecibo radio telescope (Figure 13.10), a binary coded message was transmitted toward the globular cluster in Hercules (Figure 13.11). It will arrive at the cluster in about 24,000 years. The message can be decoded by breaking the binary digits into 73 groups of 23 digits each. Both 73 and 23 are prime numbers. When the groups are arranged one under the other, a visual message results. In Figure 13.12 the zeros are white squares and the ones are black squares. Some of the information contained in the message is displayed in Figure 13.13.

We have sent four other serious messages into space. The engraved plaque shown in Figure 13.14 is on both of the Pioneer spacecraft, and similar ones are on both Voyager spacecraft. Pioneer 10 will cross the orbit of Pluto in 1987 and will leave the solar system. It is traveling

**FIGURE 13.8**
An artist's rendition of the multiple telescope array proposed for project Cyclops.

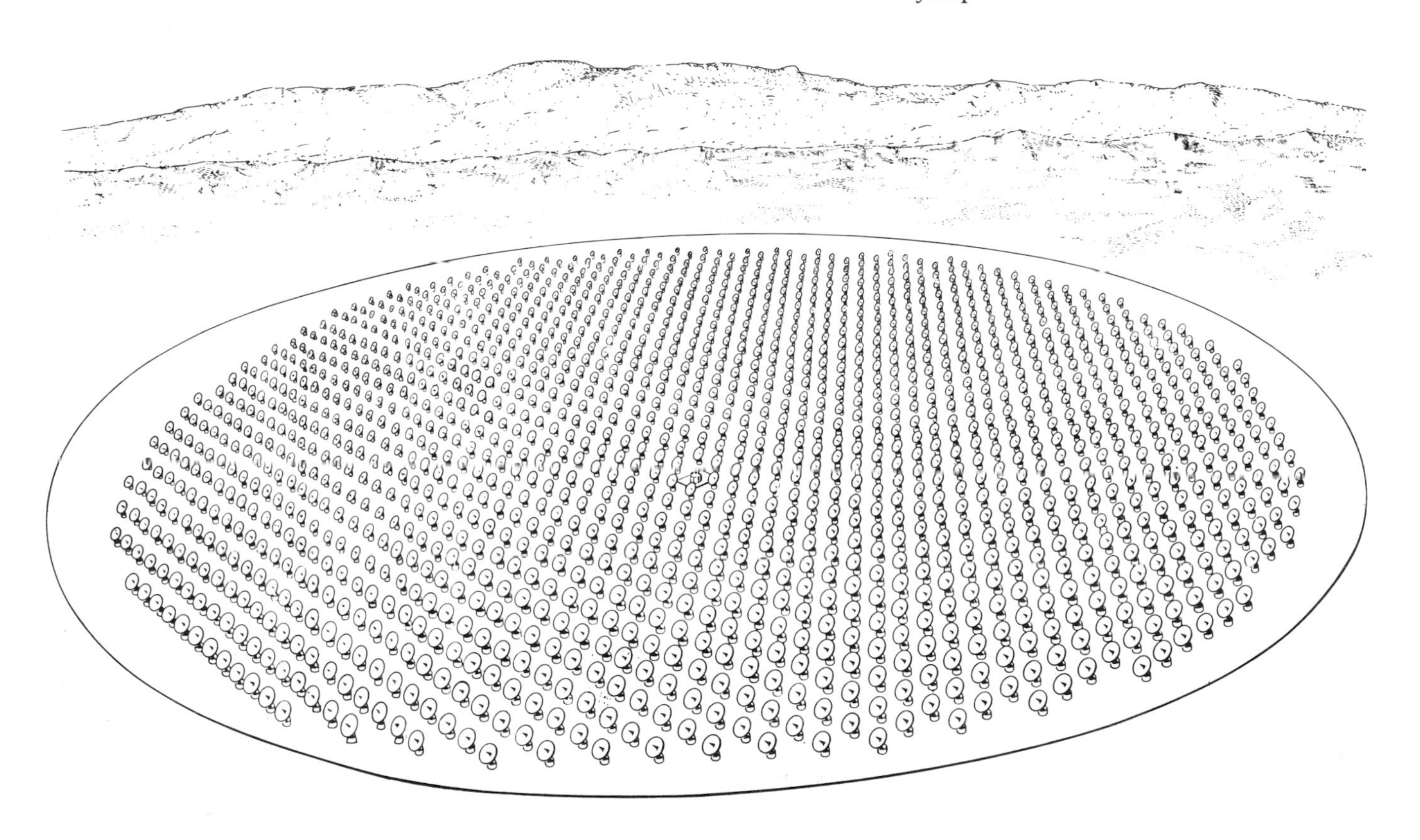

```
0000001010101000000000000101000001010
0000001001000100010001001011001010101
0101010101001001000000000000000000000
0000000000000000110000000000000000000
1101000000000000000000011010000000000
0000000010101000000000000000000111110
0000000000000000000000000000000110000
1110001100001100010000000000000110010
0001101000110001100001101011111011111
0111110111110000000000000000000000000
0100000000000000000100000000000000000
0000000000010000000000000000011111100
0000000000011111000000000000000000000
0011000011000011100011000100000001000
0000001000011010000110001110011010111
1101111101111101111100000000000000000
0000000001000000110000000001000000000
0011000000000000000100000110000000000
1111110000011000000111110000000000110
0000000000001000000001000000001000001
0000001100000001000000011000011000000
1000000000011000100001100000000000000
0110011000000000000011000100011000000
0000110000110000001000000010000001000
0000010000010000000110000000010001000
0000011000000001000100000000010000000
1000001000000010000000100000001000000
0000001100000000011000000001100000000
0100011101011000000000001000000010000
0000000000100000111110000000000001000
0101110100101101100000010011100100111
1111011100001110000011011100000000010
1000001110110010000001010000011111100
1000000101000001100000010000011011000
0000000000000000000000000000000011100
0001000000000000001110101000101010101
0100111000000000101010100000000000000
0010100000000000000111110000000000000
0001111111110000000000001110000000111
0000000001100000000000110000000110100
0000000101100000110011000000011001100
0010001010000010100010000100010010001
0010001000000001000101000100000000000
0100001000010000000000001000000000100
0000000000001001010000000000011110011
11101001111000
```

**FIGURE 13.9**
The binary message proposed by F. Drake and C. Sagan. The message is transmitted with intervals on or off, line by line like a television scan. This message is decoded in Figs. 13.12 and 13.13.

through space at about 16 km/sec. Since it must travel 80,000 years before it reaches the distance of the nearest stars and is not expected to pass even close to any star, the probability of its ever being intercepted is virtually zero. Nevertheless, it is an optimistic sign to attempt to communicate information. The symbols in the upper left corner of the plaque show how the hydrogen atom produces the 21 cm line. Twenty-one cm is adopted as the unit of length indicated by the binary "1" between the two hydrogen atoms. The size of the two people can be gotten from a comparison of the spacecraft drawn to scale behind them,

**FIGURE 13.10**
The great Arecibo telescope. It is built in a natural bowl on the island of Puerto Rico and has a diameter of 305 m. (Cornell University photograph)

**FIGURE 13.11**
The globular cluster M13 in Hercules. (Official U.S. Navy photograph)

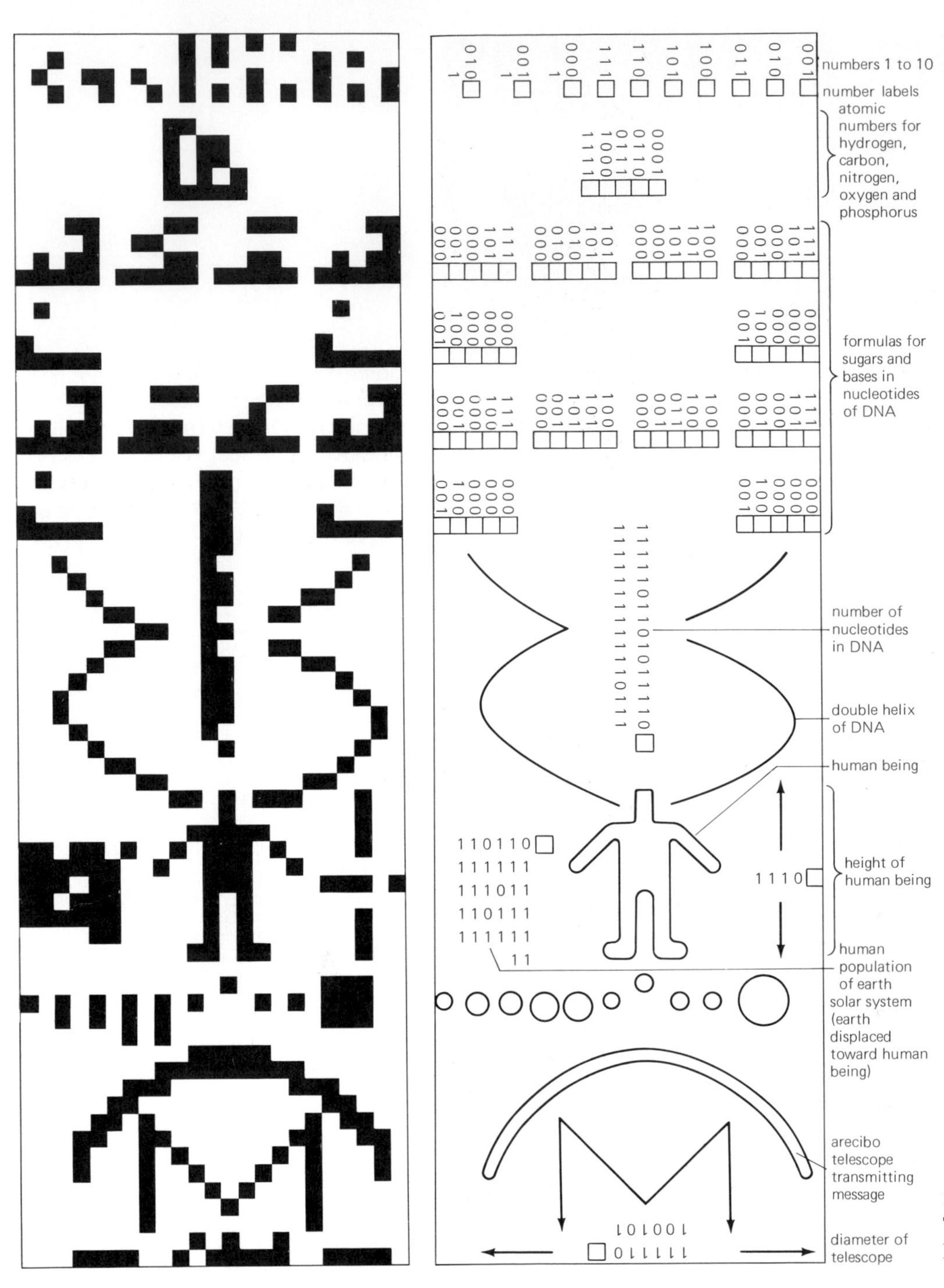


**FIGURE 13.12 (left)**
The Drake and Sagan message decoded as black and white spots. Each line has 23 spots. The message is 73 lines long.

**FIGURE 13.13 (right)**
The information in the Drake and Sagan message.

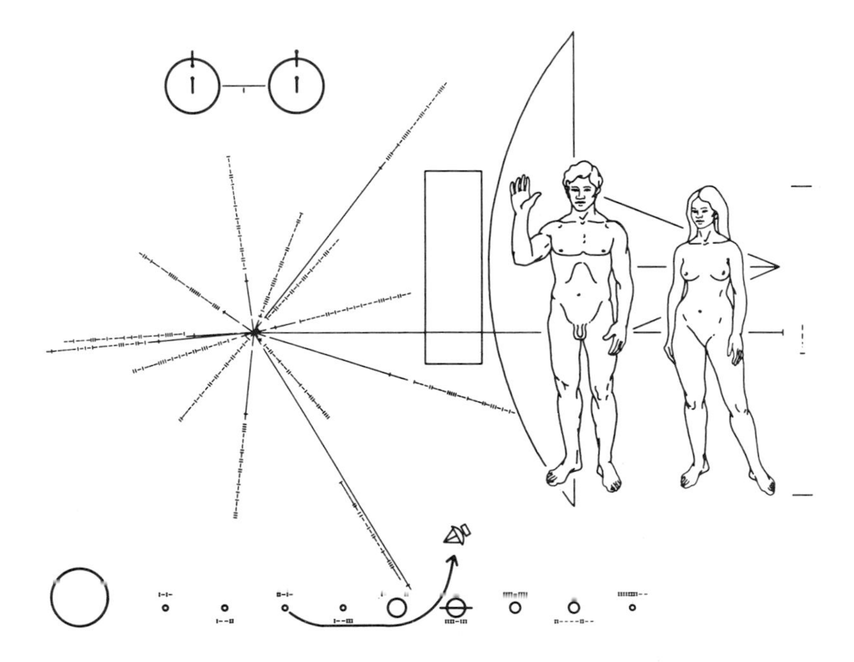

FIGURE 13.14
The plaque carried aboard Pioneers 10 and 11. If discovered by other humanoids, it is hoped that they will decipher it as explained in the text.

or from the binary coded "8" between the tote marks on the right. This means 8 × 21 cm is their height. The planets and their distances in binary numbers are located along the bottom of the plaque. The fact that the inhabitants of the third planet launched the satellite to fly by the fifth planet is also indicated.

Probably the most interesting thing on the plaque is the radial burst of 15 lines at the left center. These lines indicate both the epoch of launch and our position in the galaxy. The horizontal line to the right represents our distance from the center of the galaxy. The other 14 lines all correspond to the distances and directions of 14 individual pulsars. The binary coded number along each line gives the current period of the pulsar. Since pulsar periods are gradually slowing down, the stated period will establish the exact time or epoch of launch. The features of the man and woman are a composite representation of human characteristics.

The two Voyager spacecraft, launched in 1977, carry plaques similar to those on Pioneer 10 and 11. In addition, they have recordings of common earth sounds along with equipment for playing the recordings. After studying Jupiter and perhaps other outer planets, both Voyagers will leave the solar system with their messages. It is unlikely they will be intercepted, but we are trying.

There is no doubt that the present civilization on earth can detect extraterrestrial civilizations. If these civilizations exist, they can also detect us. Whether or not detection occurs depends a bit on luck, a bit on how long high-technology civilizations exist, a bit on how far apart they are, and a bit on desire and persistence. A million failures does not prove there are no extraterrestrial civilizations, but a single success proves that there are.

## SUMMARY

We adopt the cosmological principle as a starting point for different cosmologies. Any cosmological model must explain the fundamental observations astronomers have made about the Universe. The observations tend to favor the big bang cosmologies over the steady-state or time varying constants models. There are three variations of big bang cosmology depending on the large scale curvature of space and whether or not the Universe is open or closed. The amount of deuterium observed in the Universe presents strong evidence that the Universe is open. The age of the Universe is between 8 and 20 billion years. All cosmological models are based on simplifying assumptions. Models may change and new models may be developed as our observations improve.

Statistics suggest that there may be other planets around other stars, and on some of those planets life may have developed. Several nearby stars have unseen companions orbiting around them. If humanoid life has developed in the Universe, then a critical factor for possible communication is the length of time a technological civilization exists. To communicate over interstellar distances we can transmit binary coded messages at wavelengths of the electromagnetic spectrum that other civilizations may monitor. Most of these wavelengths are in the radio region of the spectrum. We have already sent one binary coded message into space. We have also attempted to communicate information about our civilization by sending specially designed plaques into space on the Pioneer and Voyager spacecraft. Also, our radio and television signals have been leaking into space continuously. The farthest signals are about 50 lt-yr away from the earth at this time. We have the ability to receive messages beamed directly at us as well as the technology to build antennas to eavesdrop on any civilization within several hundred light years of the earth.

### *FALLACIES AND FANTASIES*

*Straight lines are straight.*
*The Milky Way galaxy is the center of The Universe.*

## REVIEW QUESTIONS

_____ 1. What is the cosmological principle? Why is it central to cosmology?

_____ 2. Describe the difference between the cosmological principle and the perfect cosmological principle.

_____ 3. List at least three different cosmologies. Which one does the perfect cosmological principle apply to?

_____ 4. What is meant when we say the Universe is unbounded?

_____ 5. How do we know the Universe is expanding?

_____ 6. List some of the observational evidence that supports the big bang cosmology.

_____ 7. Describe a pulsating universe. Is our Universe pulsating? Describe the evidence that helps us to answer that question.

_____ 8. Give some arguments for believing life exists elsewhere in the Universe. Give some opposing arguments.

_____ 9. How is it possible to communicate over interstellar distances?

_____ 10. Suppose you receive a binary coded message from space. What kind of scientists would you select to help you decode the message? Why would you select them?

# FURTHER READINGS

Readings are given by chapter. Each article is chosen to amplify a subject or present a subject from a different point of view.

## CHAPTER 1

George Gamow, "The Evolutionary Universe." *Scientific American 195*, 136–149 (1956).
Edward R. Harrison, "Why the Sky is Dark at Night," *Physics Today* 27, 30–36 (1974).
David Layzer, "The Arrow of Time." *Scientific American 233*, 56–69 (1975).
Steven Weinberg, "The First Three Minutes." New York: Basic Books, 1977.

## CHAPTER 2

J. Darrel Mulholland, "Measures of Time in Astronomy." Publications of the Astronomical Society of the Pacific *84*, 357–364 (1972).
Michael W. Oventon, "The Origin of the Constellations." *The Philosophical Journal 3*, 1–18, (1966).
Neil L. Lark, "Astronomy in Science Fiction." *Mercury* 5, No. 3, 16–19 (1976).
Norriss S. Hetherington, "Man, Society and the Universe." *Mercury* 4, No. 6, 9–12 (1975).
Margaret M. Byard, "Poetic Responses to the Copernican Revolution." *Scientific American 236*, 120–129 (1977).

## CHAPTER 3

Stillman Drake, "Galileo's Discovery of the Law of Free Fall." *Scientific American 228,* 84–93 (1973).
S. Drake and J. Mac Lachlan, "Galileo's Discovery of the Parabolic Trajectory." *Scientific American 232,* 102–110 (1975).
C. Wilson, "How Did Kepler Discover His First Two Laws?" *Scientific American 226,* 92–96, 99–106 (1972).
E. Simonsen, "A Visit to Tycho Brahe's Observatory." *Sky and Telescope 47,* 86–88 (1974).
Joseph Ashbrook, "How Far Away are the Stars?" *Sky and Telescope 47,* 165 (1974).
Owen Gingerich, "Copernicus and Tycho." *Scientific American, 229,* 86–101 (1973).

## CHAPTER 4

William A. Fowler, "The Origin of the Elements." *Scientific American, 195,* 92–91 (1956).
David N. Schramm, "The Age of the Elements." *Scientific American, 230,* 69–77 (1974).
Stephen Boughn and Ho Jung Paik, "The Search for Gravitational Radiation." *Mercury 5,* No. 3, 9–15 (1976).
George Gamow, "Gravity." *Scientific American 204,* 94–110 (1961).

## CHAPTER 5

Robert P. Kraft, "Pulsating Stars and Cosmic Distances." *Scientific American, 201,* 48–55 (1959).
Geoffrey and Margaret Burbidge, "Stellar Populations." *Scientific American 199,* 44–65 (1958).
Icko Iben, Jr., "Globular-Cluster Stars." *Scientific American, 223,* 27–39 (1970).

## CHAPTER 6

K. I. Kellermann, "Intercontinental Radio Astronomy." *Scientific American, 226,* 72–83 (1972).
George H. Herbig, "Interstellar Smog." *American Scientist 62* 220-227 (1974).

Frank H. Shu, "Spiral Structure, Dust Clouds and Star Formation." *American Scientist 61*, 524–536 (1973).
Harold Weaver, "Steps Toward Understanding the Large Scale Structure of the Milky Way." *Mercury 4*, No. 4, 18–24; 4, No. 6, 18–19; 5, No. 1, 19–30 (1975–1976).
Bart J. Bok, "Updating Galactic Spiral Structure." *American Scientist, 60*, 709–722 (1972).

## CHAPTER 7

Barry E. Turner, "Interstellar Molecules." *Scientific American, 228*, 51–69 (1973).
Bart J. Bok, "The Birth of Stars." *Scientific American, 227*, 49–61 (1972).
Robert L. Dickman, "Bok Globules." *Scientific American 236*, 66–81 (1977).

## CHAPTER 8

William K. Hartmann, "Cratering in the Solar System." *Scientific American, 236* 84–99 (1977).
H. T. Simmons, "Mighty Jupiter." *Smithsonian* (1974).
Bruce C. Murray, "Mars from Mariner 9." *Scientific American, 228*, 49–69 (1973).
Kerry Joels, "Planetary Meteorology: A New Perspective on the Earth's Weather." *Mercury* 5, No. 4, 16–19 (1976).
John S. Lewis, "The Chemistry of the Solar System." *Scientific American 230*, 50–65 (1974).
"The Solar System." *Scientific American* 233, No. 3. (1975).

## CHAPTER 9

Peter M. Millman, "The Meteoritic Hazard of Interplanetary Travel." *American Scientist, 59*, 700–705 (1971).
Fred L. Whipple, "The Nature of Comets." *Scientific American 230* 48–57 (1974).
Clark R. Chapman, "The Nature of the Asteroids." *Scientific American 232*, 24–33 (1975).
Joseph Veverka, "Phobos and Deimos." *Scientific American 236*, 30–37 (1977).

## CHAPTER 10

M. F. Ingham, "The Spectrum of Airglow." *Scientific American 226*, 79–85 (1972).
F. R. Stephenson, "The Earliest Known Record of a Solar Eclipse." *Nature, 228*, 651–652 (1970).
John C. Brandt, "The Solar Wind Blows Some Good for Astronomy." *Smithsonian, 3*, 30–35 (1973).
Everly Driscoll, "Peeking Beneath the Sun's Skirts." *Science News, 104*, (1973).
Philip Morrison, "Neutrino Astronomy." *Scientific American 207*, 90–99 (1962).
John A. Eddy, "The Case of the Missing Sunspots." *Scientific American 236*, 80–92 (1977).

## CHAPTER 11

Philip A. Charles and J. Leonard Culhane, "X Rays from Supernova Remnants." *Scientific American 233*, 38–55 (1975).
Harold Gursky and P. J. Van de Heuvel, "X Ray Emitting Double Stars." *Scientific American 232*, 24–35 (1975).
Helmut A. Abt, "The Companions of Sunlike Stars." *Scientific American 236*, 96–105 (1977).
Robert A. Wilson, "Binary Stars: A Look at Some Interesting Developments." *Mercury 3*, No. 5, 4–12 (1974).
Lawrence Anderson, "X-rays from Degenerate Stars." *Mercury 5*, No. 5, 6–11; 5, No. 6, 2–6 (1976).
Louis C. Green, "Ordinary Stars, White Dwarfs, and Neutron Stars." *Sky and Telescope, 41*, 18–20 (1971).
Kip Thorne, "The Search for Black Holes." *Scientific American 231*, 32–43 (1974).
Ben Bova, "Obituary of Stars: Tale of Red Giants, White Dwarfs and Black Holes." *Smithsonian 4*, 54–63, (1973).
Stephen P. Maran, "In Nine Centuries, Search Unravels Many Parts of the Crab Nebula Mystery." *Smithsonian 1*, 50–57 (1970).
David J. Helfand, "Pulsars: Physics Laboratories in Our Galaxy." *Mercury 6*, 2–7 (1977).
F. Richard Stephenson and David H. Clark, "Historical Supernovas." *Scientific American 234*, 100–107 (1976).

## CHAPTER 12

Robert P. Kirshner, "Supernovas in Other Galaxies." *Scientific American 235*, 88–101 (1976).

Allan Sandage, "The Red-Shift." *Scientific American 195*, 170–186 (1956).

Alar Toomre and Juri Toomre, "Violent Tides Between Galaxies." *Scientific American, 229*, 38–48 (1973).

David S. Heeschen, "Radio Galaxies." *Scientific American 206*, 41–49 (1962).

Richard B. Larsen, "The Origin of the Galaxies." *American Scientist 65*, 188–196 (1977).

## CHAPTER 13

J. J. Callahan, "The Curvature of Space in a Finite Universe." *Scientific American 235*, 90–100 (1976).

Thomas C. Van Flandern, "Is Gravity Getting Weaker?" *Scientific American 234*, 44–59 (1976).

Ralph A. Alpher, "Large Numbers, Cosmology, and Gamow." *American Scientist 61*, 52–58 (1973).

Jay M. Pasachoff and William A. Fowler, "Deuterium in the Universe." *Scientific American 230*, 108–118 (1974).

Carl Sagan, "On the Detectivity of Advanced Galactic Civilizations." *Icarus 19*, 350–352 (1973).

Adrian Webster, "The Cosmic Background Radiation." *Scientific American 231*, 26–33 (1974).

Carl Sagan and Frank Drake, "The Search for Extraterrestrial Intelligence." *Scientific American 232*, 80–89 (1975).

J. Richard Gott III, James E. Gunn, David N. Schramm and Beatrice M. Tinsley, "Will the Universe Expand Forever?" *Scientific American 234*, 62–79 (1976).

Peter van de Kamp, "Barnard's Star: The Search for Other Solar Systems." *Natural History, 79*, 38–43 (1970).

Carl Sagan, Linda S. Sagan, and Frank Drake, "A Message from Earth." *Science, 175*, 881–884 (1972).

Georges Lemaitre, *The Primeval Atom*. New York: D. Van Nostrand, 1950.

William H. McCrea, "Cosmology Today." *American Scientist, 58*, 521–527 (1970).

Ivan R. King, "Man in the Universe." *Mercury* 5, 7–8 (1976).
George H. Herbig, "A Universe Teeming with Planetary Systems—Perhaps." *Mercury* 5, No. 2, 2–7, 31 (1976).
Ralph I. Palsson, "The Challenge of Viking: Is There Life on Mars?" *Mercury* 5, No. 2, 14–18 (1976).
Fred Hoyle, "The Steady-State Universe." *Scientific American* 195, 157–169 (1956).
Virginia Trimble, "Cosmology: Man's Place in the Universe." *American Scientist* 65, 76–86 (1977).
Sidney van den Bergh, "Are We Alone?" *Mercury* 4, No. 4, 8–10 (1975).
Don Albert, "The Meaning of Curved Space." *Mercury* 4, No. 4, 16–19 1975).

# APPENDIX



## BRIEF TABLE OF DECIMAL MULTIPLES

| *Decimal* | *Power Notation* | *Prefix* | *Symbol* |
|---|---|---|---|
| 0.001 | $10^{-3}$ | milli- | m |
| 0.01 | $10^{-2}$ | centi- | c |
| 0.1 | $10^{-1}$ | deci- | d |
| 1 | $10^{0}$ | | |
| 10 | $10^{1}$ | deca- | dk |
| 100 | $10^{2}$ | hecto- | h |
| 1,000 | $10^{3}$ | kilo- | k |
| 1,000,000 | $10^{6}$ | mega- | M |
| 1,000,000,000 | $10^{9}$ | giga- | G |

Powers-of-ten notation, also called scientific notation, is extremely convenient for writing very large or very small numbers. The rules for multiplication and division in this notation are simple. For multiplication we have $10^a \times 10^b = 10^{a+b}$, thus $10^2 \times 10^5 = 10^7$. For division we have $10^c \div 10^d = 10^{c-d}$, thus $10^2 \div 10^5 = 10^{-3}$.

## ENGLISH-METRIC CONVERSION UNITS

The principal advantage of the metric system over the English system is that the metric system is based upon powers of ten. Any powers-of-ten system is as good as any other, but the metric system has the advantage that it has been adopted by more people than any other single system.
1 inch = 2.54 centimeters
1 foot = 30.48 centimeters = 0.3048 meter
1 yard = 91.44 centimeters = 0.9144 meter
1 mile = 160934 centimeters = 1609.3 meters = 1.6093 kilometers
1 ounce = 28.3495 grams = 0.0283 kilogram
1 pound = 453.6 grams = 0.4536 kilogram
1 pint (fluid) = 47.32 centiliters = 0.4732 liter
1 quart (fluid) = 94.64 centiliters = 0.9464 liter
1 kilometer = 0.6214 miles
1 meter = 1.0936 yards
1 centimeter = 0.3937 inch
1 liter = 2.1134 pints (fluid)
1 gram = 0.0353 ounce

## TEMPERATURE SCALES

The three common temperature scales are absolute (Kelvin), centigrade (Celsius), and Fahrenheit. The absolute and centigrade scales are directly related by the constant 273. To convert absolute temperatures (°K) to centigrade temperatures (°C), subtract 273 from the absolute temperature (°K), as follows,

$$(°K) - 273 = (°C)$$

To convert centigrade to absolute, add 273 to the centigrade temperature (°C) as follows,

$$(°C) + 273 = (°K)$$

The conversion from Fahrenheit to centigrade (and vice versa) is not as simple. There are 180 degrees Fahrenheit between the freezing and boiling points of water and only 100 degrees centigrade. Further, the Fahrenheit scale defines the freezing point of water as 32° whereas on the centigrade scale it is zero. Thus to convert Fahrenheit (°F) to centigrade (°C), we must subtract 32 and multiply by 100/180 or 5/9 as follows,

$$5/9\,[(°F) - 32] = (°C)$$

Conversely,

$$(°C)\,9/5 + 32 = (°F)$$

To convert Fahrenheit to absolute, convert to centigrade and add 273.

## SOME COMMON TEMPERATURES

| | °F | °C | °K |
|---|---|---|---|
| Absolute zero | −459 | −273 | 0 |
| Freezing point of water | +32 | 0 | +273 |
| Room temperature | +68 | +20 | +293 |
| Body temperature | +98.6 | +37 | +310 |
| Boiling point of water | +212 | +100 | +373 |

## BASIC ASTRONOMICAL DATA

| | |
|---|---|
| Earth's equatorial radius | 6378 km |
| Earth's mass | $5.98 \times 10^{27}$g |
| Velocity of light | 299,793 km/sec |
| Moon's mass | 1/81.33 of Earth's |
| Moon's radius | 1738 km |
| Moon's mean distance | 384,404 km |
| Astronomical unit (AU) | 149,598,000 km |
| Gravitational constant* | $6.6730 \times 10^{-8}$dyn cm$^2$g$^{-2}$ |
| Sun's mass | $1.99 \times 10^{33}$g |
| Sun's radius | 696,000 km |
| Light year (lt-yr) | $9.46 \times 10^{12}$ km |

*Latest value by J. W. Beams, 1975.

## CHARACTERISTICS OF MEMBERS OF THE SOLAR SYSTEM

| Characteristics | Sun | Mercury | Venus | Earth | Moon | Mars | Ceres | Jupiter | Saturn | Uranus | Neptune | Pluto |
|---|---|---|---|---|---|---|---|---|---|---|---|---|
| Symbol | ⊙ | ☿ | ♀ | ⊕ | ☾ | ♂ | ① | ♃ | ♄ | ♅ | ♆ | ♇ |
| Distance from Sun (Astronomical Units) | — | 0.387 | 0.723 | 1.000 | 1.000 | 1.524 | 2.767 | 5.203 | 9.539 | 19.182 | 30.058 | 39.439 |
| Distance from Sun (million's of kilometers) | — | 57.91 | 108.20 | 149.60 | 149.60 | 227.94 | 413.98 | 778.33 | 1429.99 | 2869.57 | 4496.60 | 5900.00 |
| Period of revolution (years) | — | 0.241 | 0.615 | 1.000 | 0.075 | 1.881 | 4.604 | 11.862 | 29.458 | 84.013 | 164.794 | 247.686 |
| Obital inclination to ecliptic | — | 7°0′ | 3°24′ | 0.0 | 5°9′ | 1°51′ | 10°37′ | 1°18′ | 2°29′ | 0°46′ | 1°46′ | 17°10′ |
| Period of rotation (days) | 25.38 | 58.65 | −243 | 0.997 | 27.322 | 1.026 | — | 0.413 | 0.426 | 1.042 | 0.917 | 6.39 |
| Inclination of equator to orbit | 7°15′ | 0° | 3°18′ | 23°27′ | 6°41′ | 23°59′ | — | 3°4′ | 26°44′ | 97°53′ | 28°48′ | — |
| Diameter (kilometers) | 1,392,000 | 4868 | 12,112 | 12,756 | 3476 | 6787 | 780 | 143,200 | 120,000 | 50,800 | 49,500 | 5800: |
| Mass (earth = 1) | 332,960 | 0.05 | 0.82 | 1.0 | 0.012 | 0.11 | — | 317.9 | 95.12 | 14.6 | 17.2 | 0.11: |
| Density (water = 1) | 1.41 | 5.44 | 5.26 | 5.52 | 3.34 | 3.94 | — | 1.314 | 0.704 | 1.31 | 1.66 | 4.9: |
| Surface temperature (daytime °K) | 6000 | 700 | 740 | 295 | 400 | 250 | — | 123 | 93 | 63 | 53 | 40 |
| Main atmospheric constituent | Hydrogen Helium | — | Carbon dioxide | Nitrogen Oxygen | — — | Carbon dioxide | — | Hydrogen Helium | Hydrogen Helium | Hydrogen Helium | Hydrogen Helium | |
| Number of moons | — | 0 | 0 | 1 | — | 2 | — | 14 | 11 | 5 | 2 | — |

: denotes uncertain values.

## THE TWENTY-FIVE BRIGHTEST STARS

| Name | Spectrum | Apparent Magnitude | Parallax | Absolute Visual Magnitude |
|---|---|---|---|---|
| 1. αCMa, Sirius | A1 V | −1.43 | 0″.377 | +1.5 |
| 2. αCar, Canopus | F0 Ia | −0.73 | .018 | −4.4 |
| 3. Alpha Centauri, d | G2 V | −0.27 | .760 | +4.1 |
| 4. αBoo, Arcturus | K2 IIIp | −0.06 | .090 | −0.3 |
| 5. αLyr, Vega | A0 V | +0.04 | .123 | +0.5 |
| 6. αAur, Capella | (G0) | +0.09 | .073 | −0.6 |
| 7. βOri, Rigel | B8 Ia | +0.15 | .005 | −6.4 |
| 8. αCMi, Procyon | F5 IV-V | +0.37 | .287 | +2.7 |
| 9. αEri, Achernar | B3 V | +0.53 | .023 | −2.7 |
| 10. Beta Centauri, d | B0.5 V | +0.66 | .016 | −3.3 |
| 11. αOri, Betelgeuse, v | M2 Iab | +0.7 | .017 | −2.9 |
| 12. αAql, Altair | A7 IV, V | +0.80 | .196 | +2.3 |
| 13. αTau, Aldebaran, v | K5 III | +0.85 | .048 | −0.7 |
| 14. Alpha Crucis, d | B0.5 V | +0.87 | .015 | −3.2 |
| 15. αSco, Antares, v, d | M1 Ib | +0.98 | .019 | −2.6 |
| 16. αVir, Spica, d | B1 V | +1.00 | .021 | −2.4 |
| 17. αPis A, Formalhaut | A3 V | +1.16 | .144 | +2.0 |
| 18. βGem, Pollux | K0 III | +1.16 | .093 | +1.0 |
| 19. αCyg, Deneb | A2 Ia | +1.26 | .006 | −4.8 |
| 20. Beta Crucis | B0.5 IV | +1.31 | (.011) | −3.5 |
| 21. αLeo, Regulus | B7 V | +1.36 | .039 | −0.7 |
| 22. εCMa, Adhara | B2 II | +1.49 | (.012) | −3.1 |
| 23. αGem, Castor, d | (A0) | +1.59 | .072 | +1.0 |
| 24. λSco, Shaula | B2 IV | +1.62 | (.026) | −1.3 |
| 25. γOri, Bellatrix | B2 III | +1.64 | (.026) | −1.3 |

d—indicates a double star with a magnitude difference less than 5; combined magnitudes are given.
v—indicates variable star.
()—indicates estimated values.

## THE TWENTY-FIVE NEAREST STARS

| No. | Name | Proper Motion per Year | Parallax | Distance in Light-Years | Visual Apparent Magnitude and Spectrum: Primary Star | Companion Star |
|---|---|---|---|---|---|---|
| 1 | Sun | — | — | — | −26.8 G2 | — |
| 2 | α Centauri | 3″.68 | 0.760 | 4.3 | 0.1 G2 | 1.5 K5† |
| 3 | Barnard's star | 10.30 | .552 | 5.9 | 9.5 M5 | * |
| 4 | Wolf 359 | 4 .84 | .431 | 7.6 | 13.5 M6e | — |
| 5 | Lalande 21185 | 4 .78 | .402 | 8.1 | 7.5 M2 | * |
| 6 | Sirius | 1 .32 | .377 | 8.6 | − 1.5 A1 | 7.2 wd‡ |
| 7 | Luyten 726-8 | 3 .35 | .365 | 8.9 | 12.5 M6e | 13.0 M6 |
| 8 | Ross 154 | 0 .74 | .345 | 9.4 | 10.6 M5e | — |
| 9 | Ross 248 | 1 .82 | .317 | 10.3 | 12.2 M6e | — |
| 10 | ϵ Eridani | 0 .97 | .305 | 10.7 | 3.7 K2 | * |
| 11 | Luyten 789-6 | 3 .27 | .302 | 10.8 | 12.2 M6 | — |
| 12 | Ross 128 | 1 .40 | .301 | 10.8 | 11.1 M5 | — |
| 13 | 61 Cygni | 5 .22 | .292 | 11.2 | 5.2 K5 | 6.0 K7* |
| 14 | ϵ Indi | 4 .67 | .291 | 11.2 | 4.7 K5 | — |
| 15 | Procyon | 1 .25 | .287 | 11.4 | 0.3 F5 | 10.8 wd‡ |
| 16 | Σ 2398 | 2 .29 | .284 | 11.5 | 8.9 M3.5 | 9.7 M4 |
| 17 | Groombridge 34 | 2 .91 | .282 | 11.6 | 8.1 M1 | 11.0 M6 |
| 18 | Lacaille 9352 | 6 .87 | .279 | 11.7 | 7.4 M2 | — |
| 19 | τ Ceti | 1 .92 | .273 | 11.9 | 3.5 G8 | — |
| 20 | BD +5°1668 | 3 .73 | .266 | 12.2 | 9.8 M4 | * |
| 21 | Lacaille 8760 | 3 .46 | .260 | 12.5 | 6.7 M1 | — |
| 22 | Kapteyn's star | 8 .79 | .256 | 12.7 | 8.8 M0 | — |
| 23 | Krüger 60 | 0 .87 | .254 | 12.8 | 9.7 M4 | 11.2 M6 |
| 24 | Ross 614 | 0 .97 | .249 | 13.1 | 11.3 M5e | 14.8 — |
| 25 | BD −12°4523 | 1 .18 | .249 | 13.1 | 10.0 M5 | — |

*Unseen components.

†Alpha Centauri is a triple star system. The third star, Proxima, has an apparent magnitude of 11 and an M5 spectrum.

‡wd—indicates white dwarf.

## A BRIEF CHRONOLOGY OF ASTRONOMY

| | |
|---|---|
| ca. 3000 B.C. | The earliest known recorded observations are made in Babylonia |
| ca. 1900–1600 B.C. | Stonehenge under construction |
| ca. 1400 B.C. | Earliest known Chinese calendar |
| ca. 1000 B.C. | Earliest recorded Chinese, Hindu observations |
| ca. 800 B.C. | Earliest preserved sundial (Egyptian) |
| ca. 500 B.C. | Pythagorean school advances concept of celestial motions on concentric spheres |
| ca. 430 B.C. | Anaxagoras explains eclipses and phases of the moon |

| | |
|---|---|
| ca. 400 B.C. | Philolaus speculates that the earth moves |
| ca. 400–300 B.C. | Several cosmological systems involving moving concentric spheres proposed by Plato, Eudoxus, and others |
| ca. 350 B.C. | Earliest known star catalog (Chinese) |
| ca. 250 B.C. | Aristarchus advances arguments favoring a heliocentric cosmology |
| ca. 200 B.C. | Eratosthenes measures earth's diameter |
| 160–127 B.C. | Hipparchus develops trigonometry, analyzes generations of observational data, obtains highly accurate celestial observations |
| ca. A.D. 140 | Ptolemy measures distance to moon; proposes geocentric cosmology involving epicycles |
| 1054 | Chinese observe supernova in Taurus |
| 1543 | Copernicus publishes *De Revolutionibus* with the heliocentric theory |
| 1572 | Tycho Brahe observes supernova; immutability of celestial sphere cast in doubt |
| 1546–1601 | Tycho accurately measures motions of the planets |
| 1608 | Lippershey invents the telescope |
| 1609 | Kepler, using Tycho's measurements, shows planets move in ellipses |
| 1609 | Galileo uses telescope to observe moons of Jupiter and crescent phase of Venus, thus lending support to Copernican hypothesis; Galileo conducts experiments in dynamics |
| 1675 | Romer measures the velocity of light |
| 1686–1687 | Newton's *Principia:* Newton combines the results of terrestrial and celestial natural philosophy to obtain the fundamental laws of motion and gravity |
| ca. 1690 | Halley shows the great comets observed every 75 years are one and the same comet in an elliptical orbit; he discovers proper motions of stars |
| ca. 1690 | Huygens makes estimate for distance to stars based upon assumption that the sun is a typical star |
| 1727 | Bradley observes aberration of starlight, conclusive proof of Copernican theory |
| ca. 1750 | Wright proposes disk model for Milky Way |
| 1755 | Kant proposes nebulae are "island universes"; proposes solar system formed from rotating cloud of gas |
| 1738–1822 | W. Herschel constructs large telescopes; discovers Uranus (1781); observes gaseous nebulae |
| 1801 | First asteroid discovered |

| | |
|---|---|
| 1802 | Solar spectrum first viewed |
| 1803 | W. Hershel discovers binary stars |
| 1837 | First stellar parallax measured (by F.G.W. Struve) |
| 1840 | Draper produces first astronomical photograph |
| 1843 | The effect of motion on light spectra is explained by Doppler |
| 1845 | Earl of Rosse discovers spiral structure of some "nebulae" |
| 1846 | Neptune predicted independently by Leverrier and Adams and discovered by Galle |
| 1850–1900 | Development of spectrum analysis; stellar spectra used for first time to obtain temperatures and compositions of stars |
| 1877 | Schiaparelli sees "canals" on Mars |
| 1905 | Einstein's special theory of relativity |
| 1905–1920 | Einstein develops his theory of gravitation; general relativity |
| 1914 | Slipher discovers that spiral nebulae are receding from us; Lemaitre, DeSitter, and Eddington explain this phenomenon using general relativity |
| 1915 | Hooker reflecting telescope (2.5m) constructed at Mount Wilson |
| 1917 | Shapley shows that solar system is more than 30,000 lt-yr from the center of the Milky Way galaxy |
| 1924 | Hubble measures distances to spirals and confirms the viewpoint that they are galaxies in their own right |
| 1910–1930 | Russell, Eddington, and others develop the theory of stellar structure |
| 1920–1930 | Shapley, Oort, Linblad investigate rotation of Milky Way galaxy |
| 1930–1960 | Nuclear physics develops; used to explain the energy source of the stars |
| 1930 | Pluto discovered by Tombaugh |
| 1931 | Jansky discovers extra-terrestrial radio radiation |
| 1937 | Discovery of first interstellar molecule |
| 1947–1960 | Astronomical instruments sent by rocket above earth's atmosphere |
| 1949 | Great Hale reflector (5m) went into routine operation |
| 1951 | Observation of neutral hydrogen at 21 cm wavelength |
| 1957 | Sputnik I orbits earth |
| 1959 | Space probe hits the moon |
| 1961 | Yuri Gagarin becomes first person in space |

1963 Discovery of quasi-stellar objects
1965 Discovery of 3°K background radiation
1965 First close photographs of Mars by Mariner 4
1968 Discovery of pulsars
1969 Apollo 11 lands first men on the moon
1969 Discovery of first complex organic interstellar molecule (formaldehyde)
1973 First close photographs of Jupiter by Pioneer 10
1974 First close photographs of Venus and Mercury by Mariner 10
1974 Confirmation that some X-ray binary components are black holes
1975 Discovery of fourteenth satellite of Jupiter
1975 Venera 9 lands on Venus
1975 Radio "pictures" depict certain radio galaxies having emitting regions 17,000,000 lt-yr across
1976 Viking 1 & 2 place landers on Mars
1977 Discovery of eleventh satellite of Saturn
1977 Discovery of rings of Uranus

## GREAT REFRACTING TELESCOPES (>65CM)

| *Year* | *Optician* | *Observatory and Location* | *Objective (cm)* | *Focal Length (cm)* |
|---|---|---|---|---|
| 1897 | Alvan Clark | Yerkes Observatory, Williams Bay, Wisconsin | 102 | 1935 |
| 1888 | Alvan Clark | Lick Observatory, Mt. Hamilton, California | 91 | 1760 |
| 1893 | Henry Brothers | Observatorie de Paris, Meudon, France | 83 | 1615 |
| 1899 | Steinheil | Astrophysikalisches Observatory, Potsdam, Germany | 80 | 1200 |
| 1886 | Henry Brothers | Bischottsheim Observatory, University of Paris, at Nice, France | 76 | 1600 |
| 1914 | Brashear | Allegheny Observatory, Pittsburgh, Pennsylvania | 76 | 1411 |
| 1894 | Howard Grubb | Royal Greenwich Observatory, Herstmonceux, England | 71 | 850 |
| 1878 | Howard Grubb | Universitäts-Sternwarte, Vienna, Austria | 67 | 1050 |
| 1925 | Howard Grubb | Union Observatory, Johannesburg, South Africa | 67 | 1070 |
| 1883 | Alvan Clark | Leander McCormick Observatory, Charlottesville, Virginia | 66 | 1000 |
| 1873 | Alvan Clark | U. S. Naval Observatory, Washington, D. C. | 66 | 990 |
| 1953* | McDowell | Mount Stromlo, Canberra, Australia | 66 | 1100 |
| 1897 | Howard Grubb | Royal Greenwich Observatory, Herstmonceux, England | 66 | 680 |

*First used in Johannesburg, South Africa, 1926.

## GREAT REFLECTING TELESCOPES (> 3 METERS)

| Year | Observatory and Location | Objective (meters) |
|---|---|---|
| 1976 | Zelenchukskaya Astrophysical Observatory, U.S.S.R. | 6.0 |
| 1948 | Hale Observatory, Mt. Palomar, California | 5.1 |
| 1973 | Kitt Peak National Observatory, Kitt Peak, Arizona | 4.0 |
| 1974 | Cerro Tololo Inter-American Observatory, Cerro Tololo, Chile | 4.0 |
| 1975 | Australian National Observatory, Siding Spring Mtn., Australia | 3.8 |
| 1975 | European Southern Observatories, Cerro La Silla, Chile | 3.6 |
| U.C.* | French, Canadian, Hawaiian Observatory, Mauna Kea, Hawaii | 3.5 |
| 1959 | Lick Observatory, Mount Hamilton, California | 3.0 |

*Under construction.

## NORTH AMERICAN OBSERVATORIES

| | |
|---|---|
| Algonquin Radio Observatory | Lake Traverse, Ontario |
| Allegheny Observatory | Pittsburgh, Pennsylvania |
| Amherst College Observatory | Amherst, Massachusetts |
| Arthur J. Dyer Observatory | Nashville, Tennessee |
| Arecibo Observatory | Arecibo, Puerto Rico<br>Ithaca, New York |
| Bell Telephone Laboratories | Newstead, New York<br>Holmdel, New Jersey |
| Big Bear Solar Observatory | Big Bear City, California |
| Boston University Observatory | Boston, Massachusetts |
| Bowdoin College Observatory | Brunswick, Maine |
| Bradley Observatory | Decatur, Georgia |
| California Institute of Technology | Big Pine, California<br>Goldstone, California |
| Capilla Peak Observatory | Capilla Peak, New Mexico |
| Carnegie Institution of Washington | Derwood, Maryland |
| Catalina Station | Tucson, Arizona |
| Chabot Observatory | Oakland, California |
| Chamberlain Observatory | Denver, Colorado |
| Cincinnati Observatory | Cincinnati, Ohio |
| Columbia University Observatory | New York, New York |
| Corralitos Observatory | Las Cruces, New Mexico |
| Crawford Hill Laboratory | Holmdel, New Jersey |
| David Dunlap Observatory | Richmond Hill, Ontario |
| Dearborn Observatory | Evanston, Illinois |
| Dominion Astrophysical Observatory | Victoria, British Columbia |
| Dominion Observatory | Ottawa, Ontario |

| Observatory | Location |
|---|---|
| Dominion Radio Astrophysical Observatory | Penticton, British Columbia |
| Drake University Municipal Observatory | Des Moines, Iowa |
| Dudley Observatory | Albany, New York |
| Erwin W. Fick Observatory | Ames, Iowa |
| Feather Ridge Observatory | Cedar Rapids, Iowa |
| Fernbank Observatory | Atlanta, Georgia |
| Five College Observatories | Amherst, Massachusetts<br>New Salem, Massachusetts |
| Florida State University Radio Observatory | Tallahassee, Florida |
| Flower and Cook Observatory | Philadelphia, Pennsylvania |
| Franklin Institute Observatory | Philadelphia, Pennsylvania |
| Fuertes Observatory | Ithaca, New York |
| George R. Wallace Jr. Astrophysical Observatory | Cambridge, Massachusetts |
| Georgetown College Observatory | Washington, D.C. |
| Goddard Space Flight Center | Greenbelt, Maryland |
| Goethe Link Observatory | Bloomington, Indiana |
| Goodsell Observatory | Northfield, Minnesota |
| Griffith Observatory | Los Angeles, California |
| Hale Observatories | Pasadena, California<br>Mt. Palomar, California<br>Mt. Wilson, California |
| Harvard College Observatory | Cambridge, Massachusetts |
| Harvard Radio Astronomy Station | Fort Davis, Texas |
| Haystack Lincoln Laboratory | Tyngsboro, Massachusetts |
| High Altitude Observatory | Boulder, Colorado |
| Hopkins Observatory | Williamstown, Massachusetts |
| Institute for Astronomy | Honolulu, Hawaii |
| Institute of Astronomy and Space Science | Vancouver, British Columbia |
| Joint Observatory for Cometary Research | Socorro, New Mexico |
| Kansas University Observatory | Lawrence, Kansas |
| Kenneth Mees Observatory | Rochester, New York |
| Kitt Peak National Observatory | Tucson, Arizona |
| Kutztown State College Observatory | Kutztown, Pennsylvania |
| Ladd Observatory | Providence, Rhode Island |
| Leander McCormick Observatory<br>Fan Mountain Observing Station | Charlottesville, Virginia |
| Leuschner Observatory | Berkeley, California |
| Lick Observatory | Santa Cruz, California |

| | |
|---|---|
| Lindheimer Astronomical Research Center | Evanston, Illinois |
| Lockheed Solar Observatory | Saugus, California |
| Louisiana State University Observatory | Baton Rouge, Louisiana |
| Louisville University Observatory | Louisville, Kentucky |
| Lowell Observatory | Flagstaff, Arizona |
| Lunar and Planetary Laboratory | Tucson, Arizona |
| Maria Mitchell Observatory | Nantucket, Massachusetts |
| McDonald Observatory | Fort Davis, Texas |
| McGraw-Hill Observatory | Kitt Peak, Arizona |
| McKim Observatory | Greencastle, Indiana |
| McMath Hulbert Observatory | Pontiac, Michigan |
| Melton Memorial Observatory | Columbia, South Carolina |
| Michigan State University Observatory | East Lansing, Michigan |
| Morrison Observatory | Gayette, Missouri |
| Mount Cuba Observatory | Wilmington, Delaware |
| Mummy Mountain Observatory | Scottsdale, Arizona |
| Nassau Astronomical Station | Montville, Ohio |
| National Observatory | Tacubaya, Mexico |
| National Research Council | Ottawa, Ontario |
| National Radio Astronomy Observatory | Charlottesville, Virginia<br>Green Bank, West Virginia<br>Socorro, New Mexico<br>Tucson, Arizona |
| New Mexico State University Observatory | Las Cruces, New Mexico |
| Observatories of the University of Western Ontario | London, Canada |
| Ohio State University Radio Observatory | Columbus, Ohio |
| Ottawa River Solar Observatory | Ottawa, Ontario |
| Owens Valley Radio Observatory | Big Pine, California |
| Pan American College Observatory | Edinburg, Texas |
| Perkins Observatory | Delaware, Ohio |
| Portage Lake Observatory | Portage Lake, Michigan |
| Princeton University Observatory | Princeton, New Jersey |
| Quebec Observatory | Quebec, Canada |
| Radio Observatory | Kingston, Ontario |
| Radio Observatory (U. of CA) | Hat Creek, California |
| Rensselaer Observatory Sampson Station | Troy, New York |
| Ritter Observatory | Toledo, Ohio |

| | |
|---|---|
| Rosemary Hill Observatory | Gainesville, Florida |
| Sacramento Peak | Sunspot, New Mexico |
| Sagamore Hill Radio Observatory | Bedford, Massachusetts |
| San Diego State University Observatory | San Diego, California |
| San Fernando Observatory | El Segundo, California |
| Santa Clara Observatory | Santa Clara, California |
| Sayre Observatory | South Bethlehem, Pennsylvania |
| Shattuck Observatory | Hanover, New Hampshire |
| Smith College Observatory | Northhampton, Massachusetts |
| Smith Observatory | Geneva, New York |
| Smithsonian Astrophysical Observatory | Cambridge, Massachusetts |
| | Tucson, Arizona |
| Space Radio Research Facility | El Segundo, California |
| Sproul Observatory | Swarthmore, Pennsylvania |
| Stanford Research Institute | Stanford, California |
| | College, Alaska |
| | Palo Alto, California |
| Strawbridge Observatory | Haverford, Pennsylvania |
| Steward Observatory | Tucson, Arizona |
| Syracuse University Observatory | Syracuse, New York |
| Table Mountain Observatory | Wrightwood, California |
| Thompson Observatory | Beloit, Wisconsin |
| Tiara Observatory | South Park, Colorado |
| National Astrophysical Observatory | Tonantzintla, Mexico |
| United States Naval Observatory | Washington, D.C. |
| | Flagstaff, Arizona |
| | Richmond, Florida |
| University of Alabama Observatory | University, Alabama |
| University of Florida Radio Observatory | Gainesville, Florida |
| University of Illinois Observatory | Urbana, Illinois |
| | Danville, Illinois |
| | Oakland, Illinois |
| University of Iowa Observatory | Riverside, Iowa |
| University of Maine Observatory | Orono, Maine |
| University of Maryland Observatory | College Park, Maryland |
| | Borrego Springs, California |
| University of Michigan Observatory | Ann Arbor, Michigan |
| | Tucson, Arizona |
| University of Minnesota Observatory | Minneapolis, Minnesota |
| University of Mississippi Observatory | Oxford, Mississippi |

| | |
|---|---|
| University of Oklahoma Observatory | Norman, Oklahoma |
| University of South Florida Observatory | Tampa, Florida |
| Van Norman Solar Observatory | San Fernando, California |
| Van Vleck Observatory | Middletown, Connecticut |
| Vassar College Observatory | Poughkeepsie, New York |
| Wallace Astrophysical Observatory | Cambridge, Massachusetts |
| Warner and Swasey Observatory | Cleveland, Ohio |
| Washburn Observatory | Madison, Wisconsin |
| Washington University Observatory | St. Louis, Missouri |
| Western State College Observatory | Danbury, Connecticut |
| Whitin Observatory | Wellesley, Massachusetts |
| Williston Observatory | South Hadley, Massachusetts |
| Yale University Observatory | New Haven, Connecticut |
| Yerkes Observatory | Williams Bay, Wisconsin |

# GLOSSARY

## A

**aberration** (1) of starlight—the apparent shift in direction of the observed object due to the tangential motion of the observer. (2) of optics—defects in an optical image more or less correctable by careful design.

**absolute magnitude** the brightness a star or object would have at the arbitrary distance of 32.6 lt-yr designated by M. Sometimes subscripts indicate the region of the spectrum referred to.

**absolute zero** by definition, the lowest possible temperature, 0°K or −273°C; almost all molecular motion ceases at this temperature.

**absorption spectrum** a dark-line spectrum superposed on a continuous spectrum and caused by a cool gas between the observer and the continuous source.

**acceleration** rate of change in speed or direction of motion.

**achromat** a compound lens of two or more simple lenses corrected to remove chromatic aberration over a broad region of the visual spectrum.

**aerolite** a stoney meteorite.

**albedo** the reflectivity of a body compared to the reflectivity of a perfect reflector of the same size, shape, orientation, and distance. Usually given as a percentage.

**almanac** a yearly calendar containing astronomical data and tables of useful information.

**amplitude** the range of variability.

**angular diameter** the angle subtended by the diameter of an object.

**angular distance** the angle on the celestial sphere along a great circle between two objects.

**anomalistic month** the month measured by the moon's motion from perigee to perigee.

**antimatter** postulated matter composed of the counterparts of ordinary matter.

**apastron** see apo

**apex** usually the solar apex, the direction toward which the sun is moving. The apex of a cluster's motion is either the convergent point or the divergent point.

**aphelion** see apo

**apo** a prefix meaning the farthest point in an orbit, thus apogee is the farthest point in the orbit of an object orbiting the earth. Occasionally reduced to ap- (as in aphelion).

**apogee** see apo

**apparent magnitude** the brightness of an object in magnitudes as observed; designated by m and sometimes a subscript denoting the spectral region referred to. The apparent magnitude equals the absolute magnitude at a distance of 32.6 lt-yr.

**apparent sun** the true sun, i.e., the sun we see.

**apparition** (1) the period during which a celestial object is visible. (2) the sudden appearance of a celestial object such as a comet.

**appulse** a penumbral lunar eclipse.

**ascendant** the sign of the zodiac on the eastern horizon at any given time.

**aspect** the position of one celestial object with respect to another.

**asterism** a named grouping of stars within a constellation of a different name.

**asteroid** a minor planet.

**astrology** (1) a primitive religion originating in the Near East. (2) a pseudoscience.

**astrometry** the branch of astronomy dealing with the positions, distances, and motions of the planets and stars; it includes the determination of time and position.

**astronautics** application of the physical laws of motion to space flight.

**astrophysics** the branch of astronomy that applies the theories and laws of physics to celestial bodies.

**aurora** a lighting of the northern sky, the Aurora Borealis, or the southern sky, Aurora Australis, centered on the geomagnetic poles caused by electrons and protons that are released by solar upheavals and trapped by the earth's magnetic field.

**autumnal equinox** that point on the celestial equator where the sun passes from above to below the equator; approximately 23 September.

**azimuth** the angle from the north point eastward along the horizon to the vertical circle passing through the object in question.

# B

**Bailey's beads** in a total solar eclipse the sun's chromosphere shows between the mountains and ridges of the moon giving a beaded appearance.

**Balmer continuum** a continuous absorption spectrum produced when energy absorbed by the electrons of hydrogen at the second energy level is in excess of that necessary to remove the electrons from the atoms.

**Balmer lines** transitions in the hydrogen atom from upper levels to the second level (emission lines) or from the second level to upper levels (absorption lines).

**binary (binary star)** a pair of stars orbiting each other by their mutual gravitational attraction.

**bipolar** having two poles. A magnetic rod having both a plus pole and a minus pole is said to be a bipolar magnet.

**blackbody** a hypothetically perfect radiator that absorbs and reemits all energy falling on it.

**black dwarf** the end product of stellar evolution.

**black hole** the region around a gravitationally collapsed mass from which no signal or light can escape.

**blink comparator** a measuring machine that alternately allows the measurer to see first one photograph and then a second. Used to discover moving objects or variable stars. It is often referred to simply as the blink.

**bolide** a meteor of extreme brightness; a fireball that breaks up.

**breccia** a stone made up of sharp fragments held together by clay, lime, or sand.

**bremsstrahlung** energy radiated by an electron being accelerated usually by a nucleus or another charged particle.

**brightness** loosely and variously used by astronomers to indicate luminosity of a body or intensity of radiation.

# C

**cardinal points** the four primary points of the compass: north, east, south, west.

**catalyst** an agent that assists or speeds a reaction with no change in itself in the end product. Carbon is a catalyst in the carbon-nitrogen–oxygen cycle.

**celestial horizon** the ideal horizon, 90° from the zenith; the horizon as seen by the eye at sea level.

**celestial mechanics** the branch of astronomy dealing with the motions of multiple bodied systems.

**chondrule** rounded (spherical) granules found to make up stony meteorites.

**chromatic aberration** one of the major aberrations of an optical system where the system fails to bring all wavelengths to the same focus.

**chronograph** an accurate timekeeping device.

**chronometer** an exceedingly accurate clock.

**coelostat** a stationary telescope looking at a flat mirror that is in a polar mount. It is arranged to keep the sun or a star field fixed in the telescope.

**collimator** optical lens (lenses) and/or mirror (mirrors) designed to render light in parallel rays. A telescope is a reverse collimator.

**coma** (1) one of the major aberrations of an optical system giving a comet-like appearance to a star image. (2) the hazy material around the nucleus of a comet. (3) when capitalized, the great cluster of galaxies in the constellation Coma Berenices.

**conjunction** two celestial bodies on the same longitude. (1) Inferior conjunction occurs when a planet or body passes between the earth and sun. (2) Superior conjunction occurs when a planet is on the opposite side of the sun from the earth.

**constellation** a historical grouping of the bright stars. Constellations are useful for indicating specific portions of the celestial sphere.

**contacts** the four distinct moments in an eclipse (a) the moment the objects first touch, (b) the moment one becomes covered by the other, (c) the moment of uncovering, and (d) the moment the objects last touch.

**continental drift** the gradual shifting of whole sections of the earth's mantle during eons of time.

**coronagraph** an instrument designed for studying the corona of the sun without a total eclipse by blocking the bright disk of the sun.

**cosmic rays** particles impinging on the earth's atmosphere at extremely high energies; mostly high-energy protons.

**crescent** a phase of the moon or planets where less than half of the visible surface is illuminated.

**Cytherean** of Venus

# D

**declination** the astronomical coordinate measuring the position of a body above (+ or N) and below (− or S) the celestial equator. It is always given in circular measure (e.g., degrees).

**deferent** a circle on which another circle containing an object moves.

**density** (1) the mass of a body divided by its volume (units are usually grams per cubic centimeter). (2) the number of objects per unit volume, such as the number of G stars per cubic lt-yr.

**differential rotation** motion where the outer parts of a system rotate at a different velocity than those nearer the center.

**diffraction** in physical optics the spreading of light as it passes an opaque edge.

**dipole field** like the magnetic field of a bar magnet.

**dispersion** the separation of electromagnetic radiation by wavelength due to a characteristic of the medium.

**diurnal** daily or happening each day.

**diurnal circle** the path of a star in the celestial sphere due to the daily rotation of the earth.

**diurnal parallax** the apparent change in direction to an object due to the baseline created for an observer by the daily rotation of the earth.

**Doppler effect** the displacement of the lines of a spectrum toward shorter wavelengths if the source and observer are approaching, or toward longer wavelengths if the source and observer are receding.

**draconitic month** a lunar month as measured from a given node of the moon's orbit and back to the same node.

# E

**earthlight** sunlight reflected from the earth.

**eccentric** motion on a circle whose axis of rotation does not coincide with its center.

**eccentricity** the measure of the degree of flattening of an ellipse. The letter "e" is generally used in astronomy to denote this parameter.

**eclipse** cutting off from view (either partially or totally) one body by the interposition of another.

**electromagnetic radiation** radiant energy from $\gamma$ rays through visible light to the longest radio waves caused by an oscillating electric or magnetic charge.

**elongation** the angular distance from the center of motion; generally a planet's angular distance from the sun as viewed from the earth.

**encounter** a close chance passing of two or more bodies resulting in a gravitational perturbation of the motions.

**energy** (1) kinetic—the energy of a particle or object due to its motion. (2) potential—the energy of a particle or object due to its position.

**ephemeris** tables of computed future positions of astronomical objects.

**epicycle** (1) a circle moving on a circle; used in early cosmologies to explain the retrograde motion of the planets. (2) in astrophysics, any small perturbation on a circular orbit.

**epoch** an arbitrary date to which observations are referred or reduced.

**equant** a point in the plane of an orbit about which an epicycle or body revolves with uniform angular velocity.

**equatorial system** the astronomical coordinate system based upon the celestial equator as its fundamental plane.

**equinox** the points where the ecliptic crosses the celestial equator.

**ether** the medium once believed to be required to transmit electromagnetic radiation through space.

**exosphere** the outermost region of the atmosphere.

# F

**faculae** bright regions near the limb of the sun.

**fireball** (1) a meteor brighter than the planet Jupiter. (2) the explosion of the primeval atom in certain cosmologies.

**fission** in physics, the splitting and breaking up of a nucleus into a lighter element or elements.

**flare** a sudden, temporary brightening of a small region on the sun.

**flocculi** see plages.

**fluorescence** the absorption of radiation of one wavelength and its reemission at another wavelength that ceases when the source of radiation ceases.

**focal length** the distance between the objective and its focus when looking at parallel incoming light.

**focus** the point where incoming radiation is imaged by an optical system.

**forbidden lines** spectral lines not normally observed in the laboratory because the transitions causing them are improbable.

**Fraunhofer lines** in the visible solar (stellar) spectrum, the many dark lines that indicate wavelengths abstracted from sunlight by gases in the sun's atmosphere.

**fusion** (1) the nuclear process joining atoms together to form heavier elements and releasing energy. (2) the point of change from liquid to solid.

# G

**galactic dynamics** the study of stellar orbits in a galaxy.

**gamma ray** a very high-energy photon emitted during nuclear fission, fusion, or transitions of excited states of certain nuclei.

**gas pressure** pressure exerted by a gas due to the motion of the molecules. It is a function of the temperature.

**geo** prefix meaning earth.

**geodesy** the science dealing with measuring the size and shape of the earth.

**gibbous phase** the phases of the moon or planets where more than half, but not all, the visible surface is illuminated.

**granules** the small scale, convective cells in the solar photosphere giving it the mottled appearance.

**gravitation** the property of matter to attract other matter.

**great circle** a circle on a sphere whose center coincides with the center of the sphere.

**greatest elongation** generally the largest difference in longitude between the sun and an inferior planet.

**greenhouse effect** the heating effect from the trapping of radiation by the atmosphere.

**Greenwich meridian** the origin for terrestral longitude by common usage and agreement.

# H

**halo** (1) a ring around the sun or moon caused by refraction by water droplets or ice crystals in the earth's atmosphere. (2) the large spherical distribution of stars encompassing a spiral galaxy, usually the regions above and below the disk of a galaxy.

**harvest moon** the full moon nearest the autumnal equinox.

**heliacal rising** the first rising of a star after being invisible due to its proximity to the sun. The similar condition for setting is referred to as the heliacal setting of a star.

**helio** prefix referring to the sun.

**heliostat** a coelostat used to observe the sun.

**Hertz** unit of frequency equal to one cycle per second. A megahertz is thus a million cycles per second.

**horizon** the place where the sky and earth seem to meet; see celestial horizon.

**horoscope** a chart of the heavens based upon one's birthdate and used by astrologers who claim to foretell the future.

**hunter's moon** the first full moon after the harvest moon.

**hydrocarbon** compounds of carbon and hydrogen.

**hypothesis** an unproven theory advanced to explain certain facts.

# I

**igneous rocks** rocks formed from molten materials.

**image** optical formation of an object as seen through an optical system.

**implosion** an inward explosion

**inertia** the tendency of a body at rest to remain at rest or when in motion to remain in motion with the same direction and speed.

**inferior conjunction** when an inferior planet passes between the sun and the earth.

**inferior planet** any of the planets closer than the earth to the sun.

**insolation** a contraction of *in*coming *so*lar radi*ation*; the sun's radiation received at the surface of the earth.

**International Date Line** by common usage and consent, an arbitrary line roughly 180° from the Greenwich meridian across which the date changes by a day.

**interplanetary medium** the distribution of dust and gas in interplanetary space.

**ionosphere** the shell of atmosphere lying between 70 and 320 kilometers above the surface of the earth. It contains a high density of ionized atoms and particles.

**island universe** a galaxy.

**isotropic** the same in all directions.

# J

**Jovian planet** one of the large gaseous planets of the solar system; Jupiter, Saturn, Uranus, or Neptune.

**Julian day** a running number for days beginning in 4713 B.C. The starting point is arbitrary and was chosen to be early enough that all astronomical events on record can be given a Julian day number regardless of the calendar in use.

# L

**leap year** that year (divisible by four) every four years when an extra day is added to February to keep the civil calendar in step with the sun. Century years (1800, 1900) except those divisible by 400, are not leap years.

**limb** the edge of the sun, moon, or planet.

**limiting magnitude** the faintest magnitude observable with a given instrument under given conditions.

**longitude** (1) celestial coordinate based upon the ecliptic. (2) the angular distances of an object measured along the ecliptic from the vernal equinox.

**luminescence** a visible glow from a material induced by radiation falling on it and persisting after the radiation is removed; a phosphor glows by luminescence.

**Lyman lines** the same as Balmer lines except that the first energy level of the hydrogen atom is involved.

# M

**magnetic field** the region of influence of a magnet.

**magnifying power** the apparent size of an object seen through the telescope compared to its size with the unaided eye.

**mass** the property of a body that resists a change in motion. It is a unique property of a body and, in the body's own reference frame, remains constant regardless of where the body is located.

**mean sun** the hypothetical sun that moves uniformly along the celestial equator.

**mesosphere** the earth's atmosphere lying above 400 kilometers.

**Messier Catalogue** an early catalogue of nebulae and galaxies designed to aid comet hunters; objects in the catalogue are designated by M and a running number such as M31.

**micrometeorite** an extremely small solid particle in space.

**molecule** two or more atoms chemically united forming the smallest unit possessing the properties of a compound.

**monochromatic** literally a single color. Usually light of a sufficiently narrow wavelength region that it can be treated as having a single wavelength.

**monopole (magnetic)** a proposed particle having a single magnetic pole.

# N

**nadir** the direction opposite the zenith.

**neap tide** the lowest tide in a given month.

**new moon** when the moon and the sun are lined up on the same side of the earth; if this occurs near a node of the moon's orbit, a solar eclipse can occur.

**node** the intersection of two great circles or of the orbit of one body with a specified plane.

**north point** the point on the horizon directly under the north pole of the sky. It is located by drawing a great circle through the zenith and the celestial north pole.

**nucleus** the central part of an atom, a comet, or a galaxy.

**nutation** the nodding of the earth's axis due to the precession of the axis and the lunar attraction on the earth's tidal bulge.

# O

**objective** the main lens of a refracting telescope; or the primary mirror of a reflecting telescope.

**oblate** a circle or sphere flattened at the poles.

**observatory** the astronomer's laboratory where telescopes are mounted.

**occultation** eclipse of a celestial object by another celestial object whose apparent size is larger.

**ocular** an eyepiece for a telescope.

**opposition** the moon or a planet opposite the sun and having an elongation of 180° as seen from the earth.

# P

**Palomar Sky Survey** more correctly The National Geographic-Palomar Sky Survey, a monumental photographic atlas of the heavens in two different colors covering the celestial sphere from the north pole to about −30° declination using the 1.2-m Palomar Schmidt telescope.

**parallax** the difference in direction of an object when the object is viewed from two places (i.e., the two ends of a base line).

**partial eclipse** an eclipse in which the eclipsing body does not completely cover the body being eclipsed.

**penumbra** (1) the inverted cone of a shadow from which the light from a source is only partially excluded. (2) the gray outer region of a sunspot.

**penumbral eclipse** an eclipse of the moon when the moon only passes through the earth's penumbral shadow.

**periastron** the closest point to the center of the primary star for an object in a stellar orbit.

**perigee** the closest point to the center of the earth for an object in an earth orbit.

**perihelion** the closest point to the center of the sun for an object in a solar orbit.

**perturbation** the disturbance of the motion of one body by another.

**phase** the particular aspect of the moon; the fractional part of a periodically occuring phenomenon.

**photometer** an electronic device for measuring the brightness of a celestial object.

**photometry** the measurement of the intensity of light.

**plages** bright regions on the surface of the sun as seen in monochromatic light.

**planet** a satellite of a star, in nearly circular orbit, which was never capable of shining by a self-sustained energy reaction.

**planetarium** a complex device for projecting a representation of the night sky and its phenomenon onto a dome; loosely, the institution housing and using such a projector.

**planetesimal hypothesis** a theory holding that the planets formed from material pulled out of the sun.

**planetoid** a minor planet (asteroid).

**platonic year** the time required for the earth to complete its precessional cycle; approximately 25,800 years.

**polar mount** a telescope mount where one axis points to the poles parallel to the earth's axis.

**postulate** an unproved assumption.

**precession** (1) the slow drift of the poles of a spinning body, due to a force or set of forces acting to try to tip its poles. (2) in astronomy, when unqualified, it usually means the change in coordinates due to the precession of the earth's poles.

**primary mirror** the main mirror of a reflecting telescope.

**prime meridian** see Greenwich meridian.

**proper motion** the cross motion of a star as seen projected on the sky and always given in seconds of arc (usually per year but occasionally per century).

**protostar** the central region of a condensed cloud that is about to become a self-luminous star.

**pyrheliometer** a device for measuring incident solar radiation.

# Q

**quadrature** a lunar or planetary elongation 90° east or west of the sun.

**quantum mechanics** a general theory dealing with the interactions of matter and radiation in terms of transferring energy by finite quantities.

**quarter phase** the phase of the moon or a planet where half the visible surface is illuminated.

**quasar** a contraction for quasi stellar objects, thought to be the most distant objects in the Universe.

# R

**radial velocity** the line of sight velocity of an object given in kilometers per second.

**radiant** the point on the celestial sphere from which a meteor shower seems to radiate.

**radiation pressure** pressure exerted by light or other radiation upon an object.

**radioactivity** the spontaneous decay of an atomic nucleus into a lighter nucleus or isotope by the emission of radiation.

**radiometer** an instrument for measuring the intensity of radiant energy.

**refraction** the bending of light as it passes from a medium of one density to a medium of another density. Lenses refract light.

**regolith** rock-like rubble overlying solid rock or soil.

**relativity** generally one of two theories by Albert Einstein. The special theory treats time and distance as depending upon the motion of the object and the observer. The general theory relates the structure of space to gravitation.

**resolution** the ability of an optical system to resolve details.

**retrograde motion** (1) the apparent backward motion of the planets with respect to the stars, as viewed from the earth. (2) motion of rotation or revolution opposite to that common for the sun and planets.

**retroreflector** a specially shaped mirror or shield designed to reflect radiation back in the same direction from which it came.

**right ascension** the astronomical coordinate measured along the celestial equator eastward from the vernal equinox. It is usually given in units of time but occasionally in angular measure.

**Roche's limit** the smallest distance from the primary at which a satellite can hold together under the tidal forces created by the primary.

# S

**satellite** (1) any body that orbits a planet. (2) also, any inferior body that orbits a larger body.

**secondary mirror** usually the mirror at the front end of a reflecting telescope with the purpose of changing the direction of the light reflected to it by the primary mirror.

**secular** nonperiodic.

**sedimentary rocks** rocks formed by deposition, either by settling out of water or by precipitation out of a solution.

**seeing** the atmospheric effects on an image as seen through a telescope.

**seleno** prefix from Greek, meaning the moon.

**separation** the angular distance between two objects.

**sidereal period** (1) sidereal revolution—the interval between two successive returns of a planet to the same point in the heavens as seen from the sun. (2) Sidereal rotation—the interval between two successive meridan crossings of a distant sidereal object (for example, a galaxy).

**siderite** a meteorite composed primarily of iron.

**siderolite** a meteorite made up of iron and silicates.

**small circle** any circle on a sphere that is not a great circle.

**solar constant** the amount of radiation from the sun falling upon unit area perpendicular to the direction of the sun per unit time at the distance of the earth.

**solstice** the points on the ecliptic farthest from the equator.

**spectrum** the energy output of an object displayed as a function of wavelength (or frequency).

**specular reflection** reflection from a polished surface, such as a mirror.

**spherical aberration** one of the major defects in an optical system caused by the failure of light rays from different radii from the optical axis to come to a common focus.

**spicule** a narrow jet in the solar chromosphere; the jets give a grass-like appearance to the limb of the sun.

**spiral** an abbreviated term for spiral galaxy; a galaxy having spiral arms.

**spring tide** the highest tide of the month.

**starquake** hypothetical surface explosions causing a neutron star to quiver.

**steady state** (1) a state of equilibrium. (2) a cosmological theory, that the universe looks the same for all observers all the time.

**stellar model** theoretical calculations varying the chemical composition of an ideal star to produce a model whose properties closely approximate those of a real star.

**summer solstice** the point on the ecliptic where the sun is farthest north.

**superior conjunction** when an inferior planet and the sun have the same longitude, with the sun being between the planet and earth.

**superior planet** any of the planets farther than the earth from the sun.

**synchrotron radiation** radiation from high-velocity electrons (velocities approaching the velocity of light) spiraling in a magnetic field.

**synodic month** the period of revolution of the moon about the earth with respect to the sun.

**synodic period** the time between successive planetary oppositions or conjunctions of the same kind (e.g., superior conjunction to superior conjunction) as seen from the earth.

**syzygy** a straight line configuration of three celestial bodies, especially when the earth, moon, and sun lie on the same line; more correctly stated when the centers of the earth, moon, and sun lie in a plane vertical to the plane of the earth's orbit. An excellent word-game word.

# T

**tangential motion** motion on the sky across the line of sight usually given in km/sec.

**tectonics** the study of land structure.

**tektites** small glassy objects found scattered on the earth and believed by some to be of cosmic origin.

**telescope** a somewhat misnamed instrument used by astronomers to collect radiation from celestial objects.

**telluric** terrestrial in origin.

**terminator** the line of sunset or sunrise on the earth, moon, or planets.

**theory** one or more hypotheses that along with established laws explains a phenomenon or class of phenomena.

**thermodynamics** a branch of physics dealing with the study of heat and heat transfer.

**tidal hypotheses** a theory explaining the origin of the planetary system, currently rejected.

**torque** the force tending to produce a rotating or twisting motion.

**totality** the period of complete light cut-off during an eclipse.

**transit** (1) a smaller astronomical body passing between the observer and a larger body. (2) a celestial body crossing the observer's meridian. (3) a telescope mounted in such a way that it can observe an object only when that object crosses the local meridian.

**triple alpha process** a nuclear process, first discussed by E. Salpeter, where three helium nuclei combine directly to form a carbon nucleus.

**tropical year** the interval of the earth's revolution about the sun with respect to the vernal equinox.

**turbulence** irregular, random motions in a fluid or gas.

**Tychonic system** a cosmology proposed by Tycho Brahe.

# U

**umbra** (1) the central part of a shadow where, geometrically, the light from the source is cut off. (2) the dark central area of a sunspot.

**universal time** the local time of the $0^h0^m0^s$ meridian (Greenwich, England).

**Universe** the totality of physical reality.

# V

**variation of latitude** changes in latitude of places on the earth due to the shift of the earth's axis within the earth.

**vernal equinox** the point on the celestial equator where the sun passes from below to above the equator, approximately 21 March.

**vertical circle** a great circle passing through the zenith and perpendicular to the horizon.

**VLA** initials for the very large array of radio telescopes under construction near Socorro, New Mexico.

**Vulcan** a hypothetical planet nearer the sun than Mercury.

# W

**wandering of the poles** the shifting of the earth with respect to its axis.

**weight** a measure of the attraction of a body on a given mass.

**winter solstice** the point on the ecliptic where the sun is farthest south from the equator.

**World Calendar** a calendar proposed to have the same dates fall on the same day every year.

## Z

**Zeeman effect** the splitting of the lines of a spectrum caused by the light source being in a magnetic field.

**zenith** the point on the celestial sphere that is directly overhead.

**zodiac** the band centered on the ecliptic divided into 12 equal sections. Each section was assigned a constellation and sign by the ancients.

# PERSONAL OBSERVING LOG

Observing may or may not play a role in your astronomy course. In any event, observing can be a fun experience. You only need to be dressed warmly enough for a night outdoors and have (1) a star chart(s) and (2) a flashlight with a piece of red paper over the front of it. The red light will enable you to read your star chart and look directly back at the sky without losing your night vision. A folding chair and binoculars are nice but are not really recessary.

Unless otherwise indicated, all of the objects listed below are visible to the unaided eye at various times during the year. Check your star charts to determine the best season for observing a particular object. Check an almanac to see when the planets will be in the night sky. In addition to the date, time, and place, you may want to comment on the weather or cloud conditions. For the clusters and nebulae, rather than writing a description, you might try sketching their appearance.

| *Objects* | *Date and time* | *Place* | *How observed (unaided eye binoculars, telescope) and description* |
|---|---|---|---|
| *SOLAR SYSTEM* | | | |
| *Mercury* | | | |
| *Venus* | | | |
| *Mars* | | | |
| *Jupiter* | | | |
| *Saturn* | | | |
| *Jupiter's four bright satellites (binoculars)* | | | |

| *Objects* | *Date and time* | *Place* | *How observed (unaided eye binoculars, telescope) and description* |
|---|---|---|---|
| *Meteor shower* | | | |
| *Comet (must be a bright one)* | | | |
| *Partial eclipse of the moon* | | | |
| *Total eclipse of the moon* | | | |
| *Partial eclipse of the sun (use standard precautions)* | | | |
| *Total eclipse of the sun (use standard precautions)* | | | |

*CONSTELLATIONS THAT ARE USEFUL IN FINDING OTHER OBJECTS*

| | | | |
|---|---|---|---|
| Big Dipper in Ursa Major | | | |
| Cassiopeia | | | |
| Scorpius | | | |
| Pegasus | | | |
| Orion | | | |
| Leo | | | |

| *Objects* | *Date and time* | *Place* | *How observed (unaided eye binoculars, telescope) and description* |
|---|---|---|---|

*BRIGHT STARS*

These are often easier to spot than whole constellations. Notice colors.

| Objects | Date and time | Place | How observed (unaided eye binoculars, telescope) and description |
|---|---|---|---|
| Polaris | | | |
| Antares | | | |
| Vega | | | |
| Altair | | | |
| Deneb | | | |
| Fomalhaut | | | |
| Capella | | | |
| Aldebaran | | | |
| Rigel | | | |
| Betelgeuse | | | |
| Sirius | | | |
| Procyon | | | |

| *Objects* | *Date and time* | *Place* | *How observed (unaided eye binoculars, telescope) and description* |
|---|---|---|---|
| Pollux | | | |
| Castor | | | |
| Regulus | | | |
| Denebola | | | |
| Spica | | | |
| Arcturus | | | |

*ZODIACAL CONSTELLATIONS*

| | | | |
|---|---|---|---|
| Pisces | | | |
| Aries | | | |
| Taurus | | | |
| Gemini | | | |
| Cancer | | | |
| Leo | | | |

| *Objects* | *Date and time* | *Place* | *How observed (unaided eye binoculars, telescope) and description* |
|---|---|---|---|
| Virgo | | | |
| Libra | | | |
| Scorpius | | | |
| Sagittarius | | | |
| Capricornus | | | |
| Aquarius | | | |

*OTHER CONSTELLATIONS*

| | | | |
|---|---|---|---|
| Little Dipper in Ursa Minor | | | |
| Auriga | | | |
| Canis Major | | | |
| Bootes | | | |
| Hercules | | | |
| Cygnus | | | |

| Objects | Date and time | Place | How observed (unaided eye binoculars, telescope) and description |
|---|---|---|---|

INTERESTING STARS, CLUSTERS, AND NEBULAE
These observations take very clear, dark nights.

| | | | |
|---|---|---|---|
| Stars Alcor and Mizar | | | |
| h and $\chi$ Persei (Double cluster in Perseus) | | | |
| M31 (Great Nebula in Andromeda) | | | |
| M13 (Cluster in Hercules) | | | |
| M45 (Pleiades) | | | |
| M44 (Praesepe cluster in Cancer) | | | |
| Hyades (Cluster in Taurus) | | | |
| Band of the Milky Way | | | |
| M42 (Orion nebula—binoculars) | | | |

| *Objects* | *Date and time* | *Place* | *How observed (unaided eye binoculars, telescope) and description* |
| --- | --- | --- | --- |

*SOME SOUTHERN SKY OBSERVATIONS*

| | | | |
| --- | --- | --- | --- |
| Canopus | | | |
| $\alpha$ Centaurus | | | |
| $\beta$ Centaurus | | | |
| Southern Cross | | | |
| Large Magellanic Cloud | | | |
| Small Magellanic Cloud | | | |

# INDEX

## T



## U



## V



## W



## X



## Y



## Z